Lecture Notes in Computer Science　　10495

Commenced Publication in 1973
Founding and Former Series Editors:
Gerhard Goos, Juris Hartmanis, and Jan van Leeuwen

More information about this series at http://www.springer.com/series/7407

Ángela I. Barbero · Vitaly Skachek
Øyvind Ytrehus (Eds.)

Coding Theory
and Applications

5th International Castle Meeting, ICMCTA 2017
Vihula, Estonia, August 28–31, 2017
Proceedings

 Springer

Editors
Ángela I. Barbero 🆔
University of Valladolid
Valladolid
Spain

Vitaly Skachek 🆔
University of Tartu
Tartu
Estonia

Øyvind Ytrehus 🆔
Simula@UiB and University of Bergen
Bergen
Norway

ISSN 0302-9743   ISSN 1611-3349   (electronic)
Lecture Notes in Computer Science
ISBN 978-3-319-66277-0   ISBN 978-3-319-66278-7   (eBook)
DOI 10.1007/978-3-319-66278-7

Library of Congress Control Number: 2017951310

LNCS Sublibrary: SL1 – Theoretical Computer Science and General Issues

Printed on acid-free paper

This Springer imprint is published by Springer Nature
The registered company is Springer International Publishing AG
The registered company address is: Gewerbestrasse 11, 6330 Cham, Switzerland

Preface

It is a pleasure to welcome you to the 5th International Castle Meeting on Coding Theory and Applications (5ICMCTA), held in Vihula Manor, Estonia, during the period of August 28–31, 2017. This volume contains the extended abstracts of the 24 submitted papers that were accepted for presentation at 5ICMCTA.

Previous workshops in the ICMCTA series were organized in La Mota Castle, Medina del Campo, Spain (1999 and 2008), Parador de Cardona, Spain (2011), and Pousada de Palmela, Portugal (2014). In the spirit of ICMCTA, we aim to bring together, in a beautiful and historical environment, young researchers as well as leading experts in order to present and discuss problems and results at the research frontier, related to the theoretical issues and practical applications of coding and information theory, and to identify new challenges and foster new collaborations. 5ICMCTA is highly international: the submitted papers were co-authored by researchers from about 20 countries, and when we include the program committees, people from almost 30 countries are involved.

The program contains presentations on relevant research areas in modern Coding Theory, including codes and combinatorial structures, algebraic geometric codes, group codes, convolutional codes, network coding, other applications to communications, and applications of coding theory in cryptography. All papers have passed through a thorough review process, which assures the high quality of the program.

In addition to the submitted talks, we are proud to present talks by our distinguished invited speakers: Gérard Cohen, Tor Helleseth, Camilla Hollanti, Raquel Pinto, and Paul Siegel.

The 5ICMCTA is being organized by the University of Tartu. In particular we wish to acknowledge the efforts of Yauhen Yakimenka, Anneli Vainumäe, Eva Pruusapuu, and Raquel Pinto.

We are very grateful to:

- the authors and participants who make up the program and who, we are convinced, will make 5ICMCTA a memorable event,
- the Scientific Committee and the sub-reviewers who performed excellently during the reviewing process,
- Easychair.org who provided the tools for efficient submission, management, and reviewing,
- The Norwegian-Estonian research cooperation programme and the project EMP133, which facilitated the organization of 5ICMCTA,
- Springer's LNCS for publishing this volume of extended abstracts,
- Springer's Cryptography and Communications for agreeing to publish a special volume of selected papers after 5ICMCTA,

- University of Tartu ASTRA project PER ASPERA Doctoral School of Information and Communication Technologies for providing funding for the student workshop.

We wish you a wonderful Castle Meeting!

July 2017

Ángela Barbero
Vitaly Skachek
Øyvind Ytrehus

Organization

General Chair

Vitaly Skachek   Institute of Computer Science, University of Tartu, Estonia

Scientific Committee

Co-chairs

Ángela I. Barbero   Department of Applied Mathematics, Universidad de
                               Valladolid, Spain
Øyvind Ytrehus   Simula@UiB and University of Bergen, Norway

Alexander Barg
Irina E. Bocharova
Eimear Byrne
Joan-Josep Climent
Gérard Cohen
Olav Geil
Marcus Greferath
Tor Helleseth
Tom Høholdt
Camilla Hollanti
Kees A. Schouhamer Immink
Frank R. Kschischang
Boris D. Kudryashov
San Ling
Daniel E. Lucani
Gary McGuire
Sihem Mesnager
Muriel Médard

Diego Napp
Frédérique E. Oggier
Patric R.J. Östergård
Raquel Pinto
Josep Rifà
Paula Rocha
Joachim Rosenthal
Eirik Rosnes
Moshe Schwartz
Vladimir Sidorenko
Patrick Sole
Leo Storme
Rüdiger Urbanke
Mercè Villanueva
Pascal O. Vontobel
Dejan Vukobratovic
Jos Weber
Gilles Zémor

Administration

Anneli Vainumäe
Eva Pruusapuu

Publicity

Yauhen Yakimenka

Contents

New Bounds for Linear Codes
of Covering Radius 2

Daniele Bartoli[1], Alexander A. Davydov[2(✉)], Massimo Giulietti[1],
Stefano Marcugini[1], and Fernanda Pambianco[1]

[1] Department of Mathematics and Computer Science,
Perugia University, Perugia, Italy
{daniele.bartoli,massimo.giulietti,
stefano.marcugini,fernanda.pambianco}@unipg.it
[2] Institute for Information Transmission Problems (Kharkevich Institute),
Russian Academy of Sciences, Moscow, Russian Federation
adav@iitp.ru

Abstract. The length function $\ell_q(r, R)$ is the smallest length of a q-ary
linear code of covering radius R and codimension r. New upper bounds
on $\ell_q(r, 2)$ are obtained for odd $r \geq 3$. In particular, using the one-to-one
correspondence between linear codes of covering radius 2 and saturating
sets in the projective planes over finite fields, we prove that

$$\ell_q(3, 2) \leq \sqrt{q(3 \ln q + \ln \ln q)} + \sqrt{\frac{q}{3 \ln q}} + 3$$

and then obtain estimations of $\ell_q(r, 2)$ for all odd $r \geq 5$. The new upper
bounds are smaller than the previously known ones. Also, the new bounds
hold for all q, not necessary large, whereas the previously best known
estimations are proved only for q large enough.

Keywords: Covering codes · Saturating sets · The length function ·
Upper bounds · Projective spaces

1 Introduction

Let F_q be the Galois field with q elements. Let F_q^n be the n-dimensional vector
space over F_q. Denote by $[n, n-r]_q$ a q-ary linear code of length n and codimension (redundancy) r, that is, a subspace of F_q^n of dimension $n - r$. The sphere
of radius R with center c in F_q^n is the set $\{v : v \in F_q^n, d(v, c) \leq R\}$ where $d(v, c)$
is the Hamming distance between vectors v and c.

Definition 1. *(i) The* covering radius *of a linear $[n, n - r]_q$ code is the least
integer R such that the space F_q^n is covered by spheres of radius R centered
at codewords.*

(ii) A linear $[n, n-r]_q$ code has covering radius R *if every column of F_q^r is equal
to a linear combination of at most R columns of a parity check matrix of the
code, and R is the smallest value with such property.*

© Springer International Publishing AG 2017
A.I. Barbero et al. (Eds.): ICMCTA 2017, LNCS 10495, pp. 1–10, 2017.
DOI: 10.1007/978-3-319-66278-7_1

Definition 1(i) and (ii) are equivalent. Let an $[n, n-r]_q R$ code be an $[n, n-r]_q$ code with covering radius R. For an introduction to coverings of vector Hamming spaces over finite fields, see [3,4].

The covering density μ of an $[n, n-r]_q R$-code is defined as

$$\mu = \frac{1}{q^r} \sum_{i=0}^{R} (q-1)^i \binom{n}{i} \geq 1.$$

The covering quality of a code is better if its covering density is smaller. For fixed q, r, and R the covering density of an $[n, n-r]_q R$ code decreases with decreasing n.

Definition 2 ([3,4]). *The* length function $\ell_q(r, R)$ *is the smallest length of a q-ary linear code with covering radius R and codimension r.*

Codes investigated from the point view of the covering quality are usually called *covering codes*; see an online bibliography in [13].

In this paper we consider covering codes with radius $R = 2$.

The known lower bound on $\ell_q(r, 2)$, based on Definition 1(ii), is

$$\ell_q(r, 2) > \sqrt{2} q^{(r-2)/2}. \tag{1}$$

Really, in a parity check matrix of an $[n, n-r]_q 2$ code, one can take $\binom{n}{2}$ distinct pair of columns and then form q^2 linear combinations from every pair. By Definition 1(ii), it holds that $\binom{n}{2} q^2 \geq q^r$ whence (1) follows.

For arbitrary q, covering codes of length close to this lower bound are known only for r even [5,7,9,10]. In particular, the following bounds are obtained by algebraic constructions [7, Sect. 4.3, Eq. (4.6)], [9, Theorem 9]:

$$\ell_q(r, 2) \leq 2q^{(r-2)/2} + q^{(r-4)/2}, \ q \geq 7, \ q \neq 9, \ r = 2t \geq 4, \ t = 2, 3, 5, \ \text{and} \ t \geq 7.$$

$$\ell_q(r, 2) \leq 2q^{(r-2)/2} + q^{(r-4)/2} + q^{(r-6)/2} + q^{(r-8)/2}, \ q \geq 7, \ q \neq 9, \ r = 8, 12.$$

If r is *odd*, covering codes of length close to lower bound (1) are known only when q is an *even power of a prime*, i.e. more exactly when $q = (q')^2$ and $q = (q')^4$, where q' is a prime power, and when $q = p^6$ with prime $p \leq 73$ [5–7,10,12]. In particular, the following bounds are obtained by algebraic constructions, see [5, Example 6, Eq. (33)], [6], [7, Sect. 4.4, Eqs. (4.12), (4.13), (4.15)], [12], and the references therein:

$$\ell_q(r, 2) \leq \left(3 - \frac{1}{\sqrt{q}}\right) q^{(r-2)/2} + \left\lfloor q^{(r-5)/2} \right\rfloor, \ q = (q')^2 \geq 16, \ r = 2t + 1 \geq 3.$$

$$\ell_q(r, 2) \leq \left(2 + \frac{2}{\sqrt[4]{q}} + \frac{2}{\sqrt{q}}\right) q^{(r-2)/2} + \left\lfloor q^{(r-5)/2} \right\rfloor, \ q = (q')^4, \ r = 2t + 1 \geq 3.$$

$$\ell_q(r, 2) \leq \left(2 + \frac{2}{\sqrt[6]{q}} + \frac{2}{\sqrt[3]{q}} + \frac{2}{\sqrt{q}}\right) q^{(r-2)/2} + 2\left\lfloor q^{(r-5)/2} \right\rfloor, \ q = (q')^6,$$

$q' \leq 73$ prime, $r = 2t + 1 \geq 3, \ r \neq 9, 13.$

The *goal of this work* is to obtain new upper bounds on the length function $\ell_q(r,2)$ with r *odd* and *arbitrary* q, not necessarily having the form $q = (q')^2$ where q' is a prime power. It is a *hard open problem*. The first and the most important step in this problem is finding of upper bounds on $\ell_q(3,2)$. It is usually considered as a separate open problem.

Let $\mathrm{PG}(N,q)$, $N \geq 2$, be the N-dimensional projective space over the field F_q; see [11] for an introduction to the projective spaces over finite fields. Effective methods obtaining upper bounds on $\ell_q(r,2)$ with r odd, in particular on $\ell_q(3,2)$, are connected with saturating sets in $\mathrm{PG}(N,q)$, $N \geq 2$.

Definition 3. *A point set $\mathcal{S} \subset \mathrm{PG}(N,q)$ is* saturating *if any point of $\mathrm{PG}(N,q)\backslash\mathcal{S}$ is collinear with two points in $\mathcal{S}$.*

Saturating sets are considered in [5–10,12,14,15], see also the references therein. In the literature, saturating sets are also called "saturated sets" [5,15], "spanning sets", "dense sets", and "1-saturating sets" [6–8,12].

Let $s(N,q)$ be *the smallest size of a saturating set* in $\mathrm{PG}(N,q)$.

If q-ary positions of a column of an $r \times n$ parity check matrix of an $[n,n-r]_q2$ code are treated as homogeneous coordinates of a point in $\mathrm{PG}(r-1,q)$ then this parity check matrix defines a saturating set of size n in $\mathrm{PG}(r-1,q)$ [5–7]. So, there is *the one-to-one correspondence between $[n,n-r]_q2$ codes and saturating sets in $\mathrm{PG}(r-1,q)$*. Therefore,

$$\ell_q(r,2) = s(r-1,q), \quad \text{in particular, } \ell_q(3,2) = s(2,q).$$

In [1,2], by probabilistic methods the following upper bound is obtained in the geometrical language.

$$s(2,q) \leq 2\sqrt{(q+1)\ln(q+1)} + 2 \sim 2\sqrt{q\ln q}. \tag{2}$$

Also, in [1,2] one can find the previous results and the references on this topic.

In [14], the following bound is proved for the projective plane $\mathrm{PG}(2,q)$.

$$s(2,q) \leq (\sqrt{3} + o(1))\sqrt{q\ln q}. \tag{3}$$

The proof of (3) is given in [14] by two approaches: probabilistic and algorithmic. In both the approaches, starting with some stage of the proof, it is assumed (by the context) that q *is large enough*. As the result of the algorithmic proof of [14], the following form of the bound can be derived.

$$s(2,q) \leq \left\lceil \sqrt{3q\ln q} \right\rceil + \left\lceil \frac{1}{2}\sqrt{q} \right\rceil \leq \sqrt{3q\ln q} + \frac{1}{2}\sqrt{q} + 2, \quad q \text{ large enough.} \tag{4}$$

Note that the first steps of the algorithmic proof in [14] do not need q large enough; this allows us to use these steps in Sect. 2.

Throughout the paper we denote

$$\Upsilon(q) = \sqrt{3\ln q + \ln\ln q} + \sqrt{\frac{1}{3\ln q} + \frac{3}{\sqrt{q}}}. \tag{5}$$

Our new results are collected in Theorem 4 based on Theorems 7 and 11.

Theorem 4. *Let q be an* arbitrary *prime power. Let the value of q be not necessarily large. Let r be odd. For the length function $\ell_q(r,2)$ and for the smallest size $s(r-1,q)$ of a saturating set in the projective space $\mathrm{PG}(r-1,q)$ the following upper bounds hold.*

(i) $\ell_q(3,2) = s(2,q) \le \Upsilon(q) \cdot q^{(3-2)/2} = \Upsilon(q)\sqrt{q}.$ (6)

(ii) $\ell_q(r,2) = s(r-1,q) \le \Upsilon(q) \cdot q^{(r-2)/2} + 2q^{(r-5)/2}, \quad r = 2t+1 \ge 5,$ (7)

where $r \ne 9, 13$, $t = 2, 3, 5$, and $t \ge 7$, $q \ge 19$.

$$\ell_q(r,2) = s(r-1,q) \le \Upsilon(q) \cdot q^{(r-2)/2} + 2q^{(r-5)/2} + q^{(r-7)/2} + q^{(r-9)/2}, \quad (8)$$

where $r = 9, 13$.

These upper bounds are smaller (i.e. better) than the previously known ones, see Sect. 4.

The paper is organized as follows. In Sect. 2, a new upper bound on the length function $\ell_q(3,2)$ is obtained. In Sect. 3, upper bounds on the length function $\ell_q(r,2)$, $r \ge 5$ odd, are considered on the base of the results of Sect. 2. Finally, in Sect. 4 we compare the obtained new bounds with the previously known ones.

2   An Upper Bound on the Length Function $\ell_q(3,2)$

Assume that in $\mathrm{PG}(2,q)$ a saturating set is constructed by a step-by-step algorithm adding one new point to the set in every step.

Let $i > 0$ be an integer. Denote by $\mathcal{S}_i$ the running set obtained after the i-th step of the algorithm. A point P of $\mathrm{PG}(2,q) \setminus \mathcal{S}_i$ is covered by $\mathcal{S}_i$ if P lies on a t-secant of $\mathcal{S}_i$ with $t \ge 2$. Let $\mathcal{R}_i$ be the subset of $\mathrm{PG}(2,q) \setminus \mathcal{S}_i$ consisting of points not covered by $\mathcal{S}_i$.

In [14] the following ingenious greedy algorithm is proposed. One takes the line ℓ skew to $\mathcal{S}_i$ such that the cardinality of intersection $|\mathcal{R}_i \cap \ell|$ is the minimal among all skew lines. Then one adds to $\mathcal{S}_i$ the point on ℓ providing the greatest number of new covered points (in comparison with other points of ℓ). As a result we obtain the set $\mathcal{S}_{i+1}$ and the corresponding set $\mathcal{R}_{i+1}$.

In [14, Proposition 3.3, Proof], the following inequality is proved without requirement that q is large enough:

$$|\mathcal{R}_{i+1}| \le |\mathcal{R}_i| \cdot \left(1 - \frac{i(q-1)}{q(q+1)}\right). \quad (9)$$

The running set $\mathcal{S}_2$ contains two points; we consider the line through them. All points on this line are covered by $\mathcal{S}_2$. So, always $\mathcal{R}_2 = (q^2+q+1)-(q+1) = q^2$ where q^2+q+1 and $q+1$ are the number of points in $\mathrm{PG}(2,q)$ and in the line, respectively. Starting from $\mathcal{R}_2 = q^2$ and iteratively applying the relation (9), we obtain for some k the following:

$$|\mathcal{R}_{k+1}| \le q^2 f_q(k),$$

where

$$f_q(k) = \prod_{i=2}^{k} \left(1 - \frac{i(q-1)}{q(q+1)}\right).$$

Now we consider a *truncated iterative process*. We will stop the iterative process when $|\mathcal{R}_{k+1}| \leq \xi$ where $\xi \geq 1$ is some value that we may *assign arbitrary* to improve estimations.

By [14, Lemma 2.1] after the end of the iterative process we can add at most $\lceil |\mathcal{R}_{k+1}|/2 \rceil$ points to the running subset $\mathcal{S}_{k+1}$ in order to get the final saturating set $\mathcal{S}$. Therefore, the size s of the obtained saturating set $\mathcal{S}$ is

$$s \leq k + 1 + \left\lceil \frac{\xi}{2} \right\rceil \quad \text{under condition} \quad q^2 f_q(k) \leq \xi. \tag{10}$$

Using the inequality $1 - x \leq e^{-x}$, we obtain that

$$f_q(k) < e^{-\sum_{i=2}^{k} i(q-1)/(q^2+q)} = e^{-(k^2+k-2)(q-1)/(2q^2+2q)},$$

which implies

$$f_q(k) < e^{-(k^2+k-2)(q-1)/(2q^2+2q)} < e^{-k^2/(2q+2)}, \tag{11}$$

provided that

$$\frac{(k^2 + k - 2)(q - 1)}{q} > k^2$$

or, equivalently,

$$\frac{k^2}{k-2} < q - 1,$$

$$k < q - 4. \tag{12}$$

Lemma 5. *Let $\xi \geq 1$ be a fixed value independent of k. The value*

$$k \geq \left\lceil \sqrt{2(q+1)}\sqrt{\ln \frac{q^2}{\xi}} \right\rceil \tag{13}$$

satisfies inequality $q^2 f_q(k) \leq \xi$.

Proof. By (11), to provide $q^2 f_q(k) \leq \xi$ it is sufficient to find k such that

$$e^{-k^2/(2q+2)} < \frac{\xi}{q^2}.$$

$\square$

Theorem 6. *Let q be an arbitrary prime power. In the projective plane $\mathrm{PG}(2, q)$ it holds that*

$$s(2, q) \le \sqrt{2(q+1)}\sqrt{\ln \frac{q^2}{\xi} + \frac{\xi}{2} + 3}, \quad \xi \ge 1, \tag{14}$$

where ξ is an arbitrarily chosen value.

Proof. We substitute the value k from (13) to (10). The summand "+3" takes into account that the size of a saturating set is an integer.   □

In order to get a "good" estimation of $s(2, q)$, we are trying to reduce the right part of (14). For it, let us consider the function of ξ of the form

$$\phi(\xi) = \sqrt{2(q+1)}\sqrt{\ln \frac{q^2}{\xi} + \frac{\xi}{2} + 3}.$$

Its derivative by ξ is

$$\phi'(\xi) = \frac{1}{2} - \frac{1}{\xi}\sqrt{\frac{q+1}{2 \ln \frac{q^2}{\xi}}}.$$

It is easy to check that $\phi'(1) < 0$, $\phi'(q) > 0$, and $\phi'(\xi)$ is an increasing function of ξ. This means that for some value $\xi_0 > 1$ it holds that $\phi'(\xi_0) = 0$. Moreover, for $\xi < \xi_0$, the derivative $\phi'(\xi) < 0$ and $\phi(\xi)$ decreases, while for $\xi > \xi_0$, the derivative is positive and $\phi(\xi)$ increases. So, in the point $\xi = \xi_0$ we have the minimum of $\phi(\xi)$. Now we will find a value of ξ such that $\phi'(\xi)$ is close to 0 and, in addition, the expression of the results is relatively simple.

Put $\phi'(\xi) = 0$. Then it is easy to see that

$$\xi^2 = \frac{q+1}{\ln q - \frac{1}{2}\ln \xi}. \tag{15}$$

We find ξ in the form $\xi = \sqrt{\frac{q+1}{c \ln q}}$. By (15),

$$c = 1 - \frac{\ln(q+1)}{4\ln q} + \frac{\ln c + \ln \ln q}{4 \ln q}.$$

We choose $c \approx 1 - \frac{\ln(q+1)}{4\ln q} \approx \frac{3}{4}$ and put $\xi = \sqrt{\frac{4q}{3\ln q}}$. The value

$$\phi'\left(\sqrt{\frac{4q}{3\ln q}}\right) = \frac{1}{2} - \frac{1}{2}\sqrt{\frac{3(q+1)\ln q}{q\left(3\ln q + \ln\ln q + \ln\frac{3}{4}\right)}}$$

is close to zero for growing q. Also, see below, the expression of the results for such ξ is quite simple.

So, the choice $\xi = \sqrt{\frac{4q}{3\ln q}}$ in (14) seems to be convenient.

Theorem 7. *Let q be an arbitrary prime power.*

(i) In $\mathrm{PG}(2,q)$, there is a saturating set of size $\leq \Upsilon(q)\sqrt{q}$.
(ii) There exists an $[n, n-3]_q 2$ code with $n \leq \Upsilon(q)\sqrt{q}$.

Proof. (i) We substitute $\xi = \sqrt{\frac{4q}{3\ln q}}$ in (14) and obtain

$$s(2,q) \leq \sqrt{(q+1)\left(3\ln q + \ln\ln q + \ln\frac{3}{4}\right)} + \sqrt{\frac{q}{3\ln q}} + 3.$$

It can be shown (e.g. by considering the corresponding derivatives) that

$$\sqrt{(q+1)\left(3\ln q + \ln\ln q + \ln\frac{3}{4}\right)} + \sqrt{\frac{q}{3\ln q}} + 3 < \Upsilon(q)\sqrt{q} \text{ for } q \geq 43.$$

Also, the necessary condition (12) holds as $\Upsilon(q)\sqrt{q} < q - 4$.
So, we have proved that a saturating set of size $\leq \Upsilon(q)\sqrt{q}$ exists in $\mathrm{PG}(2,q)$ for $q \geq 43$.
Now note that in [7, Table 1], the smallest known (up to September 2010) sizes of saturating sets in $\mathrm{PG}(2,q)$, $q \leq 1217$, are given. All these sizes (including the region $q < 43$) are smaller than $\Upsilon(q)\sqrt{q}$.
The assertion **(i)** is proved.
(ii) The one-to-one correspondence between saturating sets and covering codes, see Introduction, implies the existence of an $[n, n-3]_q 2$ code with $n \leq \Upsilon(q)\sqrt{q}$.

$\square$

Theorem 7 immediately implies the estimation (6) of Theorem 4(i).

Remark 8. Let $\xi = 1$. From (14) we have

$$s(2,q) \leq 2\sqrt{(q+1)\ln q} + 3, \tag{16}$$

that practically coincides with bound (2) from [1,2].
Let $\xi = \sqrt{q}$. From (14) we obtain the estimation

$$s(2,q) \leq \sqrt{3(q+1)\ln q} + \frac{1}{2}\sqrt{q} + 3 \tag{17}$$

which practically coincides with bound (4) of [14].
However, the value $\xi = \sqrt{\frac{4q}{3\ln q}}$ gives the estimation (6) that is smaller (i.e. better) than (16) and (17), see Sect. 4.

Remark 9. In fact, the estimations (2) from [1,2], (3) and (4) of [14], and the new estimation (6), proved in this section, hold in an arbitrary finite plane of order q, not necessarily Desarguesian. But in a non-Desarguesian plane we have not the one-to-one correspondence between $[n, n-3]_q 2$ codes and saturating sets. It is why we consider here only the Desarguesian plane $\mathrm{PG}(2,q)$.

3   Upper Bounds on the Length Function $\ell_q(r,2)$, $r \geq 5$ Odd

For upper bounds on the length function $\ell_q(r,2)$, $r \geq 5$ odd, an important tool is the inductive construction of [5,7] providing the following code parameters.

Proposition 10 ([5, Example 6] [7, Theorem 4.4]). *Let an* $[n_q, n_q - 3]_q 2$ *code exist. Then the following holds.*

(i) *Under conditions* $n_q < q$ *and* $q + 1 \geq 2n_q$, *there is an infinite family of* $[n, n - r]_q 2$ *codes with the parameters*

$$n = n_q q^{(r-3)/2} + 2q^{(r-5)/2}, \ r = 2t - 1 \geq 5, \ r \neq 9,13, \ t = 3,4,6, \ and \ t \geq 8. \tag{18}$$

(ii) *Under condition* $n_q < q$ *there is an infinite family of* $[n, n - r]_q 2$ *codes with*

$$n = n_q q^{(r-3)/2} + 2q^{(r-5)/2} + q^{(r-7)/2} + q^{(r-9)/2}, \ r = 9,13. \tag{19}$$

Theorem 11. *Let* q *be an arbitrary prime power. Then there exists an infinite family of* $[n, n - r]_q 2$ *codes with the parameters*

$$n = \Upsilon(q) \cdot q^{(r-2)/2} + 2q^{(r-5)/2}, \quad r = 2t + 1 \geq 5, \ r \neq 9,13, \tag{20}$$

where $t = 2,3,5$, *and* $t \geq 7$, $q \geq 19$.
    Also there exists an infinite family of $[n, n - r]_q 2$ *codes with the parameters*

$$n = \Upsilon(q) \cdot q^{(r-2)/2} + 2q^{(r-5)/2} + q^{(r-7)/2} + q^{(r-9)/2}, \ r = 9,13. \tag{21}$$

Proof. Since $\Upsilon(q)\sqrt{q} < q$, we may put that the starting $[n_q, n_q - 3]_q 2$ code of Proposition 10 is the $[n, n - 3]_q 2$ code, $n \leq \Upsilon(q)\sqrt{q}$, of Theorem 7. It is easy to check directly that the condition $q + 1 \geq 2\Upsilon(q)\sqrt{q}$ holds for $q \geq 79$. Now, similarly to the proof of Theorem 7, we use the smallest known sizes of saturating sets in $PG(2,q)$ from [7, Table 1]. For $q < 79$, these sizes are smaller than $\Upsilon(q)\sqrt{q}$ and, moreover, for $19 \leq q < 79$ they provide the condition $q + 1 \geq 2n_q$ for Proposition 10. Now the relations (20) and (21) follow from (18) and (19), respectively.   □

Theorem 11 immediately implies the estimations (7) and (8) of Theorem 4(ii).

4   Comparison with the Previously Known Results

Surveys on the results on non-binary covering codes in [7,10] show that the inductive approach of Proposition 10 is the main tool to obtain upper bounds on the length function $\ell_q(r,2)$, $r \geq 5$ odd. Proposition 10 uses the length function $\ell_q(3,2)$ as the base for inductive estimations. Therefore upper bounds on $\ell_q(3,2)$, smaller than the known ones, provide bounds on $\ell_q(2t+1,2)$, $2t+1 \geq 5$, that are less than the corresponding known results. So, in the beginning we should compare the new bound on $\ell_q(3,2)$, see (6), with the best corresponding known bound, see (4).

First of all we should emphasize that the new bound (6) holds for all q, *not necessary large*, whereas the known bound (4) is proved only for q large enough.

Then we consider the difference $\Delta(q)$ between the bounds (4) and (6) where

$$\Delta(q) = \sqrt{3q \ln q} + \frac{1}{2}\sqrt{q} + 2 - \Upsilon(q)\sqrt{q}.$$

It can be shown (e.g. by considering the derivatives) that $\Delta(q) > 0$ for $q \geq 337$ and, moreover, $\Delta(q)$ and $\frac{\Delta(q)}{\sqrt{q}}$ are increasing functions of q. For illustration, see Fig. 1 where the top curve shows $\Delta(q)$ while the bottom one $\sqrt{q/7}$ is given for comparison.

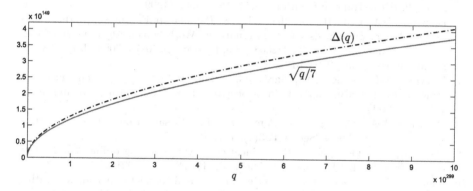

Fig. 1. The difference $\Delta(q)$ (*top dashed-dotted curve*) vs $\sqrt{q/7}$ (*bottom solid curve*)

Note also that

$$\lim_{q \to \infty} \frac{\Delta(q)}{\sqrt{q}} = \lim_{q \to \infty} \left(\sqrt{3 \ln q} + \frac{1}{2} - \sqrt{3 \ln q + \ln \ln q} - \frac{1}{\sqrt{3 \ln q}} - \frac{1}{\sqrt{q}} \right) = \frac{1}{2}.$$

Finally, if one uses Proposition 10 to estimate $\ell_q(r,2)$, $r \geq 5$ odd, then the difference between new and known results will be of order $\Delta(q)q^{(r-3)/2}$. It means that our improvements for $r = 3$ directly expand to odd $r \geq 5$.

Acknowledgements. The research of D. Bartoli, M. Giulietti, S. Marcugini, and F. Pambianco was supported in part by Ministry for Education, University and Research of Italy (MIUR) (Project "Geometrie di Galois e strutture di incidenza") and by the Italian National Group for Algebraic and Geometric Structures and their Applications (GNSAGA - INDAM). The research of A.A. Davydov was carried out at the IITP RAS at the expense of the Russian Foundation for Sciences (project 14-50-00150).

References

1. Bartoli, D., Davydov, A.A., Giulietti, M., Marcugini, S., Pambianco, F.: On upper bounds on the smallest size of a saturating set in a projective plane. [math.CO] (2015). https://arxiv.org/abs/1505.01426
2. Bartoli, D., Davydov, A.A., Giulietti, M., Marcugini, S., Pambianco, F.: New upper bounds on the smallest size of a saturating set in a projective plane. In: 2016 XV International Symposium Problems of Redundancy in Information and Control Systems (REDUNDANCY), St. Petersburg, pp. 18–22. IEEE (2016). http://ieeexplore.ieee.org/document/7779320
3. Brualdi, R.A., Litsyn, S., Pless, V.S.: Covering radius. In: Pless, V.S., Huffman, W.C., Brualdi, R.A. (eds.) Handbook of coding theory, vol. 1, pp. 755–826. Elsevier, Amsterdam (1998)
4. Cohen, G., Honkala, I., Litsyn, S., Lobstein, A.: Covering Codes. North-Holland Mathematical Library, vol. 54. Elsevier, Amsterdam (1997)
5. Davydov, A.A.: Constructions and families of nonbinary linear codes with covering radius 2. IEEE Trans. Inf. Theor. **45**(5), 1679–1686 (1999)
6. Davydov, A.A., Giulietti, M., Marcugini, S., Pambianco, F.: Linear covering codes over nonbinary finite fields. In: XI International Workshop on Algebraic and Combintorial Coding Theory (ACCT 2008), Pamporovo, pp. 70–75 (2008). http://www.moi.math.bas.bg/acct2008/b12.pdf
7. Davydov, A.A., Giulietti, M., Marcugini, S., Pambianco, F.: Linear nonbinary covering codes and saturating sets in projective spaces. Adv. Math. Commun. **5**(1), 119–147 (2011)
8. Davydov, A.A., Marcugini, S., Pambianco, F.: On saturating sets in projective spaces. J. Comb. Theor. Ser. A **103**(1), 1–15 (2003)
9. Davydov, A.A., Östergård, P.R.J.: Linear codes with covering radius $R = 2, 3$ and codimension tR. IEEE Trans. Inf. Theor. **47**(1), 416–421 (2001)
10. Giulietti, M.: The geometry of covering codes: small complete caps and saturating sets in Galois spaces. In: Blackburn, S.R., Holloway, R., Wildon, M. (eds.) Surveys in Combinatorics 2013. London Mathematical Society Lecture Note Series, vol. 409, pp. 51–90. Cambridge Univ Press, Cambridge (2013)
11. Hirschfeld, J.W.P.: Projective Geometries Over Finite Fields. Oxford Mathematical Monographs, 2nd edn. Clarendon Press, Oxford (1998)
12. Kiss, G., Kóvacs, I., Kutnar, K., Ruff, J., Šparl, P.: A note on a geometric construction of large cayley graphs of given degree and diameter. Studia Univ. Babes-Bolyai Math. **54**(3), 77–84 (2009)
13. Lobstein, A.: Covering radius, an online bibliography. http://perso.telecom-paristech.fr/~lobstein/bib-a-jour.pdf
14. Nagy, Z.L.: Saturating sets in projective planes and hypergraph covers. [math.CO] (2017). arxiv:1701.01379
15. Ughi, E.: Saturated configurations of points in projective Galois spaces. Eur. J. Comb. **8**(3), 325–334 (1987)

Multidimensional Decoding Networks for Trapping Set Analysis

Allison Beemer[(✉)] and Christine A. Kelley

University of Nebraska-Lincoln, Lincoln, NE, USA
allison.beemer@huskers.unl.edu

Abstract. We present a novel multidimensional network model as a means to analyze decoder failure and characterize trapping sets of graph-based codes. We identify a special class of these decoding networks, which we call transitive networks, and show how they may be used to identify trapping sets and inducing sets. Many codes have transitive decoding network representations. We conclude by investigating the decoding networks of codes arising from product, half-product, and protograph code constructions.

Keywords: Trapping sets · Iterative decoding · Multidimensional decoding networks · Finite state machines · LDPC codes · Redundancy

1 Introduction

Failure of message-passing decoders of low-density parity-check (LDPC) codes has been shown to be characterized by graphical (sub)structures in the code's Tanner graph, such as pseudocodewords, absorbing sets, trapping sets, and stopping sets (a subclass of trapping sets) [1–4]. In particular, these structures contribute to persistent error floors in the Bit Error Rate (BER) curves of these codes. The presence of such structures naturally depends on the choice of Tanner graph representation used in decoding. However, while absorbing sets are combinatorially-defined, trapping sets are heavily decoder dependent, making them more difficult to analyze. Nevertheless, for many practical channels, the error floor behavior is dominated by the harmful trapping sets in the graph [3].

In this work, we present a multidimensional network model for analyzing hard-decision message-passing decoders. The structure of this network is dependent on the code, as well as the choice of Tanner graph representation and decoder. Thus, our model takes into account all parameters determining the presence of harmful trapping sets. We show how these decoding networks may be used to identify trapping sets, and therefore analyze decoder behavior of LDPC codes, as well as provide insight into the optimal number of iterations to be run on a given code (and representation) with a chosen decoder. We show that this analysis is simplified for networks with a transitivity property, and discuss the connection between transitive networks and redundancy in their corresponding parity-check matrices. Finally, we relate the decoding networks of product and

© Springer International Publishing AG 2017
Á.I. Barbero et al. (Eds.): ICMCTA 2017, LNCS 10495, pp. 11–20, 2017.
DOI: 10.1007/978-3-319-66278-7_2

half-product codes to those of their underlying component codes, and examine the connection between the decoding networks of a protograph and its lift.

While the results herein present the case of binary linear codes transmitted over the binary erasure and symmetric channels (BEC and BSC), the definitions and results may be extended for linear codes over larger fields. Moreover, the decoding network concept and its applications may be extended to other channels and decoding algorithms. This paper is organized as follows. In Sect. 2, we provide notation and background. We present multidimensional decoding networks for decoder analysis in Sect. 3. In Sect. 4, we define trapping sets and describe a method for identifying them using this model. In Sect. 5, we discuss the connection between redundancy and transitivity, and apply the decoding network framework to various codes.

2   Preliminaries

In this section we introduce relevant background. Let C be a binary LDPC code of length n with associated Tanner graph G, to be decoded with some chosen hard- or soft-decision decoder. Following the notation and definitions in [5], suppose that the codeword $\mathbf{x}$ is transmitted, and $\hat{\mathbf{x}}$ is received. Let $\hat{\mathbf{x}}^\ell$ be the output after ℓ iterations of the decoder are run on G. A node $\hat{x}_i$, with $1 \leq i \leq n$, is said to be *eventually correct* if there exists $L \in \mathbb{Z}_{\geq 0}$ such that $\hat{x}_i^\ell = x_i$ for all $\ell \geq L$.

Definition 1. Let $T(\hat{\mathbf{x}})$ denote the set of variable nodes that are not eventually correct for a received word $\hat{\mathbf{x}}$, and let $G[T]$ denote the subgraph induced by $T(\hat{\mathbf{x}})$ and its neighbors in the graph G. If $G[T]$ has a variable nodes and b odd-degree check nodes, $T(\hat{\mathbf{x}})$ is said to be an (a, b)-*trapping set*. In this case, the set of variable nodes in error in the received word $\hat{\mathbf{x}}$ is called an *inducing set* for $T(\hat{\mathbf{x}})$. The *critical number* of a trapping set T, denoted $m(T)$, is the minimum number of variable nodes in an inducing set of T.

Note that the critical number may arise from an inducing set not fully contained in T. Moreover, observe that stopping sets [4] are simply trapping sets over the BEC, all of whose inducing sets contain the stopping set itself. Since a stopping set is an inducing set for itself, the critical number of a stopping set is equal to its size.

Due to their complexity, trapping sets are typically analyzed under hard-decision decoders, such as Gallager A/B [6,7], although interesting work has also been done on trapping set analysis for soft-decision decoders on the AWGN channel [8,9]. Simulation results suggest that trapping sets with respect to hard decision decoders also affect the error floor performance of soft-decision decoders [5,10]. Recall that the Gallager A algorithm operates for transmission over the BSC; messages are calculated at nodes and sent across edges of the associated Tanner graph. Variable to check ($\mu_{v \to c}$) and check to variable ($\mu_{c \to v}$) messages are calculated as follows:

$$\mu_{v \to c}(r, m_1, \ldots, m_{d(v)-1}) = \begin{cases} r+1 \text{ if } m_1 = \cdots = m_{d(v)-1} = r+1 \\ r \text{ else,} \end{cases} \qquad (1)$$

where r is the received value at variable node v, d_v is the degree of v, and $m_1, \ldots, m_{d(v)-1}$ are the most recent messages received from the check nodes which are not those to which the current message is being calculated. In the other direction,

$$\mu_{c \to v}(m_1, \ldots, m_{d(c)-1}) = \sum_{i=1}^{d(c)-1} m_i , \qquad (2)$$

where here $m_1, \ldots, m_{d(c)-1}$ are messages received from variable nodes. Note that calculations are performed in $\mathbb{F}_2$. Gallager B relaxes the unanimity condition of $\mu_{v \to c}$. Throughout the examples in this work which use Gallager decoding, we consider Gallager A decoding with the final decoder output at a variable node v given by the received value at v, unless incoming messages $\mu_{c \to v}$ for c adjacent to v unanimously disagree with this received value. Further results for Gallager B and more relaxed decoding rules will be treated in a forthcoming work.

3   Multidimensional Network Framework

Suppose we decode a code $\mathcal{C}$ using a hard-decision decoder, and consider the labeled directed graph (digraph) for a fixed $\ell \in \mathbb{Z}_{>0}$, denoted $\mathcal{D}_\ell$, with vertex and edge sets $V = \{\mathbf{x} : \mathbf{x} \in S\}$ and $E = \{(\mathbf{x}_i, \mathbf{x}_j, \ell) : \mathbf{x}_i^\ell = \mathbf{x}_j\}$, respectively, where S is the set of possible received words, $(\mathbf{x}_i, \mathbf{x}_j, \ell)$ denotes an edge from $\mathbf{x}_i$ to $\mathbf{x}_j$ with edge label ℓ, and $\mathbf{x}_i^\ell$ is the output of the decoder after ℓ iterations with input $\mathbf{x}_i$. Note that we allow loops, which are edges of the form $(\mathbf{x}_i, \mathbf{x}_i, \ell)$. For simplicity, we refer to the label of a vertex – that is, its corresponding word in S – interchangeably with the vertex itself. There will be a potentially distinct digraph on this same vertex set for each choice of $\ell \in \mathbb{Z}_{>0}$. We call the union of these digraphs for all $\ell \in \mathbb{Z}_{>0}$ the *(multidimensional) decoding network* corresponding to the code $\mathcal{C}$ and the specific choice of decoder, as we may consider the digraph which incorporates the information for all ℓ as a *multidimensional network*.

Definition 2 ([11]). A *multidimensional network* is an edge-labeled directed graph $\mathcal{D} = (V, E, D)$, where V is a set of vertices, D a set of edge labels, called *dimensions*, and E is a set of triples (u, v, d) where $u, v \in V$ and $d \in D$. We say that an edge (or vertex) *belongs to* a given dimension d if it is labeled d (or is incident to an edge labeled d).

In this framework, the decoding network is a multidimensional network with $D = \mathbb{Z}_{>0}$, and each edge labeled with the number of decoder iterations, ℓ, to which it corresponds. Notice that, in any given dimension (i.e. iteration), every vertex has outdegree equal to one. Next, we introduce an important type of decoding network.

Definition 3. We say that a decoding network is *transitive* if $(v_1, v_2, \ell) \in E$ if and only if for every choice of $1 \leq k \leq \ell - 1$, there exists $v_k \in V$ such that $(v_1, v_k, k), (v_k, v_2, \ell - k) \in E$. We say a decoder is transitive for a code $\mathcal{C}$ and a choice of representation of $\mathcal{C}$ if its resulting decoding network is transitive.

Transitivity corresponds to the case in which running i iterations of the decoder on a received word, and using that output as input to a subsequent j decoding iterations, is equivalent to running $i + j$ iterations on the original received word.

Let $\mathcal{D}_\ell$ denote the digraph corresponding to the ℓth dimension of the decoding network $\mathcal{D}$, and let $A(\mathcal{D}_\ell)$ denote the adjacency matrix of the digraph $\mathcal{D}_\ell$. Observe that a decoding network $\mathcal{D}$ is transitive if and only if $A(\mathcal{D}_\ell) = (A(\mathcal{D}_1))^\ell$ for all $\ell \geq 1$, as the product $(A(\mathcal{D}_1))^\ell$ gives directed paths of length ℓ in dimension 1 of $\mathcal{D}$.

Example 1. Consider a simple binary parity-check code of length n, with parity-check matrix given by the $1 \times n$ all-ones matrix. The Tanner graph representation of such a code is a single check node with n variable node leaves. Thus, if codewords are sent over the BSC and decoded with the Gallager A algorithm, the message the solitary check node sends to each of its adjacent variable nodes is either (a) the node's channel value, if a codeword is received, or (b) the opposite of its originally-received value, otherwise. Each variable node will always send back its channel value. For any number of iterations, codewords will decode to codewords, and any non-codeword $\hat{x}$ will be decoded to $\hat{x} + 1$. If n is odd, every received word will decode to a codeword, and the network will be transitive. If n is even, $\hat{x} + 1$ will not be a codeword, and the network will not be transitive. The cases $n = 3$ and $n = 4$ are shown in Fig. 1; in both networks, edges belonging to higher dimensions are suppressed, as all dimensions are identical.

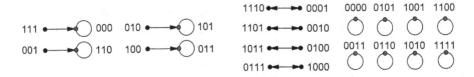

Fig. 1. The decoding networks of parity-check codes of lengths 3 (left) and 4 (right).

Example 2. Consider the binary Hamming code of length $7 = 2^3 - 1$, denoted $\mathcal{H}_3$. Recall that this code's canonical 3×7 parity-check matrix has columns consisting of all nonzero binary words of length 3. The corresponding Tanner graph may be seen in representation A of Fig. 2. However, $\mathcal{H}_3$ may also be defined by the parity-check matrix whose Tanner graph is representation B in Fig. 2. Under Gallager A decoding, representation A does not yield a transitive decoding network. However, if representation B is decoded via Gallager A, the resulting decoding network is transitive: every word decodes under a single iteration to a codeword, and decodes identically for any higher number of iterations.

If a decoder is transitive for all representations of all codes $\mathcal{C}$, we say that it is *universally transitive*. Any decoder which ignores channel values at each subsequent iteration will be universally transitive; some examples are given next.

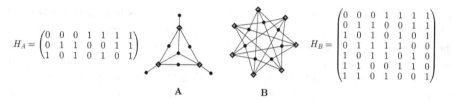

Fig. 2. Two distinct Tanner graph representations of $\mathcal{H}_3$, along with their parity-check matrices. Variable nodes are denoted by •, and check nodes are denoted by ◊.

Example 3. If codewords from a code $\mathcal{C}$ are sent over the BEC, and words are decoded using a peeling decoder which iteratively corrects erasures, then the corresponding decoding network of $\mathcal{C}$ is universally transitive. Indeed, corrections at each iteration are performed regardless of previous iterations. Similarly, iterative bit-flipping decoders over the BSC are universally transitive.

4   Trapping Set Characterization

Within this multidimensional decoding network framework, trapping sets of a code transmitted over the BSC may be determined by looking at the supports of the words corresponding to vertices in the decoding network that have nonzero indegree in an infinite number of dimensions. That is,

Theorem 1. *For each vertex* $\mathbf{x} \in V = \mathbb{F}_2^n$ *in a decoding network* $\mathcal{D} = (V, E, D)$ *for a code* $\mathcal{C}$, *let* $M_{\mathbf{x}}$ *be the set of vertices* $\mathbf{y} \in V$ *for which there is an edge* $(\mathbf{x}, \mathbf{y}, \ell) \in E$ *for infinitely many choices of* ℓ. *Then the set of variable nodes corresponding to*

$$\bigcup_{\mathbf{y} \in M_{\mathbf{x}}} \mathrm{supp}(\mathbf{y}) \ ,$$

denoted $\mathcal{T}(\mathbf{x})$, *is a trapping set with an inducing set given by the variable nodes corresponding to* $\mathrm{supp}(\mathbf{x})$. *Furthermore, the set of trapping sets of the code* $\mathcal{C}$ *is*

$$\{\mathcal{T}(\mathbf{x}) : \mathbf{x} \in \mathbb{F}_2^n\} \ ,$$

and, given a trapping set $\mathcal{T}$, *its set of inducing sets is given by the variable nodes corresponding to*

$$\{\mathrm{supp}(\mathbf{x}) : \mathcal{T}(\mathbf{x}) = \mathcal{T}\} \ ,$$

and its critical number is

$$m(\mathcal{T}) = \min\{|\mathrm{supp}(\mathbf{x})| : \mathcal{T}(\mathbf{x}) = \mathcal{T}\} \ .$$

Proof. Assuming that the all-zero codeword was sent over the BSC, any decoding errors will be given by 1's. If, during the process of decoding the received word $\mathbf{x}$, there is some word $\mathbf{y}$ such that an edge from $\mathbf{x}$ to $\mathbf{y}$ occurs in an infinite number of network dimensions, the support of $\mathbf{y}$ gives variable node positions which are not eventually correct. By definition, these variable nodes belong to a trapping set induced by the variable nodes of the support of $\mathbf{x}$. However, these may not be the only variable nodes that are not eventually correct given the received word $\mathbf{x}$. Taking the union of the supports of all such $\mathbf{y}$ gives us our expression for $\mathcal{T}(\mathbf{x})$, the trapping set induced by $\mathbf{x}$. Repeating this for each possible received word, we find all trapping sets of the code. Note that each trapping set may be induced by multiple received words.

Example 4. Let $\mathcal{C}$ be the binary repetition code of length 3, with the parity-check matrix

$$H = \begin{pmatrix} 1 & 1 & 0 \\ 0 & 1 & 1 \end{pmatrix}$$

The associated Tanner graph is a path of length 4 with variable nodes v_1, v_2, and v_3. Let $\mathcal{D}$ be the (non-transitive) decoding network of $\mathcal{C}$ under Gallager A, shown in Fig. 3. Assuming $\mathbf{0}$ was transmitted, the received word 011, for example, decodes in one iteration to the codeword 111, but decodes for any number of iterations greater than 1 to the (non-code)word 110. Thus, $\{v_1, v_2\}$ is a (2,1)-trapping set. Note that the support of 011, which induces this trapping set, is not contained in the trapping set. Similarly, $\{v_1, v_2, v_3\}$ is a trapping set corresponding to the codeword 111, with inducing sets $\{v_1, v_3\}$ and $\{v_1, v_2, v_3\}$. Other trapping sets of the code include $\{v_2, v_3\}$ (induced by $\{v_1, v_2\}$), $\{v_1\}$ (induced by $\{v_3\}$), and $\{v_3\}$ (induced by $\{v_1\}$).

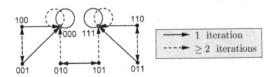

Fig. 3. The decoding network of a binary repetition code of length 3.

In the case of transitive decoding networks, trapping sets may be identified by looking only at dimension 1, as follows:

Corollary 1. *If the decoding network, $\mathcal{D}$, of a code $\mathcal{C}$ is transitive, then the trapping sets are given by (1) the sets of variable nodes corresponding to supports of vertices with loops in $\mathcal{D}_1$, and (2) the sets of variable nodes corresponding to unions of the supports of vertices forming directed cycles in $\mathcal{D}_1$. Furthermore, inducing sets of trapping sets in a transitive decoding network are given by the variable nodes corresponding to the support of any vertex which has a directed path to either a (nonzero) vertex with a loop, or to a directed cycle, regardless of where that path enters the cycle.*

Proof. In a transitive decoding network, the edges in dimension ℓ correspond to directed paths of length ℓ in $\mathcal{D}_1$. Thus, in order for a word to appear as the output of the decoder in an infinite number of dimensions, it must be the terminating vertex of infinitely many directed paths (of distinct lengths) from the received word. Because the decoding network for a code is finite, and the outdegree of every vertex in $\mathcal{D}_1$ is equal to 1, this can only occur if there is a loop at that vertex, or if it belongs to a directed cycle. The result follows from Theorem 1.

In the adjacency matrix of dimension 1 of a decoding network, nonzero diagonal entries indicate loops. There are numerous algorithms for finding directed cycles in a digraph, such as Depth-First Search (DFS) [12]. We further note that several works have addressed the computational aspects of finding and/or removing trapping sets [10,13–15].

We conclude this section by discussing the ideal number of iterations given a fixed code and decoder. Note that in Example 4, if the all-zero codeword was sent, more received words are decoded correctly if only a single iteration of the decoder is run, rather than multiple iterations. To this end, we introduce a parameter we call the *decoding diameter*.

Definition 4. Let $\mathcal{D} = (V, E, D)$ be the decoding network of a code $\mathcal{C}$ under a fixed decoder. For $\mathbf{x} \in V$, let $L_\mathbf{x}$ be the minimum nonnegative integer such that for all $\ell \geq L_\mathbf{x}$, $\mathbf{x}^\ell$, the output of the decoder after ℓ iterations, appears an infinite number of times in the sequence $\{\mathbf{x}^k\}_{k=L_\mathbf{x}}^\infty$. Then, the *decoding diameter* of $\mathcal{D}$ is given by $\Delta(\mathcal{D}) = \max_{\mathbf{x} \in V} L_\mathbf{x}$.

After running $\Delta(\mathcal{D})$ iterations, all errors will be contained in trapping sets of the code. In Example 2, $\mathcal{H}_3$ with canonical graph representation A decoded via the Gallager A algorithm has decoding diameter 3, and $\mathcal{H}_3$ with representation B has decoding diameter 1. The decoding diameter of a transitive decoding network is the maximum distance in $\mathcal{D}_1$ of any vertex to either a vertex with loop or to a directed cycle.

5   Representations Yielding Transitivity

Due to the effect of the choice of representation on a decoding network's structure, it is natural to ask which representations, if any, ensure that a code's decoding network is transitive under a fixed decoder. Recall from Example 2 that the canonical representation of the Hamming code $\mathcal{H}_3$ does not yield a transitive decoding network under Gallager A, while the decoding network arising from the representation given by the parity-check matrix including all nonzero codewords of the dual code is transitive. In fact,

Theorem 2. *Every binary linear code has a parity-check matrix representation whose corresponding decoding network is transitive under Gallager A decoding.*

In particular, adding exactly the nonzero codewords of the dual to the parity-check matrix of a code (a representation which we will refer to as the *complete representation*) will result in a transitive decoding network under Gallager

A decoding: using symmetries of the complete representation, we may show that after a single iteration of the decoder, the received word either decodes to a codeword and continues decoding to that codeword for any number of iterations, or it will decode to itself under any number of iterations. The full proof of this result will appear in the full version of this paper.

While the complete representation establishes the existence of a representation yielding a transitive network for any code, this level of redundancy is not necessarily required, and in fact may create an excess of trapping sets in the code. In Example 2, adding row seven of representation B to the canonical representation A gives a transitive network, as does adding any three additional rows to the canonical representation. However, any other combination of row additions does not yield a transitive network. Thus, it is interesting to determine the minimum level of redundancy needed to yield a transitive decoding network for a code with a fixed choice of decoder.

In the remainder of this section, we apply the decoding network model to product and half-product codes, which have been subject to renewed interest [16–18], as well as to codes constructed from protographs. By phrasing the decoding networks of these classes of codes in terms of the decoding networks of their component codes, we can identify trapping sets in the larger codes with fewer up-front computations. Proofs have been omitted for space.

Definition 5. Let C_1 and C_2 be binary linear codes of lengths n and m, respectively. Then the *product code* $C_1 \times C_2$ is the set of $m \times n$ binary arrays such that each row forms a codeword in C_1, and each column forms a codeword in C_2.

Consider a product code, $C_1 \times C_2$, with a decoder that operates by iteratively decoding one component code at a time. At each iteration, channel information is dispensed with and decoding is performed based solely on the current estimate.

Theorem 3. *Let C_1 and C_2 be codes of lengths n and m, respectively, with decoding networks $\mathcal{D}^1$ and $\mathcal{D}^2$. Let A_1 be the adjacency matrix of the directed graph product $(\mathcal{D}_1^1)^m$, and let A_2 be the adjacency matrix of $(\mathcal{D}_1^2)^n$. Then, the adjacency matrix of dimension ℓ of the transitive decoding network, $\mathcal{D}$, of the product code $C_1 \times C_2$ is given by $(A_1 A_2)^\ell$.*

Definition 6 ([17]). Let C be a binary linear code of length n, and let $\tilde{C}_H = \{X - X^T : X \in C \times C\}$. The *half-product code* with component code C, denoted C_H, is obtained from $\tilde{C}_H$ by setting the symbols below the diagonal of each element of $\tilde{C}_H$ equal to zero.

A decoder runs by iteratively decoding at each of the n constraints corresponding to "folded" codewords, as in [17,18]. Again, channel information is dispensed with at each subsequent iteration. Let A_i be the adjacency matrix of the digraph on the vertex set of the decoding network of the half-product code, $\mathcal{D}^H$, which gives the behavior of a single decoder iteration run on the ith constraint code. While decoding is performed on the ith constraint, all $(n-1)(n-2)/2$ symbols not participating in constraint i are fixed. Let $\mathcal{D}^i$ be the decoding network associated with the ith constraint code. Then, A_i is the adjacency matrix of a disjoint union of $2^{(n-1)(n-2)/2}$ copies of $\mathcal{D}^i$, corresponding to all the ways non-participating symbols may be fixed. Permute the rows and columns of the A_i's so that they all correspond to a single ordering of the vertices in $\mathcal{D}^H$, and let S_n denote the symmetric group on n elements.

Theorem 4. *The product $(A_{\sigma(1)} \cdots A_{\sigma(n)})^\ell$, where $\sigma \in S_n$, gives the adjacency matrix of $\mathcal{D}_\ell^H$, dimension ℓ of the decoding network of the half-product code $\mathcal{C}_H$, where the component constraints are decoded in the order determined by σ.*

Theorem 5. *Let $\mathcal{C}$ be a binary linear code with Tanner graph G and decoding network $\mathcal{D}$ with respect to a fixed decoder. Viewing G as a protograph, let $\hat{\mathcal{C}}$ be the code corresponding to a degree h lift of G, denoted $\hat{G}$, and (with an abuse of notation), let $\hat{\mathcal{D}}$ be the decoding network of $\hat{\mathcal{C}}$ with respect to the same decoder. Then, there exists a subgraph of $\hat{\mathcal{D}}$ that is isomorphic to $\mathcal{D}$. In particular, if $\mathcal{D}$ is not transitive, then $\hat{\mathcal{D}}$ is not transitive. However, transitivity of $\mathcal{D}$ does not necessarily imply transitivity of $\hat{\mathcal{D}}$.*

6   Conclusions

We presented a multidimensional network framework for the analysis of trapping sets under hard-decision message-passing decoders. We showed that when the decoding network of a code is transitive, trapping sets and their inducing sets can be found by examining the behavior of the decoder in a single iteration. We further showed that all codes have a transitive decoding network representation under Gallager A decoding.

This work leads to many interesting avenues for future pursuit: determining, for certain classes of codes, the minimum number of parity-check matrix rows needed to obtain a transitive decoding network, using the transitive networks of these codes to improve existing results on their trapping and inducing sets, characterizing which lifts of transitive decoding networks preserve transitivity, developing an efficient method of finding trapping sets and predicting the decoding diameter for non-transitive networks, and extending these results to Gallager B and other decoding algorithms. As a first direction, an interesting open problem is to determine a graph theoretic condition (or equivalently, a condition on the parity-check matrix) that guarantees transitivity.

Acknowledgements. We thank the anonymous reviewers for their useful comments that improved the quality of the paper.

References

1. Koetter, R., Vontobel, P.: Graph-covers and iterative decoding of finite length codes. In: Proceedings of the 3rd International Symposium on Turbo Codes (2003)
2. Dolecek, L., Zhang, Z., Anantharam, V., Wainwright, M., Nikolic, B.: Analysis of absorbing sets for array-based LDPC codes. In: IEEE International Conference on Communications, pp. 6261–6268 (2007)

3. Richardson, T.: Error-floors of LDPC codes. In: Proceedings of the 41st Allerton Conference on Communication, Control and Computing, pp. 1426–1435 (2003)
4. Di, C., Proietti, D., Telatar, I.E., Richardson, T.J., Urbanke, R.L.: Finite-length analysis of low-density parity-check codes on the binary erasure channel. IEEE Trans. Inf. Theor. **48**, 1570–1579 (2002)
5. Nguyen, D.V., Chilappagari, S.K., Marcellin, M.W., Vasic, B.: On the construction of structured LDPC codes free of small trapping sets. IEEE Trans. Inf. Theor. **58**, 2280–2302 (2012)
6. Sankaranarayanan, S., Chilappagari, S.K., Radhakrishnan, R., Vasić, B.: Failures of the Gallager B decoder: analysis and applications. In: IEEE International Conference on Communications (2006)
7. Gallager, R.: Low-density parity-check codes. IRE Trans. Inf. Theor. **8**, 21–28 (1962)
8. Sun, J.: Studies on graph-based coding systems. Ph.D. Dissertation, Department of Electrical and Computer Engineering, University of California, Columbus (2004)
9. Butler, B.K., Siegel, P.H.: Error floor approximation for LDPC codes in the AWGN channel. IEEE Trans. Inf. Theor. **60**, 7416–7441 (2014)
10. Vasic, B., Chilappagari, S.K., Nguyen, D.V., Planjery, S.K.: Trapping set ontology. In: Proceedings of the 47th Allerton Conference on Communication, Control, and Computing, pp. 1–7 (2009)
11. Berlingerio, M., Coscia, M., Giannotti, F., Monreale, A., Pedreschi, D.: Foundations of multidimensional network analysis. In: International Conference on Advances in Social Networks Analysis and Mining, pp. 485–489 (2011)
12. Cormen, T.H., Leiserson, C.E., Rivest, R.L., Stein, C.: Introduction to Algorithms, 3rd edn. MIT Press, Cambridge (2009)
13. Beemer, A., Kelley, C.A.: Avoiding trapping sets in SC-LDPC codes under windowed decoding. In: Proceedings IEEE International Symposium on Information Theory and Applications (2016)
14. Karimi, M., Banihashemi, A.H.: Efficient algorithm for finding dominant trapping sets of LDPC codes. IEEE Trans. Inf. Theor. **58**, 6942–6958 (2012)
15. Hashemi, Y., Banihashemi, A.H.: Characterization and efficient exhaustive search algorithm for elementary trapping sets of irregular LDPC codes. In: Proceedings of IEEE International Symposium on Information Theory (2017)
16. Justesen, J.: Performance of product codes and related structures with iterated decoding. IEEE Trans. Commun. **59**, 407–415 (2011)
17. Mittelholzer, T., Parnell, T., Papandreou, N., Pozidis, H.: Improving the error-floor performance of binary half-product codes. In: Proceedings of IEEE International Symposium Information Theory and Applications (2016)
18. Pfister, H.D., Emmadi, S.K., Narayanan, K.: Symmetric product codes. In: Information Theory and Applications Workshop, pp. 282–290 (2015)

Erasure Correction and Locality of Hypergraph Codes

Allison Beemer, Carolyn Mayer[(✉)], and Christine A. Kelley

University of Nebraska-Lincoln, Lincoln, NE, USA
cmayer@huskers.unl.edu

Abstract. We examine erasure correction and locality properties of regular and biregular hypergraph codes. We propose a construction of t-uniform t-partite biregular hypergraphs based on (c,d)-regular bipartite graphs. We show that for both the regular and biregular case, when the underlying hypergraph has expansion properties, the guaranteed erasure correcting capability for the resulting codes is improved. We provide bounds on the minimum stopping set size and cooperative locality of these hypergraph codes for application in distributed storage systems.

Keywords: Hypergraphs · Generalized LDPC codes · Locality · (r, ℓ)-cooperative locality · Availability · Stopping sets · Expander graphs

1 Introduction

The past decade has seen an increased interest in coding for distributed storage systems (DSS) due to the increasing amounts of data that need to be stored and accessed across many servers. A primary focus in this area is the design of codes with locality properties, where error correction of small sets of symbols may be performed efficiently without having to access all symbols or all information from accessed symbols, and where data may be protected by multiple repair groups. Coding techniques optimizing different aspects of the storage and retrieval problem yield different families of codes, such as batch codes, locally repairable codes, private information retrieval codes, and local regeneration codes, [1–3], with connections among them [4,5].

Codes from regular hypergraphs with expansion-like properties were introduced and analyzed in [6,7]. As was shown for expander codes that have underlying expander graph representations, the authors show that better expansion (referred to as ϵ-homogeneity in the hypergraph case) implies improved minimum distance and error correction. Indeed, [8] gives a construction of one of the few explicit algebraic families of asymptotically good codes known to date, and there have since been several papers addressing the design and analysis of codes using expander graphs [9,10]. In this paper, we consider codes based on regular hypergraphs (i.e. all vertices have the same degree) as well as biregular hypergraphs (i.e. vertices have one of two distinct degrees), and present bounds

© Springer International Publishing AG 2017
A.I. Barbero et al. (Eds.): ICMCTA 2017, LNCS 10495, pp. 21–29, 2017.
DOI: 10.1007/978-3-319-66278-7_3

on their error-correction capabilities in the context of distributed storage, specifically the minimum stopping set size and cooperative locality of the codes.

This paper is organized as follows. Section 2 presents relevant background. We derive bounds on the minimum stopping set size and cooperative locality of regular hypergraph codes in Sect. 3. In Sect. 4, we show how to construct a t-uniform t-partite biregular hypergraph from an underlying (c, d)-regular bipartite graph and provide similar bounds. Section 5 concludes the paper.

2   Preliminaries

A *hypergraph* $\mathcal{H}$ is a set of vertices V along with a set of edges E, where E is a set of subsets of V. A hypergraph is *t-uniform* if every edge contains exactly t vertices, and it is *t-partite* if the vertex set V can be partitioned into t vertex sets $V_1, \ldots, V_t$ such that no edge contains more than one vertex from any part. In this paper we will use $\mathcal{H} = (V_1, V_2, \ldots, V_t; E)$ to denote a t-uniform t-partite hypergraph in which each edge contains exactly one vertex from each part. A hypergraph is Δ-*regular* if every vertex belongs to Δ edges. We assume that all hypergraphs considered in this paper have no parallel edges. That is, no two distinct edges are comprised of exactly the same set of vertices.

Such t-uniform t-partite Δ-regular hypergraphs were used in [6,7] to design codes where the edges of the hypergraph represent the code symbols, and the vertices of the hypergraph represent the constraints. Specifically, when there are n vertices in each part, the block length of the corresponding *hypergraph code* is $n\Delta$ and the number of constraint vertices is nt. As in [11], each constraint node represents a linear block length Δ code (commonly referred to as a "subcode"), and is satisfied when the assignment on the edges incident to that constraint node form a codeword of the subcode. This is similar to generalized low density parity-check (LDPC) codes as defined in [11], in which the constraint nodes may be more sophisticated than simple parity-check codes.

Sipser and Spielman [8] show that the guaranteed error correction capabilites of a code may be improved when the underlying graph is a good expander. Let G be a (c, d)-regular bipartite graph with m left vertices of degree c and n right vertices of degree d, and let μ be the second largest eigenvalue (in absolute value) of the adjacency matrix of G. Then for subsets S and T of the left and right vertices, respectively, the number of edges in the subgraph induced by $S \cup T$ is at most $|E(S, T)| \leq \frac{d}{m}|S||T| + \frac{\mu}{2}(|S| + |T|)$ [12]. Loosely speaking, a graph is a good expander if small sets of vertices have large sets of neighbors; graphs with small second largest eigenvalue have this property. Bilu and Hoory introduced an analogous notion for hypergraph expansion as follows:

Definition 1. *[6] Let* $\mathcal{H} = (V_1, V_2, \ldots, V_t; E)$ *be a t-uniform t-partite Δ-regular hypergraph with n vertices in each part. Then $\mathcal{H}$ is ϵ-homogeneous if for every choice of $A_1, A_2, \ldots, A_t$ with $A_i \subseteq V_i$ and $|A_i| = \alpha_i n$,*

$$\frac{|E(A_1, A_2, \ldots, A_t)|}{n\Delta} \leq \prod_{i=1}^{t} \alpha_i + \epsilon\sqrt{\alpha_{\sigma(1)}\alpha_{\sigma(2)}},$$

where $\sigma \in S_t$ is a permutation on $[t]$ such that $\alpha_{\sigma(i)} \leq \alpha_{\sigma(i+1)}$ for each $i \in [t-1]$, and $E(A_1, \ldots, A_t)$ denotes the set of edges which intersect all of the A_i's.

Let $[N, K, D]$ denote a binary linear code with block length N, dimension K, and minimum distance D. The following bounds on the rate and minimum distance of a code $\mathcal{Z}$ from an ϵ-homogeneous t-uniform t-partite Δ-regular hypergraph with n vertices in each part and a $[\Delta, \Delta R, \Delta \delta]$ subcode C at each constraint node are given in [6]:

$$\text{rate}(\mathcal{Z}) \geq tR - (t-1) \tag{1}$$

$$d_{\min}(\mathcal{Z}) \geq n\Delta \left(\delta^{\frac{t}{t-1}} - c(\epsilon, \delta, t) \right) \tag{2}$$

where $c(\epsilon, \delta, t) \to 0$ as $\epsilon \to 0$. Note that $n\Delta$ is the blocklength of $\mathcal{Z}$.

The *locality* of a code measures how many code symbols must be used to recover an erased code symbol. While there are a variety of locality notions relevant to coding for distributed storage, in this paper we focus on (r, ℓ)-cooperative locality and (r, τ)-availability [13, 14].

Definition 2. *A code C has (r, ℓ)-cooperative locality if for any $\mathbf{y} \in C$, any set of ℓ symbols in $\mathbf{y}$ are functions of at most r other symbols. Furthermore, C has (r, τ)-availability if any symbol in $\mathbf{y}$ can be recovered by using any of τ disjoint sets of symbols each of size at most r.*

3   Bounds on Regular Hypergraph Codes

In this section, we examine the erasure correction and cooperative locality of regular hypergraph codes. First, a *stopping set* in the Tanner graph representing a generalized LDPC code is a subset S of the variable nodes such that each neighbor of S has at least $d_{\min}(C)$ neighbors in S, where C is the subcode represented by the constraint vertices [15, 16]. It was initially shown in [17] that stopping sets characterize iterative decoder failure for simple LDPC codes (i.e. when the subcodes are simple parity check codes) on the binary erasure channel. We first define stopping sets for regular hypergraph codes and give a lower bound on the minimum stopping set size.

Definition 3. *Let $\mathcal{Z}$ be a code on a hypergraph $\mathcal{H} = (V_1, \ldots, V_t; E)$, with edges representing code symbols and vertices representing the constraints of a subcode C. Then a stopping set S is a subset of the edges of $\mathcal{H}$ such that every vertex contained in an element of S is contained in at least $d_{\min}(C)$ elements of S.*

Though the size of a minimum stopping set depends on both the hypergraph representation and the choice of subcode, we denote this size by $s_{\min}(\mathcal{H})$, and assume that the subcode is clear from context.

Theorem 1. *Let $\mathcal{H}$ be a t-uniform t-partite Δ-regular hypergraph. If the vertices of $\mathcal{H}$ represent constraints of a subcode C with minimum distance $d_{\min}(C)$ and block length Δ, then the size of the minimum stopping set, $s_{\min}(\mathcal{H})$, is bounded by*

$$s_{\min}(\mathcal{H}) \geq d_{\min}(C)^{t/(t-1)}. \tag{3}$$

Proof. Let $\mathcal{H}$ be as above, and let S be a minimum stopping set. Each edge in S contains exactly one constraint node from each of the t parts of $\mathcal{H}$, so each part of $\mathcal{H}$ has exactly $|S| = s_{\min}(\mathcal{H})$ incident edges belonging to S. Each constraint node contained in an edge in S must be contained in at least $d_{\min}(C)$ edges in S. By the pigeonhole principle, the number of vertices in any part of $\mathcal{H}$ that are contained in some edge in S is bounded above by $s_{\min}(\mathcal{H})/d_{\min}(C)$. Indeed, were there more than $s_{\min}(\mathcal{H})/d_{\min}(C)$ vertices incident to S in a single part, some vertex must have fewer than $s_{\min}(\mathcal{H})/(s_{\min}(\mathcal{H})/d_{\min}(C)) = d_{\min}(C)$ incident edges from S, a contradiction. Now consider the maximum size of S: this amounts to counting the number of edges possible, given that each edge is incident to exactly one vertex of (at most) $s_{\min}(\mathcal{H})/d_{\min}(C)$ vertices in each of the t parts of $\mathcal{H}$. That is, there are at most $(s_{\min}(\mathcal{H})/d_{\min}(C))^t$ edges in S. Thus,

$$\left(\frac{s_{\min}(\mathcal{H})}{d_{\min}(C)}\right)^t \geq s_{\min}(\mathcal{H}) \Rightarrow s_{\min}(\mathcal{H}) \geq d_{\min}(C)^{t/(t-1)}.$$

The bound of Theorem 1 is tight. For example, when $\mathcal{H}$ is a complete 3-uniform 3-partite hypergraph with at least two vertices in each part and constraint code C such that $d_{\min}(C) = 4$, it is easy to show that $s_{\min}(\mathcal{H}) = 8$.

Since the errors of particular relevance to DSS are erasures (such as a server going down), we can use the stopping set bound to characterize how many errors can be corrected. Theorem 1 guarantees that we may correct any $d_{\min}(C)^{t/(t-1)} - 1$ erasures using iterative decoding. If C is a code with locality r_1, at most $(s_{\min}(\mathcal{H})/d_{\min}(C)) \cdot r_1 \cdot t$ other codeword symbols are involved in the repair of the erasures in the decoding process. This yields the following:

Corollary 1. *If the subcodes C of the regular hypergraph code $\mathcal{Z}$ have r_1 locality, then $\mathcal{Z}$ has (r, ℓ)-cooperative locality where*

$$r = r_1 t s_{\min}(\mathcal{H})/d_{\min}(C) \tag{4}$$

$$s_{\min}(\mathcal{H}) - 1 \geq \ell \geq d_{\min}(C)^{t/(t-1)} - 1. \tag{5}$$

Observe that if the subcode C has (r, τ)-availability, then the hypergraph code $\mathcal{Z}$ has at least (r, τ)-availability.

We now extend the result to codes on hypergraphs with known ϵ-homogeneity.

Theorem 2. *Let* $\mathcal{H} = (V_1, V_2, \ldots, V_t; E)$ *be a t-uniform t-partite Δ-regular ϵ-homogeneous hypergraph where there are n vertices in each of the t parts. If the subcodes C have minimum distance $d_{\min}(C)$,*

$$s_{\min}(\mathcal{H}) \geq \left(\left(1 - \frac{\epsilon \Delta}{d_{\min}(C)} \right) \frac{n^{t-1} d_{\min}(C)^t}{\Delta} \right)^{1/(t-1)}. \tag{6}$$

For $\epsilon < \frac{d_{\min}(C)(n^{t-1} - \Delta)}{\Delta n^{t-1}}$, this gives an improvement on the bound in Theorem 1.

Proof. Let S be a minimum stopping set. By Theorem 1, $s_{\min}(\mathcal{H}) \geq d_{\min}(C)^{t/(t-1)}$. Now, let $A_i \subseteq V_i$ be the set of vertices in V_i, for $i \in [t]$, contained in an edge in S. By ϵ-homogeneity,

$$s_{\min}(\mathcal{H}) = |S| \leq |E(A_1, \ldots, A_t)| \leq n\Delta \left(\prod_{i=1}^{t} \alpha_i + \epsilon \sqrt{\alpha_{\sigma(1)} \alpha_{\sigma(2)}} \right).$$

Since $|A_i| \leq s_{\min}(\mathcal{H})/d_{\min}(C)$ for all i, $\alpha_i \leq s_{\min}(\mathcal{H})/n d_{\min}(C)$. Thus, the above inequality simplifies to obtain the result:

$$s_{\min}(\mathcal{H}) \leq n\Delta \left(\left(\frac{s_{\min}(\mathcal{H})}{n d_{\min}(C)} \right)^t + \epsilon \frac{s_{\min}(\mathcal{H})}{n d_{\min}(C)} \right).$$

Observe that we have shown in general that

$$s_{\min}(\mathcal{H}) \geq \left(\left(1 - \frac{\epsilon \Delta}{d_{\min}(C)} \right) \frac{n^{t-1}}{\Delta} \right)^{1/(t-1)} d_{\min}(C)^{t/(t-1)}.$$

Then, if $\left(\left(1 - \frac{\epsilon \Delta}{d_{\min}(C)} \right) \frac{n^{t-1}}{\Delta} \right)^{1/(t-1)} > 1$, this gives a better lower bound for $s_{\min}(\mathcal{H})$ than that found in Theorem 1. Simplifying, we have our condition on ϵ.

Corollary 2. *Using iterative decoding on a code $\mathcal{Z}$ based on a t-uniform t-partite Δ-regular ϵ-homogeneous hypergraph with vertices representing constraints of a subcode C, up to*

$$\left(\left(1 - \frac{\epsilon \Delta}{d_{\min}(C)} \right) \frac{n^{t-1} d_{\min}(C)^t}{\Delta} \right)^{1/(t-1)} - 1$$

erasures may be corrected.

In other words, if δ is the relative minimum distance of C, and N is the total number of edges in the hypergraph (that is, the block length of the code $\mathcal{Z}$), we may correct up to a $\delta(\delta - \epsilon)^{1/(t-1)} - \frac{1}{N}$ fraction of erasures.

Remark 1. We may correct up to a $\delta^{t/(t-1)} - \frac{1}{N} - c(\epsilon, \delta, t)$ fraction of erasures, where $c(\epsilon, \delta, t) \to 0$ as $\epsilon \to 0$. It can be shown that the bound in Corollary 2 improves the error correction capability of

$$\binom{t-1}{t/2}^{-2/t} \left(\frac{\delta}{2}\right)^{(t+2)/t} - c'(\epsilon, \delta, t)$$

in [6] for any $0 < \delta < 1$ and $t \geq 2$ (i.e. for all relevant cases) and large block length. Note that $c'(\epsilon, \delta, t) \neq c(\epsilon, \delta, t)$, but that both vanish as $\epsilon \to 0$.

It is important to note that we are focusing solely on erasures, while [6] gives a decoding algorithm and correction capabilities for more general errors.

4   Bounds on Biregular Hypergraph Codes

A construction of t-uniform t-partite Δ-regular hypergraphs is presented in [6] based on an underlying regular bipartite expander graph. In this section we show how to obtain a t-uniform t-partite (Δ_1, Δ_2)-biregular hypergraph from a (c, d)-regular bipartite graph in a similar way. We provide bounds on the stopping set size, cooperative locality, rate, and minimum distance for the resulting hypergraph codes.

Definition 4. *We say that a t-uniform t-partite hypergraph $\mathcal{H} = (V_1, \ldots, V_t; E)$ is (Δ_1, Δ_2)-biregular if the parts can be labeled such that each vertex in an odd (resp., even) index part is contained in Δ_1 (resp., Δ_2) edges.*

Construction 1. *Let $G = V \cup W$ be a (c, d)-regular bipartite expander graph with $|V| \geq |W|$. For $t \in \mathbb{N}$, construct a t-uniform t-partite hypergraph $\mathcal{H}$ with parts $V_1, \ldots, V_t$ as follows. For odd (resp., even) i, let V_i be a copy of V (resp., W). Take $E(\mathcal{H})$ to be the set of edges corresponding to walks of length $t-1$ in G. That is $(v_1, \ldots, v_t)$ with $v_i \in V_i$ is in $E(\mathcal{H})$ if and only if $(v_1, \ldots, v_t)$ corresponds to a walk in G.*

Note that $\mathcal{H}$ is indeed t-uniform and t-partite, and has vertices of degree $\Delta_1 = c^{\lceil \frac{t}{2} \rceil} d^{\lfloor \frac{t}{2} \rfloor - 1}$ (resp., $\Delta_2 = c^{\lceil \frac{t}{2} \rceil - 1} d^{\lfloor \frac{t}{2} \rfloor}$) in odd (resp., even) index parts.

The definition of a stopping set may be extended to biregular hypergraph codes.

Definition 5. *Let $\mathcal{Z}$ be a code on a hypergraph $\mathcal{H} = (V_1, V_2, \ldots, V_t; E)$, with the edges representing the code symbols and the vertices representing the constraints of a subcode C_1 (resp., C_2) if the vertex is in an odd (resp., even) index part. Then a stopping set S is a subset of the edges of $\mathcal{H}$ such that every vertex contained in an element of S is contained in at least $d_{\min}(C_1)$ (resp., $d_{\min}(C_2)$) elements of S if the vertex is in an odd (resp., even) index part.*

We now give a bound on the minimum stopping set size and (r, ℓ)-cooperative locality of codes resulting from Construction 1. The proofs are similar to those in Sect. 3 and are thus omitted.

Theorem 3. *Let $\mathcal{H}$ be a t-uniform t-partite (Δ_1, Δ_2)-biregular hypergraph. If the vertices in an odd (resp., even) index part of $\mathcal{H}$ represent constraints of a subcode C_1 (resp., C_2) with block length Δ_1 (resp., Δ_2), then the size of the minimum stopping set, $s_{\min}(\mathcal{H})$, is bounded by*

$$s_{\min}(\mathcal{H}) \geq \left(d_{\min}(C_1)^{\lceil \frac{t}{2} \rceil} d_{\min}(C_2)^{\lfloor \frac{t}{2} \rfloor} \right)^{1/(t-1)}. \tag{7}$$

Corollary 3. *If the subcodes C_1 (resp., C_2) of the biregular hypergraph code $\mathcal{Z}$ have r_1 (resp., r_2) locality then $\mathcal{Z}$ has (r, ℓ)-cooperative locality where*

$$r = r_1 \left\lceil \frac{t}{2} \right\rceil \frac{s_{\min}(\mathcal{H})}{d_{\min}(C_1)} + r_2 \left\lfloor \frac{t}{2} \right\rfloor \frac{s_{\min}(\mathcal{H})}{d_{\min}(C_2)} \tag{8}$$

$$s_{\min}(\mathcal{H}) - 1 \geq \ell \geq \left(d_{\min}(C_1)^{\lceil \frac{t}{2} \rceil} d_{\min}(C_2)^{\lfloor \frac{t}{2} \rfloor} \right)^{1/(t-1)} - 1. \tag{9}$$

Observe that $\mathcal{Z}$ has at least the (r, τ)-availability of its subcodes.

We next extend the definition of ϵ-homogeneity to biregular hypergraphs and give an improved minimum stopping set bound for the corresponding codes.

Definition 6. *Let $\mathcal{H} = (V_1, V_2, \ldots, V_t; E)$ be a t-uniform t-partite (Δ_1, Δ_2)-biregular hypergraph with n_1 vertices in each odd index part and n_2 vertices in each even index part. We say that $\mathcal{H}$ is ϵ-homogeneous if for every choice of $A_1, A_2, \ldots, A_t$, with $A_i \subseteq V_i$,*

$$\frac{|E(A_1, A_2, \ldots A_t)|}{\Delta_1 n_1} \leq \prod_{i=1}^{t} \alpha_i + \epsilon \sqrt{\alpha_{\sigma(1)} \alpha_{\sigma(2)}},$$

where σ is a permutation on $[t]$ such that $\alpha_{\sigma(i)} \leq \alpha_{\sigma(i+1)}$ for each $i \in [t-1]$, and $|A_i| = \alpha_i n_1$ if i is odd and $|A_i| = \alpha_i n_2$ if i is even.

Theorem 4. *Let $\mathcal{H} = (V_1, \ldots, V_t; E)$ be a t-uniform t-partite (Δ_1, Δ_2)-regular ϵ-homogeneous hypergraph where there are n_1 (resp., n_2) vertices in each of the odd (resp., even) index parts. Let C_1 and C_2 be the subcodes of the odd and even index parts, respectively. Then $s_{\min}(\mathcal{H})$ is bounded below by*

$$\left(\frac{(n_1 d_{\min}(C_1))^{\lceil \frac{t}{2} \rceil} (n_2 d_{\min}(C_2))^{\lfloor \frac{t}{2} \rfloor}}{n_1 \Delta_1} \left(1 - \frac{\epsilon n_1 \Delta_1}{\min_{i=1,2}\{n_i d_{\min}(C_i)\}} \right) \right)^{\frac{1}{t-1}}. \tag{10}$$

For $\epsilon < \left(1 - \frac{n_1 \Delta_1}{n_1^{\lceil \frac{t}{2} \rceil} n_2^{\lfloor \frac{t}{2} \rfloor}} \right) \frac{\min_{i=1,2}\{n_i d_{\min}(C_i)\}}{n_1 \Delta_1}$, this improves the Theorem 3 bound.

We now give bounds on the rate and minimum distance of a length $n_1 \Delta_1$ code $\mathcal{Z}$ from an ϵ-homogeneous t-uniform t-partite (Δ_1, Δ_2)-regular hypergraph with

n_1 (resp., n_2) vertices in each odd (resp., even) index part and $[\Delta_1, \Delta_1 R_1, \Delta_1 \delta_1]$ subcodes C_1 (resp., $[\Delta_2, \Delta_2 R_2, \Delta_2 \delta_2]$ subcodes C_2).

$$\text{rate}(\mathcal{Z}) \geq R_1 \lceil \frac{t}{2} \rceil + R_2 \lfloor \frac{t}{2} \rfloor - (t-1) \tag{11}$$

$$d_{\min}(\mathcal{Z}) \geq n_1 \Delta_1 \left((\delta_1^{\lceil \frac{t}{2} \rceil} \delta_2^{\lfloor \frac{t}{2} \rfloor})^{\frac{1}{t-1}} - c(\epsilon, \delta_1, \delta_2, t) \right) \tag{12}$$

where $c(\epsilon, \delta_1, \delta_2, t) \to 0$ as $\epsilon \to 0$. The proofs are similar to those for the regular hypergraph bounds given in [6] and are thus omitted.

Ramanujan graphs have the largest possible gap between their first and second largest eigenvalues (in absolute value), and are thus the best in terms of expansion. In [18], the authors give an explicit algebraic construction of regular Ramanujan graphs. The existence of infinite families of (c, d)-regular bipartite Ramanujan graphs was shown in [19]. Moreover, the regular t-uniform t-partite hypergraphs constructed in [6] from regular expander graphs with second largest eigenvalue λ were shown to be $2(t-1)\lambda$-homogeneous. We conjecture that when Construction 1 starts with a (c, d)-regular bipartite expander graph, the resulting hypergraph will be ϵ-homogeneous, where ϵ depends on the second largest eigenvalue of the underlying expander graph.

5   Conclusions

We examined the erasure correcting capability of regular hypergraph codes as well as a new code construction based on biregular hypergraphs. We provided lower bounds on the minimum stopping set size, cooperative locality, and availability of these codes, and gave improved parameters for ϵ-homogeneous hypergraph codes. Designing explicit families of hypergraphs that are optimal with respect to their corresponding codes' locality properties is a line for future work. An interesting open problem is whether the proposed construction of biregular hypergraphs can be proven to yield biregular ϵ-homogeneous hypergraphs when starting from biregular expander graphs.

Acknowledgements. We thank the anonymous reviewers for their useful comments that improved the quality of the paper.

References

1. Ishai, Y., Kushilevitz, E., Ostrovsky, R., Sahai, A.: Batch codes and their applications. In: ACM Symposium on Theory of Computation, Chicago, IL, pp. 262–271, June 2004
2. Dimakis, A.G., Ramchandran, K., Wu, Y., Suh, C.: A survey on network codes for distributed storage. In: Proceedings of IEEE, vol. 99, no. 3, March 2011
3. Kamath, G.M., Prakash, N., Lalitha, V., Kumar, P.V.: Codes with local regeneration. In: IEEE Information Theory and Application Workshop, pp. 1–5, San Diego, CA, February 2013

4. Ernvall, T., Westerbäck, T., Freij-Hollanti, R., Hollanti, C.: A connection between locally repairable codes and exact regenerating codes. In: Proceedings of IEEE International Symposium on Information Theory, Barcelona, July 2016
5. Skachek, V.: Batch and PIR Codes and their Connections to Locally Repairable Codes. arXiv:1611.09914v2, March 2017
6. Bilu, Y., Hoory, S.: On codes from hypergraphs. Eur. J. Comb. **25**, 339–354 (2004)
7. Barg, A., Zémor, G.: Codes on hypergraphs. In: IEEE International Symposium on Information Theory, Toronto, ON, pp. 156–160, July 2008
8. Sipser, M., Spielman, D.A.: Expander codes. IEEE Trans. Inf. Theory **42**, 1710–1722 (1996)
9. Janwa, H.L., Lal, A.K.: On expander graphs: parameters and applications, preprint. arXiv:0406048
10. Zémor, G.: On expander codes. IEEE Trans. Inf. Theory **47**(2), 835–837 (2001)
11. Tanner, R.M.: A recursive approach to low complexity codes. IEEE Trans. Inf. Theory **27**, 533–547 (1981)
12. Janwa, H., Lal, A.K.: On Tanner codes: minimum distance and decoding. Proc. AAECC **13**, 335–347 (2003)
13. Rawat, A.S., Mazumdar, A., Vishwanath, S.: Cooperative local repair in distributed storage. arXiv:1409.3900
14. Rawat, A.S., Papailiopoulos, D.S., Dimakis, A.G., Vishwanath, S.: Locality and availability in distributed storage. IEEE Trans. Inf. Theory **62**(8), 4481–4493 (2016)
15. Kelley, C.A., Sridhara, D.: Eigenvalue bounds on the pseudocodeword weight of expander codes. Adv. Math. Commun **1**, 287–307 (2007)
16. Miladinovic, N., Fossorier, M.P.C.: Generalized LDPC codes and generalized stopping sets. IEEE Trans. Commun. **56**, 201–212 (2008)
17. Di, C., Proietti, D., Richardson, T., Teletar, E., Urbanke, R.: Finite-length analysis of low-density parity-check codes on the binary erasure channel. IEEE Trans. Inf. Theory **48**, 1570–1579 (2002)
18. Lubotzky, A., Phillips, R., Sarnak, P.: Ramanujan graphs. Combinatorica **8**, 261–277 (1988)
19. Marcus, A.W., Spielman, D.A., Srivastava, N.: Interlacing families I: bipartite Ramanujan graphs of all degrees. Ann. Math. **182**, 307–325 (2015)

Reed-Muller Codes: Information Sets from Defining Sets

José Joaquín Bernal[⊠] and Juan Jacobo Simón

Departamento de Matemáticas, Universidad de Murcia, 30100 Murcia, Spain
{josejoaquin.bernal,jsimon}@um.es

Abstract. As affine-invariant codes, Reed-Muller codes are extension of cyclic group codes and they have a defining set that determines them uniquely. In this paper we identify those cyclic codes with multidimensional abelian codes and we use the techniques introduced in [3] to construct information sets for first and second order Reed-Muller codes from its defining set.

1 Introduction

We are interested in finding information sets for Reed-Muller codes, a question that has been addressed for many authors earlier. From the geometric point of view several ideas for finding information sets has been presented. In [5,10] Moorhouse and Blokhuis gave bases formed by the incidence vectors of certain lines valid for a more general family of geometric codes. Later, J. D. Key, T. P. McDonough and V. C. Mavron extended these definitions in [8] in order to apply the permutation decoding algorithm. Finally, in [9], the same authors gave a description of information sets for Reed-Muller codes by using the polynomial approach.

In this work we use the fact that Reed-Muller codes can be seen as affine-invariant codes to get information sets. From this context any Reed-Muller code is a parity check extension of a cyclic group code, so any information set of that cyclic code is obviously an information set for the Reed-Muller code; moreover, there exists a direct connection between the respective defining sets. On the other hand, in [3] we introduced a method for constructing information sets for any abelian code starting from its defining set. Then, the goal of this paper is to obtain information sets for Reed-Muller codes of first and second order respectively by applying those techniques shown in [3] to the punctured cyclic group code seen as a multidimensional abelian code.

The paper is structured as follows. Section 2 include the basic notation and the necessary preliminaries about abelian codes. Then, Sect. 3 contains the explanation of how the method given in [3] works in the particular case of two-dimensional abelian codes. Section 4 shows some essential results related to the

This work was partially supported by MINECO, project MTM2016-77445-P, and Fundación Séneca of Murcia, project 19880/GERM/15.

© Springer International Publishing AG 2017
A.I. Barbero et al. (Eds.): ICMCTA 2017, LNCS 10495, pp. 30–47, 2017.
DOI: 10.1007/978-3-319-66278-7_4

case of cyclic codes seen as two-dimensional cyclic codes. In Sect. 5 we recall the definition of Reed-Muller codes as affine-invariant codes in an abelian group algebra and then we show how we apply the results in the previous sections to them. The core of this paper are Sects. 6 and 7 where we present the main results, to wit, explicit descriptions of the information sets for first and second order Reed-Muller codes respectively.

2 Preliminaries

In this paper we deal with Reed-Muller codes identified as abelian codes, so for the convenience of the reader we give an introduction to abelian codes just in the binary case. Then all throughout $\mathbb{F}$ denotes the field with two elements.

A binary abelian code is an ideal of a group algebra $\mathbb{F}G$, where G is an abelian group. It is well-known that there exist integers $r_1, \ldots, r_l$ such that G is isomorphic to the direct product $C_{r_1} \times \cdots \times C_{r_l}$, with C_{r_i} the cyclic group of order r_i, $i = 1, \ldots, l$. Moreover, this decomposition yields an isomorphism of $\mathbb{F}$-algebras from $\mathbb{F}G$ to

$$\mathbb{F}[X_1, \ldots, X_l] / \langle X_1^{r_1} - 1, \ldots, X_l^{r_l} - 1 \rangle.$$

We denote this quotient algebra by $\mathbb{A}(r_1, \ldots, r_l)$ and we identify the codewords with polynomials $P(X_1, \ldots, X_l)$ such that every monomial satisfy that the degree of the indeterminate X_i is in $\mathbb{Z}_{r_i}$, the ring of integers modulo r_i, that we always write as canonical representatives. We write the elements $P \in \mathbb{A}(r_1, \ldots, r_l)$ as $P = P(X_1, \ldots, X_l) = \sum a_{\mathbf{j}} X^{\mathbf{j}}$, where $\mathbf{j} = (j_1, \ldots, j_l) \in \mathbb{Z}_{r_1} \times \cdots \times \mathbb{Z}_{r_l}$, $X^{\mathbf{j}} = X_1^{j_1} \cdots X_l^{j_l}$ and $a_{\mathbf{j}} \in \mathbb{F}$. We always assume that r_i is odd for every $i = 1, \ldots, l$, that is, we assume that $\mathbb{A}(r_1, \ldots, r_l)$ is a semisimple algebra.

Our main tool to study the construction of information sets for abelian codes is the notion of defining set.

Definition 1. *Let $C \subseteq \mathbb{A}(r_1, \ldots, r_l)$ be an abelian code. Let R_i be the set of r_i-th roots of unity, $i = 1, \ldots, l$. Then the root set of C is given by*

$$\mathcal{Z}(C) = \left\{ (\beta_1, \ldots, \beta_l) \in \prod_{i=1}^{l} R_i \mid P(\beta_1, \ldots, \beta_l) = 0 \text{ for all } P(X_1, \ldots, X_l) \in C \right\}.$$

Then, fixed a primitive r_i-th root of unity α_i in some extension of $\mathbb{F}$, $i = 1, \ldots, l$, the defining set of C with respect to $\alpha = \{\alpha_1, \ldots, \alpha_l\}$ is

$$D_\alpha(C) = \{(a_1, \ldots, a_l) \in \mathbb{Z}_{r_1} \times \cdots \times \mathbb{Z}_{r_l} \mid (\alpha_1^{a_1}, \ldots, \alpha_l^{a_l}) \in \mathcal{Z}(C)\}.$$

It can be proved that, fixed a collection of primitive roots of unity, every abelian code is totally determined by its defining set.

In order to describe the structure of the defining set of an abelian code we need to introduce the following definitions.

Definition 2. *Let a, r and γ be integers. The 2^γ-cyclotomic coset of a modulo r is the set*

$$C_{2^\gamma, r}(a) = \left\{ a \cdot 2^{\gamma \cdot i} \mid i \in \mathbb{N} \right\} \subseteq \mathbb{Z}_r.$$

We shall write $C_r(a)$ when $\gamma = 1$.

Definition 3. *Given $(a_1, \ldots, a_l) \in \mathbb{Z}_{r_1} \times \cdots \times \mathbb{Z}_{r_l}$, its 2-orbit modulo $(r_1, \ldots, r_l)$ is the set*

$$Q(a_1, \ldots, a_l) = \left\{ (a_1 \cdot 2^i, \ldots, a_l \cdot 2^i) \mid i \in \mathbb{N} \right\} \subseteq \mathbb{Z}_{r_1} \times \cdots \times \mathbb{Z}_{r_l}.$$

It is well known that for every abelian code $\mathcal{C} \subseteq \mathbb{A}(r_1, \ldots, r_l)$, $D(\mathcal{C})$ is closed under multiplication by 2 in $\mathbb{Z}_{r_1} \times \cdots \times \mathbb{Z}_{r_l}$, and so $D(\mathcal{C})$ is a disjoint union of 2-orbits modulo $(r_1, \ldots, r_l)$. Conversely, every union of 2-orbits modulo $(r_1, \ldots, r_l)$ defines an abelian code in $\mathbb{A}(r_1, \ldots, r_l)$. From now on, we will only write 2-orbit, and the tuple of integers will always be clear by the context.

Now, we need to give the notion of information set of a code in the context of abelian codes. Let $P = \sum a_\mathbf{j} X^\mathbf{j} \in \mathbb{A}(r_1, \ldots, r_l)$ and let $\mathcal{I}$ be a subset of $\mathbb{Z}_{r_1} \times \cdots \times \mathbb{Z}_{r_l}$, we denote by $P_\mathcal{I}$ the vector $(a_\mathbf{j})_{\mathbf{j} \in \mathcal{I}} \in \mathbb{F}^{|\mathcal{I}|}$. Now, for an abelian code $\mathcal{C} \subseteq \mathbb{A}(r_1, \ldots, r_l)$ we denote by $\mathcal{C}_\mathcal{I}$ the linear code $\{P_\mathcal{I} : P \in \mathcal{C}\} \subseteq \mathbb{F}^{|\mathcal{I}|}$.

Definition 4. *An information set for an abelian code $\mathcal{C} \subseteq \mathbb{A}(r_1, \ldots, r_l)$ with dimension k is a set $\mathcal{I} \subseteq \mathbb{Z}_{r_1} \times \cdots \times \mathbb{Z}_{r_l}$ such that $|\mathcal{I}| = k$ and $\mathcal{C}_\mathcal{I} = \mathbb{F}^k$.*

The complementary set $(\mathbb{Z}_{r_1} \times \cdots \times \mathbb{Z}_{r_l}) \setminus \mathcal{I}$ is called a set of check positions for $\mathcal{C}$.

To finish this section, let us recall that, as usual, we denote by $\mathcal{C}^\perp$ the dual code of $\mathcal{C}$, that is, the set of codewords $v \in \mathbb{A}(r_1, \ldots, r_l)$ such that $v \cdot u = 0$, for all $u \in \mathcal{C}$, where "$\cdot$" denotes the usual inner product. It is easy to see that any information set for $\mathcal{C}$ is a set of check positions for $\mathcal{C}^\perp$ and vice versa.

3   Information Sets for Abelian Codes

In [3] we introduced a method for constructing information sets for any multidimensional abelian code just in terms of its defining set. Although one may apply this construction for any value of l, for the shake of simplicity, in this section we only recall the two-dimensional construction. So, from now on we take $l = 2$ and the ambient space will be $\mathbb{A}(r_1, r_2)$.

Let $e = (e_1, e_2) \in \mathbb{Z}_{r_1} \times \mathbb{Z}_{r_2}$. We define

$$m(e_1) = |C_{r_1}(e_1)|$$

and

$$m(e) = m(e_1, e_2) = \left| C_{2^{m(e_1)}, r_2}(e_2) \right|. \tag{1}$$

The construction of information sets is based on the computation of the parameters defined in (1) on a special subset of the defining set of the given abelian code. Specifically this set has to satisfy the conditions described in the following definition. For any subset $A \subset \mathbb{Z}_{r_1} \times \mathbb{Z}_{r_2}$ we denote its projection onto the first coordinate by A_1.

Definition 5. *Let D be a union of 2-orbits modulo (r_1, r_2) and $\overline{D} \subset D$ a complete set of representatives. Then $\overline{D}$ is called a set of restricted representatives if $\overline{D}_1$ is a complete set of representatives of the 2-cyclotomic cosets modulo r_1 in D_1.*

Example 1. Consider $r_1 = 3, r_2 = 5$ and let $\mathcal{C} \subseteq \mathbb{A}(3, 5)$ be the abelian code with defining set $\mathcal{D}(\mathcal{C}) = Q(1, 1) \cup Q(0, 1) \cup Q(2, 0)$, where $Q(1, 1) = \{(1, 1), (2, 2), (1, 4), (2, 3)\}$, $Q(0, 1) = \{(0, 1), (0, 2), (0, 3), (0, 4)\}$ and $Q(2, 0) = \{(1, 0), (2, 0)\}$. Then, according to Definition 5, the set of representatives $\{(1, 1), (0, 1), (2, 0)\}$ is not restricted, because $C_3(1) = C_3(2)$, while $\{(1, 1), (0, 1), (1, 0)\}$ is indeed restricted.

Now, let $\mathcal{C} \subseteq \mathbb{A}(r_1, r_2)$ be an abelian code with defining set $D_\alpha \subseteq \mathbb{Z}_{r_1} \times \mathbb{Z}_{r_2}$, with respect to $\alpha = \{\alpha_1, \alpha_2\}$. Take $\overline{D} \subset D_\alpha$ a set of restricted representatives. Given $e_1 \in \overline{D}_1$, let

$$R(e_1) = \{e_2 \in \mathbb{Z}_{r_2} \mid (e_1, e_2) \in \overline{D}\}.$$

For each $e_1 \in \overline{D}_1$, we define

$$M(e_1) = \sum_{e_2 \in R(e_1)} m(e_1, e_2) \tag{2}$$

and we consider the values $\{M(e_1)\}_{e_1 \in \overline{D}_1}$. Then we denote

$$f_1 = \max_{e_1 \in \overline{D}_1} \{M(e_1)\} \qquad \text{and}$$

$$f_i = \max_{e_1 \in \overline{D}_1} \{M(e_1) \mid M(e_1) < f_{i-1}\}.$$

So, we obtain the sequence

$$f_1 > \cdots > f_s > 0 = f_{s+1}, \tag{3}$$

that is, we denote by f_s the minimum value of the parameters $M(\cdot)$ and we set $f_{s+1} = 0$ by convention. Note that $M(e_1) > 0$, for all $e_1 \in \overline{D}_1$, by definition.

From the previous values $f_\bullet$ we define for $i = 1, \ldots, s$

$$g_i = \sum_{M(e_1) \geq f_i} m(e_1) \tag{4}$$

and then we obtain the sequence

$$g_1 < g_2 < \cdots < g_s. \tag{5}$$

Finally, we define the set

$$\Gamma(\mathcal{C}) = \{(i_1, i_2) \in \mathbb{Z}_{r_1} \times \mathbb{Z}_{r_2} \mid \text{ there exists} \tag{6}$$
$$1 \leq j \leq s \text{ with } f_{j+1} \leq i_2 < f_j, \text{ and } 0 \leq i_1 < g_j\}.$$

The following theorem, proved in [3] for any abelian code, establishes that $\Gamma(\mathcal{C})$ is a set of check positions for $\mathcal{C}$, and consequently $\Gamma(\mathcal{C})$ defines an information set for $\mathcal{C}^\perp$.

Theorem 1. *Let r_1, r_2 be odd integers and let C be an abelian code in $\mathbb{A}(r_1, r_2)$ with defining set $\mathcal{D}_\alpha(C)$ with respect to $\alpha = \{\alpha_1, \alpha_2\}$. Then $\Gamma(C)$ is a set of check positions for C.*

Let us observe that given an abelian code $C \subseteq \mathbb{A}(r_1, r_2)$ with defining set $D_\alpha(C)$ if one chooses different primitive roots of unity, say $\gamma = \{\gamma_1, \gamma_2\}$, then the *structure* of the q-orbits in $D_\gamma(C)$ is the same as in $D_\alpha(C)$, that is, we obtain the same values for the parameters (1) and (2), so we get the same set of check positions $\Gamma(C)$ (see [4, p. 100]). That is why we do not use any reference to the roots of unity taken in the notation of the set of check positions. So, in the rest of the paper, for any abelian code C we denote its defining set by $D(C)$ and the corresponding set of check positions by $\Gamma(C)$ without any mention to the primitive roots.

Example 2. We continue with the code C considered in Example 1. Then $\mathcal{D}(C) = Q(1,1) \cup Q(0,1) \cup Q(2,0)$ and we take $\overline{\mathcal{D}} = \{(1,1), (0,1), (1,0)\}$ as set of restricted representatives. From (2) we have that $M(1) = m(1,1) + m(1,0) = 2 + 1 = 3$ and $M(0) = m(0,1) = 4$. So $f_1 = 4 > f_2 = 3 > f_3 = 0$. On the other hand, from (4) we obtain $g_1 = m(0) = 1, g_2 = g_1 + m(1) = 3$. Therefore,

$$\Gamma(C) = \{(i_1, i_2) \in \mathbb{Z}_{r_1} \times \mathbb{Z}_{r_2} \mid$$
$$(1 \le i_2 < 4 \text{ and } 0 \le i_1 < 1) \text{ or } (0 \le i_2 < 3 \text{ and } 0 \le i_1 < 3)\}$$
$$= \{(0,0), (0,1), (0,2), (0,3), (1,0), (1,1), (1,2), (2,0), (2,1), (2,2)\}$$

is a set of check positions for C.

Remark 1. In [3] we showed how to construct the previous set of check positions for any multidimensional abelian code. As we have already mentioned, in this paper we only need to use the two-dimensional case which yields the same information set that was introduced by H. Imai in [7].

4   Cyclic Codes as Two-Dimensional Cyclic Codes

We are going to construct an information set for the punctured cyclic code of a Reed-Muller code viewed as a multidimensional abelian code. So, we are interested in applying the results of the previous section when the original abelian code is in fact cyclic.

Let C^* be a binary cyclic code with length $n = r_1 \cdot r_2$. All throughout this section we assume that $\gcd(r_1, r_2) = 1$ and r_1, r_2 are odd. Let $T : \mathbb{Z}_n \to \mathbb{Z}_{r_1} \times \mathbb{Z}_{r_2}$ be the isomorphism given by the Chinese Remainder Theorem, and let us denote $T = (T_1, T_2)$; that is, $T_i(e)$ is the projection of $T(e)$ onto $\mathbb{Z}_{r_i}$, for $i = 1, 2$ and any $e \in \mathbb{Z}_n$.

Lemma 1. *For every $e \in \mathbb{Z}_n$ the equality $T(C_n(e)) = Q(T(e))$ holds; in particular $|C_n(e)| = |Q(T(e))|$. Moreover, one has that the projection of $Q(T(e))$ onto $\mathbb{Z}_{r_1}$ is equal to $C_{r_1}(T_1(e))$.*

Let $\mathcal{D}^* = D_\alpha(\mathcal{C}^*) \subseteq \mathbb{Z}_n$ be the defining set of $\mathcal{C}^*$ with respect to an arbitrary primitive n-th root of unity α. Then, since there exist integers η_1, η_2 such that $\eta_1 r_1 + \eta_2 r_2 = 1$, we have that $\alpha_1 = \alpha^{\eta_2 r_2}$ and $\alpha_2 = \alpha^{\eta_1 r_1}$ are primitive r_1-th and r_2-th roots of unity respectively. Set $T(1) = (\delta_1, \delta_2)$; observe that $\gcd(\delta_1, r_1) = 1$ and $\gcd(\delta_2, r_2) = 1$. We define the abelian code $\mathcal{C} = \mathcal{C}_{\mathcal{C}^*} \subseteq \mathbb{A}(r_1, r_2)$ as the code with defining set $\mathcal{D} = D(\mathcal{C}) = T(\mathcal{D}^*)$, with respect to $(\beta_1, \beta_2) = (\alpha_1^{\delta_1^{-1}}, \alpha_2^{\delta_2^{-1}})$. In this situation, we have that $\mathcal{C}$ is the image of $\mathcal{C}^*$ by the map

$$\mathbb{A}(n) \longrightarrow \mathbb{A}(r_1, r_2)$$
$$\sum a_i X^i \hookrightarrow \sum b_{jl} X^j Y^l,$$

where $b_{jl} = a_i$ if and only if $T(i) = (j, l)$. Therefore, $\mathcal{I}$ is an information set for $\mathcal{C}$ if and only if $T^{-1}(\mathcal{I})$ is an information set for $\mathcal{C}^*$. Usually, we omit the reference to the original cyclic code in the notation of the new abelian code, and we will write $\mathcal{C}$ instead of $\mathcal{C}_{\mathcal{C}^*}$; that reference will by clear by the context. For the rest of this section we assume that we have fixed an election of α and consequently, the roots β_1, β_2 are also fixed.

Remark 2. Let us observe that, since the defining set $\mathcal{D}^* = D_\alpha(\mathcal{C}^*)$ depends on the election of α, if we fix another one β, we get a new defining set $D_\beta(\mathcal{C}^*)$ and consequently we obtain a different abelian code $\mathcal{C}$. However, this new abelian code has a defining set with the same structure of q-orbits so it yields the same set $\Gamma(\mathcal{C})$ (see paragraph after Theorem 1).

On the other hand, it is easy to prove that any change of the isomorphism T is equivalent to a change of the primitive root α in order to get the same abelian code in $\mathbb{A}(r_1, r_2)$. Nevertheless, this implies that, since we get the same set $\Gamma(\mathcal{C})$, we could obtain a different set of check positions $T^{-1}(\Gamma(\mathcal{C}))$ at the end. Therefore, by the method we are describing in this section we could get at most as many information sets as isomorphisms from $\mathbb{Z}_n$ to $\mathbb{Z}_{r_1} \times \mathbb{Z}_{r_2}$ exist. To simplify the development of this extended abstract we continue assuming that we always manage the isomorphism given by the Chinese Remainder Theorem.

Then, the goal of this section is to describe the values of the parameters defined in (2), used to get a set of check positions for $\mathcal{C}$ (6), just in terms of the defining set of $\mathcal{C}^*$. This will allow us to define an information set for the original cyclic code $\mathcal{C}^*$ (via T) without any mention to the abelian code $\mathcal{C}$.

Let $\overline{\mathcal{D}^*} \subseteq \mathcal{D}^* \subseteq \mathbb{Z}_n$ be a complete set of representatives of the 2-cyclotomic cosets modulo n in $\mathcal{D}^*$. As we have noted in Lemma 1, for any $e \in \overline{\mathcal{D}^*}$, $T(C_n(e)) = Q(T(e))$, so $T(\overline{\mathcal{D}^*})$ is a complete set of representatives of the 2-orbits modulo (r_1, r_2) in $T(\mathcal{D}^*)$. However, it might not be a set of restricted representatives according to Definition 5. Then we introduce the following definition.

Definition 6. *Let $\overline{\mathcal{D}^*} \subseteq \mathcal{D}^*$ be a complete set of representatives of the 2-cyclotomic cosets modulo n in $\mathcal{D}^*$. Then $\overline{\mathcal{D}^*}$ is said to be a suitable set of representatives if $T(\overline{\mathcal{D}^*})$ is a set of restricted representatives of the 2-orbits in $T(\mathcal{D}^*)$.*

The next lemma says that we will always be able to take a suitable set of representatives in $\mathcal{D}^*$.

Lemma 2. *Let $\mathcal{D}^*$ be the defining set of a cyclic code $\mathcal{C}^* \subseteq \mathbb{A}(n)$. Let $\mathcal{C} \subseteq \mathbb{A}(r_1, r_2)$ be the abelian code with defining set $T(\mathcal{D}^*)$. Then, there always exists a suitable set of representatives $\overline{\mathcal{D}^*} \subseteq \mathcal{D}^*$.*

Example 3. Let us consider the binary cyclic code $\mathcal{C}^* \subseteq \mathbb{A}(15)$ whose defining set, with respect to a 15-th primitive root of unity α, is the following union of 2-cyclotomic cosets modulo 15, $\mathcal{D}^* = \{1, 2, 4, 8\} \cup \{3, 6, 12, 9\} \cup \{5, 10\}$. Take $T :$ $\mathbb{Z}_{15} \to \mathbb{Z}_3 \times \mathbb{Z}_5$ the isomorphism given by the Chinese Remainder Theorem. Then $\mathcal{C}$ denotes the abelian code in $\mathbb{A}(3, 5)$ with defining set $\mathcal{D} = T(\mathcal{D}^*)$. The reader may check that $\mathcal{C}$ is the code of Examples 1 and 2. The set $B = \{1, 3, 5\} \subseteq \mathcal{D}^*$ is a complete set of representatives of the 2-cyclotomic cosets in $\mathcal{D}^*$; however it is not a suitable set of representatives since $T(B) = \{(1, 1), (0, 3), (2, 0)\}$ is not a restricted set of representatives in $\mathcal{D}$ (note that $C_3(1) = C_3(2)$). We may solve this problem by substituting 5 by 10. Then $\overline{\mathcal{D}^*} = \{1, 3, 10\}$ is a suitable set of representatives because $T(\overline{\mathcal{D}^*}) = \{(1, 1), (0, 3), (1, 0)\}$ is a restricted set of representatives in $\mathcal{D}$.

Now we deal with the construction of a set of check positions for the abelian code $\mathcal{C} = \mathcal{C}_{\mathcal{C}^*} \subseteq \mathbb{A}(r_1, r_2)$ just in terms of $\mathcal{D}^*$, the defining set of the cyclic code $\mathcal{C}^*$.

Given $\overline{\mathcal{D}^*} \subseteq \mathcal{D}^*$ a suitable set of representatives, we consider $\sim$ the equivalence relation on $\overline{\mathcal{D}^*}$ given by the rule

$$a \sim b \in \overline{\mathcal{D}^*} \text{ if and only if } a \equiv b \mod r_1. \tag{7}$$

From now on, we denote by $\overline{\mathcal{D}^*}$ a suitable set of representatives and $\mathcal{U} \subseteq \overline{\mathcal{D}^*}$ a complete set of representatives of the equivalence classes related to $\sim$. In addition, for any $u \in \mathcal{U}$ we write

$$\mathcal{O}(u) = \{a \in \overline{\mathcal{D}^*} \mid a \sim u\}.$$

Observe that if $\overline{\mathcal{D}} = T(\overline{\mathcal{D}^*})$ then $\overline{\mathcal{D}}_1 = T_1(\mathcal{U})$. Furthermore, for any $e \in \overline{\mathcal{D}^*}$ there exists a unique $u \in \mathcal{U}$ such that $T_1(u) = T_1(e)$. By abuse of notation, we will write

$$C_{r_1}(e) = C_{r_1}(T_1(e)) = C_{r_1}(T_1(u)) = C_{r_1}(u).$$

Example 4. Following Example 3, from the suitable set of representatives $\overline{\mathcal{D}^*} = \{1, 3, 10\}$ we may define, for instance, $\mathcal{U} = \{1, 3\}$. Note that $\mathcal{O}(1) = \{1, 10\}$.

The following theorem says how to obtain the values $M(\cdot)$ (defined in (2)) corresponding with the set $\overline{\mathcal{D}} = T(\overline{\mathcal{D}^*}) \subset \mathbb{Z}_{r_1} \times \mathbb{Z}_{r_2}$, in terms of the elements in $\mathcal{U} \subseteq \mathbb{Z}_n$.

Theorem 2. *For each $(e_1, e_2) \in \overline{\mathcal{D}} = T(\overline{\mathcal{D}^*})$, there exists a unique $u \in \mathcal{U}$ such that $T_1(u) = e_1$ and*

$$M(e_1) = \frac{1}{|C_{r_1}(u)|} \sum_{v \in \mathcal{O}(u)} |C_n(v)| = \sum_{v \in \mathcal{O}(u)} |C_{2^{|C_{r_1}(u)|}, r_2}(T_2(v))|.$$

Therefore, fixed a n-th primitive root of unity α, for a given cyclic code C^* in $\mathbb{A}(n)$ with defining set with respect to α, $\mathcal{D}^*$, we consider $C \subseteq \mathbb{A}(r_1, r_2)$ the abelian code with defining set $\overline{\mathcal{D}} = T(\mathcal{D}^*)$ (taking as reference the suitable primitive roots of unity mentioned at the beginning of this section). Then, we have obtained that we can take a suitable set of representatives $\overline{\mathcal{D}^*} \subseteq \mathcal{D}^*$ and a set $\mathcal{U} \subseteq \overline{\mathcal{D}^*}$, with $T_1(\mathcal{U}) = \overline{\mathcal{D}}_1$, in such a way that we are able to construct the set of check positions given in (6) for C in terms of $\mathcal{U}$ as follows:

Since for any $e_1 \in \overline{\mathcal{D}}_1$ there exists a unique element $u \in \mathcal{U}$ that satisfies $T_1(u) = e_1$, by abuse of notation, we may write $M(u) = M(T_1(u)) = M(e_1)$. Then, the set of values (2) can be described as

$$\left\{ M(u) = \frac{1}{|C_{r_1}(u)|} \sum_{v \in \mathcal{O}(u)} |C_n(v)| \mid u \in \mathcal{U} \right\} \tag{8}$$

which yields the sequence $f_1 > \cdots > f_s > 0 = f_{s+1}$ (see (3)). On the other hand, for any $k = 1, \ldots, s$ the values (4) can be computed as

$$g_k = \sum_{\substack{u \in \mathcal{U} \\ M(u) \geq f_k}} |C_{r_1}(u)|$$

which give us the sequence $g_1 < g_2 < \cdots < g_s$ (see (5)) and hence the set $\Gamma(C)$. Finally, $T^{-1}(\Gamma(C))$ is a set of check positions for C^* described in terms of $\mathcal{U}$ uniquely.

Example 5. We apply the results in this section to the code C^* given in Example 3. Recall that its defining set is $\mathcal{D}^* = \{1, 2, 4, 8\} \cup \{3, 6, 12, 9\} \cup \{5, 10\}$. We have chosen $\overline{\mathcal{D}^*} = \{1, 3, 10\}$ as suitable set of representatives and $\mathcal{U} = \{1, 3\}$. Then, from (8) we have

$$M(1) = \frac{1}{|C_3(1)|} (|C_{15}(1)| + |C_{15}(10)|) = \frac{1}{2}(4 + 2) = 3,$$

$$M(3) = \frac{1}{|C_3(3)|} \cdot |C_{15}(3)| = 4,$$

and so $f_1 = 4 > f_2 = 3 > f_3 = 0$. On the other hand, $g_1 = m(0) = 1 < g_2 = 1 + m(1) = 3$. These sequences yield the set of check positions for C

$$\Gamma(C) = \{(0,0), (0,1), (0,2), (0,3), (1,0), (1,1), (1,2), (2,0), (2,1), (2,2)\}.$$

Finally, we have that $T^{-1}(\Gamma(C)) = \{0, 1, 2, 3, 5, 6, 7, 10, 11, 12\}$ is a set of check positions for C^*.

5   Reed-Muller Codes

In this section we shall introduce Reed-Muller codes from the group-algebra point of view. Specifically, we are going to present Reed-Muller codes as a type of codes contained in the family of so-called affine-invariant extended cyclic codes (see, for instance, [1,2] or [6]).

Recall that $\mathbb{F}$ denotes the binary field. Let G be the additive subgroup of the field of 2^m elements. So G is an elementary abelian group of order $|G| = 2^m$ and $G^* = G \setminus \{0\}$ is a cyclic group. From now on we denote $n = 2^m - 1$. We consider the group algebra $\mathbb{F}G$ which will be the ambient space for Reed-Muller codes. We denote the elements in $\mathbb{F}G$ as $\sum_{g \in G} a_g X^g$. Notice that X^0 is the unit element in $\mathbb{F}G$.

The following definitions introduce the family of affine-invariant extended cyclic codes. All throughout this section we fix α, a generator of the cyclic group G^*, that is, a primitive n-th root of unity.

Definition 7. *Let α be a generator of G^*. A code $\mathcal{C} \subseteq \mathbb{F}G$ is an extended cyclic code if for any $\sum_{g \in G} a_g X^g \in \mathcal{C}$ one has that $\sum_{g \in G} a_g X^{\alpha g} \in \mathcal{C}$ and $\sum_{g \in G} a_g = 0$.*

Definition 8. *We say that an extended cyclic code $\mathcal{C} \subseteq \mathbb{F}G$ is affine-invariant if for any $\sum_{g \in G} a_g X^g \in \mathcal{C}$ one has that $\sum_{g \in G} a_g X^{hg+k}$ belongs to $\mathcal{C}$ for all $h, k \in G, h \neq 0$.*

It is clear that if $\mathcal{C} \subseteq \mathbb{F}G$ is affine-invariant then $\mathcal{C}$ is an ideal in $\mathbb{F}G$ and $\mathcal{C}^* \subseteq \mathbb{F}G^*$, the punctured code at the position X^0, is cyclic in the sense that it is the projection to $\mathbb{F}G^*$ of the image of a cyclic code via the map

$$\mathbb{F}[X]/\langle X^n - 1 \rangle \longrightarrow \mathbb{F}G$$

$$\sum_{i=0}^{n-1} a_i X^i \longmapsto \left(-\sum_{i=0}^{n-1} a_i \right) X^0 + \sum_{i=0}^{n-1} a_i X^{\alpha^i}, \qquad (9)$$

where α is the fixed n-th root of unity.

Now, for any $s \in \{0, \ldots, n = 2^m - 1\}$ we consider the $\mathbb{F}$-linear map $\phi_s : \mathbb{F}G \to G$ given by

$$\phi_s \left(\sum_{g \in G} a_g X^g \right) = \sum_{g \in G} a_g g^s$$

where we assume $0^0 = 1 \in \mathbb{F}$ by convention.

Definition 9. *Let $\mathcal{C} \subseteq \mathbb{F}G$ be an affine-invariant code. The set*

$$D(\mathcal{C}) = \{i \mid \phi_i(x) = 0 \text{ for all } x \in \mathcal{C}\}$$

is called the defining set of $\mathcal{C}$.

Note first that since $\mathcal{C}$ is an extended-cyclic code, one has that $0 \in D(\mathcal{C})$ because $\phi_0 \left(\sum_{g \in G} a_g X^g \right) = \sum_{g \in G} a_g g^0 = \sum_{g \in G} a_g$. Furthermore, it follows from the equality $\phi_{2s}(x) = (\phi_s(x))^2$ $(x \in \mathcal{C})$ that $D(\mathcal{C})$ is a union of 2-cyclotomic cosets modulo n. On the other hand, keeping in mind the map (9), one has that the zeros of the cyclic code $\mathcal{C}^*$ are $\{\alpha^s \mid s \in D(\mathcal{C}), s \neq 0\}$; that is, the defining set of $\mathcal{C}^*$, according to Definition 1 (and with respect to that root of unity), is $D(\mathcal{C}^*) = D(\mathcal{C}) \setminus \{0\}$.

It is easy to prove that any affine-invariant code is totally determined by its defining set. Conversely, any subset of $\{0, \ldots, n\}$ which is a union of 2-cyclotomic cosets and contains 0 defines an affine-invariant code in $\mathbb{F}G$.

Remark 3. It may occur that $n, 0 \in D(\mathcal{C})$ which could yield confusion in order to consider the 2-cyclotomic cosets modulo n. Those elements are considered distinct and they indicate different properties of the code $\mathcal{C}$; to wit, 0 always belongs to $D(\mathcal{C})$ because $\mathcal{C}$ is an extended cyclic code, while if n belongs to $D(\mathcal{C})$ then the cyclic code $\mathcal{C}^*$ is even-like which implies that $\mathcal{C}$ is a trivial extension.

Finally, to introduce the family of Reed-Muller codes as affine-invariant codes we need to recall the notions of binary expansion and 2-weight. For any natural number k its binary expansion is the sum $\sum_{r\geq 0} k_r 2^r = k$ with $k_r = 0, 1$. The 2-weight or simply weight of k is $\mathrm{wt}(k) = \sum_{r\geq 0} k_r$.

Definition 10. *Let $0 < \rho < m$. The Reed-Muller code of order ρ and length 2^m, denoted by $R(\rho, m)$, is the affine-invariant code in $\mathbb{F}G$ with defining set*

$$D(R(\rho, m)) = \{i \mid 0 \leq i < 2^m - 1 \text{ and } \mathrm{wt}(i) < m - \rho\}.$$

From the classical point of view, Reed-Muller codes are also defined for the cases $\rho = 0, m$, but they correspond with the trivial cases $R(m, m) = \mathbb{F}G$ and $R(0, m) = \langle \sum_{g\in G} g \rangle$ (the repetition code) which do not have interest in the context of this paper. On the other hand, it is well known that $R(\rho, m)^\perp = R(m - \rho - 1, m)$ and that $R(m - 1, m)$ is the code of all even weight vectors in $\mathbb{F}G$ (the reader may see [2]). Then $R(m - 1, m)^\perp = R(0, m)$ and therefore the problem of searching for information sets has no interest in the case $\rho = m - 1$, so in the rest of the paper we will assume that $\rho < m - 1$.

As a consequence of Definition 10 we have that the defining set of the punctured code $R^*(\rho, m)$, at the position X^0 and with respect to the fixed n-th root of unity α, is given by the following union of cyclotomic cosets modulo n

$$D(R^*(\rho, m)) = \bigcup_{i \in D(R(\rho,m))\setminus\{0\}} C_n(i).$$

Remark 4. Note that, fixed any primitive root of unity α, following the notation used to describe the elements in $\mathbb{F}G$, $\left(-\sum_{i=0}^{n-1} a_i \right) X^0 + \sum_{i=0}^{n-1} a_i X^{\alpha^i} \in \mathbb{F}G$, a set $\mathcal{I} \subseteq \{0, \alpha^0, \ldots, \alpha^{n-1}\}$ is an information set for a code $\mathcal{C} \subseteq \mathbb{F}G$, with dimension k, if $|\mathcal{I}| = k$ and $\mathcal{C}_{\mathcal{I}} = \mathbb{F}^{|\mathcal{I}|}$.

On the other hand, since $R^*(\rho, m)$ is contained in $\mathbb{F}G^*$ we have that any information set for it will be a subset of $\{\alpha^0, \ldots, \alpha^{n-1}\}$. Obviously, an information set for $R^*(\rho, m)$ is an information set for $R(\rho, m)$ too. However, note that if Γ is a set of check positions for $R^*(\rho, m)$ then a set of check positions for $R(\rho, m)$ is $\Gamma \cup \{0\}$.

In the following sections we will deal with the application of the results contained in Sect. 4 to the cyclic code $R^*(\rho, m)$ in order to obtain an information set for $R(\rho, m)$. To use a notation congruent with that used in that section we write $\mathcal{D}^* = D(R^*(\rho, m))$. We assume that there exist integers r_1, r_2 such that $n = 2^m - 1 = r_1 \cdot r_2$ with r_1, r_2 odd and $\gcd(r_1, r_2) = 1$. Now, we fix some notation and introduce some definitions and basic results that will be needed.

For any integer $0 < K < m - 1$ we define

$$\Omega(K) = \{0 < j < 2^m - 1 \mid \mathrm{wt}(j) = K\}$$
$$= \{2^{t_1} + \cdots + 2^{t_K} \mid 0 \le t_1 < \cdots < t_K < m\}$$

and hence,

$$\mathcal{D}^* = \bigcup_{K=1}^{m-\rho-1} \{0 < j < 2^m - 1 \mid \mathrm{wt}(j) = K\} = \bigcup_{K=1}^{m-\rho-1} \Omega(K).$$

Example 6. Take $m = 4$. Then $n = 2^4 - 1 = 15, r_1 = 3, r_2 = 5$. For these parameters one has that

$$\Omega(1) = \{1, 2, 4, 8\}, \Omega(2) = \{3, 5, 6, 9, 10, 12\} \text{ and } \Omega(3) = \{7, 11, 13, 14\}.$$

The code $R(1, 4)$ has defining set $\mathcal{D}(R(1, 4)) = \{0, 1, 2, 3, 4, 5, 6, 8, 9, 10, 12\}$ and so $\mathcal{D}^* = D(R^*(1, 4)) = \Omega(1) \cup \Omega(2) = \{1, 2, 3, 4, 5, 6, 8, 9, 10, 12\}$. The reader may check that $\mathcal{D}^*$ corresponds to the code of Examples 1 to 5.

Finally, the code $R(2, 4)$ has defining set $\mathcal{D}(R(2, 4)) = \{0, 1, 2, 4, 8\}$, which implies $\mathcal{D}^* = \mathcal{D}(R^*(2, 4)) = \Omega(1)$.

Now, we take $\overline{\mathcal{D}^*}$ a suitable set of representatives in $\mathcal{D}^*$ and a set $\mathcal{U}$ as in the previous section (see Definition 6 and subsequent paragraphs).

We are interested in handling convenient elements in the fixed suitable set of representatives in order to compute the necessary cyclotomic cosets. It is clear that if $e = 2^s \in \Omega(1)$, with $0 \le s < m$, then $C_n(e) = C_n(1)$. On the other hand, it is easy to check that given any element $e \in \Omega(2)$ there always exists

$$e' = 1 + 2^t, \tag{10}$$

with $t \le \lfloor \frac{m}{2} \rfloor$ such that $C_n(e) = C_n(e')$.

As we will see in the next section, this fact becomes essential for our purposes. In addition, it is important to note that the elements in a given suitable set of representatives might not verify condition (10); we include the next example to show that case.

Example 7. Let us consider the Reed-Muller code $R(3, 6)$. So $m = 6, n = 2^6 - 1 = 63, r_1 = 7, r_2 = 9$. For these parameters we have

$$\Omega(1) = C_{63}(1) = \{1, 2, 4, 8, 16, 32\}$$

and

$$\Omega(2) = C_{63}(3) \cup C_{63}(5) \cup C_{63}(9) = \{3, 5, 6, 9, 10, 12, 17, 18, 20, 24, 33, 34, 36, 40, 48\}.$$

The defining set of $R^*(3,6)$ is $\mathcal{D}^* = \Omega(1) \cup \Omega(2)$. The set $\overline{\mathcal{D}^*} = \{1, 3, 5, 36\}$ is a suitable set of representatives according to Definition 6, however while $1, 3$ and 5 satisfy condition (10) the element $36 = 2^2 + 2^5$ does not satisfy it. There exists a unique element in $C_{63}(36) = C_{63}(9)$ that satisfies that condition, to wit, 9.

6   Information Sets for First-Order Reed-Muller Codes

In this section we are applying the results of Sect. 4 to the punctured codes of first-order Reed-Muller codes. Specifically, we shall construct an information set for the punctured code $R^*(1, m)$ by using those techniques and then we shall give an information set for the Reed-Muller code $R(1, m)$. To use the mentioned results we need to assume that there exist odd integers r_1, r_2 such that $n = 2^m - 1 = r_1 \cdot r_2$ and $\gcd(r_1, r_2) = 1, r_1, r_2 > 1$.

All throughout this section we fix a primitive n-th root of unity α; recall that $T : \mathbb{Z}_n \to \mathbb{Z}_{r_1} \times \mathbb{Z}_{r_2}$ is the isomorphism of groups given by the Chinese Remainder Theorem.

We have two different possibilities to get an information set for the code $R(1, m)$; to wit, from an information set of $R^*(1, m)$, which comes from the defining set of $R^*(1, m)$, and from a check of set positions of $R(1, m)^{\perp} = R(m - 2, m)$, which depends on the defining set of $R^*(m - 2, m)$. In general it is more convenient to develop the results in terms of $R^*(m - 2, m)$ because its defining set is much smaller than that of $R^*(1, m)$.

Let us denote $\mathcal{C}^* = R^*(m - 2, m)$. Then, following the notation fixed in the previous section, we have that

$$\mathcal{D}^* = \Omega(1)$$

where $\Omega(1) = \{2^t \mid 0 \le t < m\}$. Observe that $\Omega(1) = C_n(1)$ so we take $\overline{\mathcal{D}^*} = \{1\}$ as our suitable set of representatives (see Definition 6). Then $\mathcal{U} = \{1\}$.

The following theorem gives us the description of an information set for the code $R(1, m)$. We denote by $\mathrm{Ord}_{r_1}(2)$ the order of 2 modulo r_1, that is, the smallest integer such that $2^{\mathrm{Ord}_{r_1}(2)} \equiv 1$ modulo r_1.

Theorem 3. *Suppose that $n = 2^m - 1 = r_1 \cdot r_2$, where r_1, r_2 are odd integers such that $\gcd(r_1, r_2) = 1, r_1, r_2 > 1$. Let $\mathcal{C}^* = R^*(m - 2, m)$ and $a = \mathrm{Ord}_{r_1}(2)$. Let $\mathcal{D}^* \subseteq \mathbb{Z}_n$ be the defining set of $\mathcal{C}^*$ with respect to α, a primitive n-th root of unity, and let $\mathcal{C} \subseteq \mathbb{A}(r_1, r_2)$ be the abelian code with defining set $D(\mathcal{C}) = T(\mathcal{D}^*)$. Then, the set $T^{-1}(\Gamma)$ where*

$$\Gamma = \Gamma(\mathcal{C}) = \left\{ (i_1, i_2) \in \mathbb{Z}_{r_1} \times \mathbb{Z}_{r_2} \mid 0 \le i_1 < a, 0 \le i_2 < \frac{m}{a} \right\}$$

is a set of check positions for $R^*(m - 2, m)$. Furthermore, $\{0, \alpha^i \mid i \in T^{-1}(\Gamma)\}$ is an information set for $R(1, m)$ and $\{\alpha^i \mid i \notin T^{-1}(\Gamma)\}$ is an information set for $R(m - 2, m)$.

Example 8. The first value for m that satisfies the required conditions is $m = 4$. In this case, $n = 2^4 - 1 = 15, r_1 = 3, r_2 = 5, \mathrm{Ord}_3(2) = 2$. So, let us give an information set for $R(1, 4)$. By Theorem 3 we have that

$$\Gamma = \{(i_1, i_2) \in \mathbb{Z}_{r_1} \times \mathbb{Z}_{r_2} \mid 0 \le i_1 < 2, 0 \le i_2 < 2\} = \{(0, 0), (1, 0), (0, 1), (1, 1)\}.$$

Then, we have that $T^{-1}(\Gamma) = \{0, 1, 6, 10\}$ and then $\{0, 1, \alpha, \alpha^6, \alpha^{10}\}$ is an information set for $R(1, 4)$. Moreover $\{\alpha^i \mid i \ne 0, 1, 6, 10\}$ is an information set for $R(2, 4)$.

The next value for m is $m = 6$. In this case, $n = 2^6 - 1 = 63, r_1 = 7$, $r_2 = 9, a = 3$. So

$$\begin{aligned} \Gamma &= \{(i_1, i_2) \in \mathbb{Z}_{r_1} \times \mathbb{Z}_{r_2} \text{ such that } 0 \le i_1 < 3, 0 \le i_2 < 2\} \\ &= \{(0, 0), (1, 0), (2, 0), (0, 1), (1, 1), (2, 1)\}. \end{aligned}$$

Since $T^{-1}(\Gamma) = \{0, 1, 9, 28, 36, 37\}$ one has that $\{0, 1, \alpha, \alpha^9, \alpha^{28}, \alpha^{36}, \alpha^{37}\}$ is an information set for $R(1, 6)$ and $\{\alpha^i \mid i \ne 0, 1, 9, 28, 36, 37\}$ is an information set for $R(4, 6)$.

Finally, let us see the case $m = 8$, that is, the Reed-Muller codes of length 256. We consider the decomposition of $n = 2^8 - 1 = 255$ defined as $(r_1 = 3, r_2 = 85)$. Then $T^{-1}(\Gamma) = \{0, 1, 3, 85, 87, 88, 171, 172\}$.

7 Information Sets for Second-Order Reed-Muller Codes

In this section we deal with second-order Reed-Muller codes. As in the previous section, we shall apply the results of Sect. 4, so, again, we need to assume that there exist integers r_1, r_2 such that $n = 2^m - 1 = r_1 \cdot r_2$ and $\gcd(r_1, r_2) = 1, r_1, r_2 > 1$. Again, we fix a primitive n-th root of unity α.

In a similar way to the case of $R(1, m)$, we have two possibilities to get an information set for the code $R^*(2, m)$, to wit, $(\mathbb{Z}_{r_1} \times \mathbb{Z}_{r_2}) \setminus \Gamma(R^*(2, m))$ and $\Gamma(R^*(2, m)^\perp)$. Once more, we consider more convenient to develop the results in terms of $R(m - 3, m) = R^*(2, m)^\perp$.

Throughout this section we denote $\mathcal{C}^* = R^*(m - 3, m)$. In this case we have that

$$\mathcal{D}^* = \Omega(1) \cup \Omega(2)$$

where $\Omega(1) = \{2^t \mid 0 \le t < m\}$ and $\Omega(2) = \{2^{t_1} + 2^{t_2} \mid 0 \le t_1 < t_2 < m\}$. Let $\overline{\mathcal{D}^*} \subseteq \mathcal{D}^*$ be a suitable set of representatives and take $\mathcal{U} \subseteq \overline{\mathcal{D}^*}$ a complete set of representatives of the equivalence classes modulo r_1 (see (7)). We will always assume that $1 \in \mathcal{U} \subseteq \overline{\mathcal{D}^*}$ (recall that $\Omega(1) = C_n(1)$).

To get the expressions (8) we need to compute the cardinalities $|C_n(e)|$, for all $e \in \overline{\mathcal{D}^*}$, and $|C_{r_1}(u)|$ for any element u in the fixed set $\mathcal{U}$. As we have seen in the previous section, these computations can be made directly in the case of elements in $\Omega(1)$; however, they turn to be much more complicated when we work with the set $\Omega(2)$. So, we impose some conditions on r_1 in order to make the mentioned computations avaliable, to wit, all throughout we also assume that $r_1 = 2^a - 1$, with a an integer. Then, we can sum up the restrictions on the paremeters as follows

$$n = 2^m - 1 = r_1 \cdot r_2, \text{ where } r_1 = 2^a - 1 \text{ and } \gcd(r_1, r_2) = 1, r_1, r_2 > 1. \quad (11)$$

Note that from these conditions one follows that $a = \mathrm{Ord}_{r_1}(2)$ and so the notation is consistent with that used in Theorem 3. In what follows we write $m = ab$.

The first lemma, probably a well-known result, shows that the value of the cardinalities $|C_n(e)|$ and $|C_{r_1}(e)|$ can be easily computed for the elements in $\Omega(2)$ of the form $1 + 2^t$.

Lemma 3. *Let μ, ν integers such that $\mu = 2^\nu - 1$. Then $|C_\mu(1)| = \nu$ and for any natural number $0 < t \leq \nu - 1$*

$$|C_\mu(1 + 2^t)| = \begin{cases} \dfrac{\nu}{2} & \text{in case} \quad t = \dfrac{\nu}{2} \\[2ex] \nu & \text{otherwise} \end{cases}$$

As we have observed in the previous section, the elements in the suitable set $\overline{\mathcal{D}}^*$ might not verify condition (10). However, we can relate any element in $\overline{\mathcal{D}}^*$ with another one in the same 2-cyclotomic coset modulo n that satisfies that condition. The following result details this relationship.

Proposition 1. *Let $\overline{\mathcal{D}}^*$ be a suitable set of representatives. There exists a bijection $\varepsilon : \overline{\mathcal{D}}^* \setminus \{1\} \longrightarrow \{1, \ldots, \lfloor \frac{m}{2} \rfloor\}$ where $\varepsilon(e)$ is the unique integer such that $1 + 2^{\varepsilon(e)}$ belongs to $C_n(e)$ and verifies condition (10).*

Remark 5. Observe that for all $e = 2^{t_1} + 2^{t_2} \in \Omega(2)$ one has that $1 + 2^\delta$, with $\delta = \min\{t_2 - t_1, m - t_2 + t_1\}$, belongs to $C_n(e)$ and verifies (10), therefore $\varepsilon(e) = \min\{t_2 - t_1, m - t_2 + t_1\}$.

Under the notation introduced by Proposition 1 we can say that for any $e \in \Omega(2)$ the element $1 + 2^{\varepsilon(e)}$ is the unique element in $C_n(e)$ satisfying condition (10).

Now, from Proposition 1 and Lemma 3 we obtain the next result which gives us the desired cardinalities.

Proposition 2. *Let $n = 2^m - 1 = r_1 \cdot r_2$ satisfying (11), that is, $m = ab$, with a such that $r_1 = 2^a - 1$ and $(r_1, n/r_1) = 1$. For any $e \in \Omega(2)$ one has that*

1. $|C_n(e)| = \begin{cases} \dfrac{m}{2} & \text{in case} \quad \varepsilon(e) = \dfrac{m}{2} \\[2ex] m & \text{otherwise} \end{cases}$

2. $|C_{r_1}(e)| = \begin{cases} \dfrac{a}{2} & \text{in case} \quad \varepsilon(e) \equiv \dfrac{a}{2} \bmod a \\[2ex] a & \text{otherwise} \end{cases}$

Now, all that is required to obtain the expressions (8) is the description of the sets $\mathcal{O}(u)$ with $u \in \mathcal{U}$, that is, $\mathcal{O}(1)$ and $\mathcal{O}(e)$ with $e \in \Omega(2) \cap \mathcal{U}$. The next results solve this problem. The first one gives the information we need about $\mathcal{O}(1)$. Recall that all throughout $\overline{\mathcal{D}^*}$ denotes a suitable set of representatives.

Proposition 3. *1.* $\mathcal{O}(1) = \{1\} \cup \{e \in \overline{\mathcal{D}^*} \mid \varepsilon(e) = \lambda a \text{ with } 1 \leq \lambda \leq \lfloor \frac{b}{2} \rfloor\}$.
2. $|\mathcal{O}(1)| = 1 + \lfloor b/2 \rfloor$.
3. $|C_{r_1}(1)| = a$.
4. *If b is even then* $\varepsilon^{-1}(m/2) \in \mathcal{O}(1)$, $|C_n(\varepsilon^{-1}(m/2))| = m/2$, *and* $|C_n(e)| = m$
for any $e \in \mathcal{O}(1) \setminus \{\varepsilon^{-1}(m/2)\}$.
5. *If b is odd then* $|C_n(e)| = m$ *for any* $e \in \mathcal{O}(1)$.

Remark 6. As we have already seen we assume that $1 \in \mathcal{U} \subseteq \overline{\mathcal{D}^*}$. Furthermore, depending on the parity of a and b we will take the following elements in $\mathcal{U}$ by convention:

1. If m is even we also assume that $\varepsilon^{-1}(m/2) \in \mathcal{U}$ unless $\varepsilon^{-1}(m/2) \in \mathcal{O}(1)$.
2. If a is even we assume that $\varepsilon^{-1}(a/2) \in \mathcal{U}$ unless $\varepsilon^{-1}(a/2) \in \mathcal{O}(\varepsilon^{-1}(m/2))$.
   Observe that $\varepsilon^{-1}(a/2)$ never belongs to $\mathcal{O}(1)$.

The next result yields the description of the set $\mathcal{O}(e)$ for any $e \in \mathcal{U} \setminus \{1\}$. Let us observe that $e \in \mathcal{U} \setminus \{1\}$ implies $e \in \mathcal{U} \cap \Omega(2)$.

Proposition 4. *Let* $e \in \mathcal{U} \setminus \{1\}$. *Then*

$$\mathcal{O}(e) = \{e' \in \overline{\mathcal{D}^*} \mid \varepsilon(e') \equiv \varepsilon(e) \bmod a \text{ or } \varepsilon(e') \equiv a - \varepsilon(e) \bmod a\}.$$

The following propositions complete the information about the sets $\mathcal{O}(e)$, with $e \in \mathcal{U} \setminus \{1\}$, by making a distinction between the cases b even and b odd ($m = ab$).

Proposition 5. *Let* $e \in \mathcal{U} \setminus \{1\}$ *and suppose that b is even. Then*

1. *For any* $e' \in \mathcal{O}(e)$ *one has that* $|C_n(e')| = m$.
2. *If a is odd then* $|\mathcal{O}(e)| = b$ *and* $|C_{r_1}(e)| = a$.
3. *If a is even then in case* $e = \varepsilon^{-1}(a/2)$ *one has that* $|\mathcal{O}(e)| = b/2$, $|C_{r_1}(e)| = a/2$ *and otherwise* $|\mathcal{O}(e)| = b$, $|C_{r_1}(e)| = a$.

Proposition 6. *Let* $e \in \mathcal{U} \setminus \{1\}$ *and suppose that b is odd. Then* $|C_n(e')| = m$ *for all* $e' \in \mathcal{O}(e) \setminus \{e\}$ *and*

1. *If a is odd then* $|\mathcal{O}(e)| = b$, $|C_{r_1}(e)| = a$ *and* $|C_n(e)| = m$.
2. *If a is even then*
   (a) If $e = \varepsilon^{-1}(m/2)$ *then* $|\mathcal{O}(e)| = (b+1)/2$, $|C_n(e)| = m/2$ *and* $|C_{r_1}(e)| = a/2$.
   (b) If $e \neq \varepsilon^{-1}(m/2)$ *then* $|\mathcal{O}(e)| = b$, $|C_n(e')| = m$ *and* $|C_{r_1}(e)| = a$.

The last result we need before enunciating our main theorem gives us the cardinality of any election of the set $\mathcal{U}$. It uses the mentioned concept of 2-weight of an integer (see paragraph before Definition 10). We talk about the 2-weight of a 2-cyclotomic coset when one of its elements (and so all of them) has that 2-weight.

Lemma 4. *For any election of the set of representatives $\mathcal{U}$ one has that*

$$|\mathcal{U}| = 1 + |\{C_{r_1}(e) \mid e \in \mathcal{U} \cap \Omega(2)\}| = 1 + \lfloor a/2 \rfloor.$$

Now, we can present the main result for second-order Reed-Muller codes.

Theorem 4. *Let $C^* = R^*(m-3, m)$. Suppose that $n = 2^m - 1 = r_1 \cdot r_2$, where $m = ab$, $r_1 = 2^a - 1$ and $\gcd(r_1, r_2) = 1, r_1, r_2 > 1$. Let $\mathcal{D}^* \subseteq \mathbb{Z}_n$ be the defining set of C^*, with respect to α, a primitive n-th root of unity, and $C \subseteq \mathbb{A}(r_1, r_2)$ the abelian code with defining set $D(C) = T(\mathcal{D}^*)$. Then, the values that appear in (3) and (5) are*

$$f_1 = b^2, f_2 = \frac{b(b+1)}{2} \quad and \quad g_1 = \frac{a(a-1)}{2}, g_2 = \frac{a(a+1)}{2},$$

respectively.

Therefore the set $T^{-1}(\Gamma)$ where

$$\Gamma = \Gamma(C) = \{(i_1, i_2) \in \mathbb{Z}_{r_1} \times \mathbb{Z}_{r_2} \text{ such that}$$

$$\left(0 \leq i_1 < \frac{a(a-1)}{2} \quad and \quad 0 \leq i_2 < b^2 \right) \text{ or}$$

$$\left(\frac{a(a-1)}{2} \leq i_1 < \frac{a(a+1)}{2} \quad and \quad 0 \leq i_2 < \frac{b(b+1)}{2} \right) \}$$

is a set of check positions for $R^*(m-3, m)$. Furthermore, $\{0, \alpha^i \mid i \in T^{-1}(\Gamma)\}$ is an information set for $R(2, m)$ and $\{\alpha^i \mid i \notin T^{-1}(\Gamma)\}$ is an information set for $R(m-3, m)$.

Example 9. The first value for m that satisfies conditions (11) is $m = 4$. In this case, $n = 2^4 - 1 = 15, r_1 = 3, r_2 = 5, a = b = 2$. So, let us give an information set for $R(2, 4)$. By Theorem 4 we have that

$$\Gamma = \{(0,0), (1,0), (2,0), (0,1), (1,1), (2,1), (0,2), (1,2), (2,2), (0,3)\}.$$

Then, we have $T^{-1}(\Gamma) = \{0, 1, 2, 3, 5, 6, 7, 10, 11, 12\}$. So $\{0, \alpha^i \mid i \in T^{-1}(\Gamma)\}$ is an information set for $R(2, 4)$ and $\{\alpha^i \mid i \notin T^{-1}(\Gamma)\}$ is an information set for $R(1, 4)$.

The next value for m is $m = 6$. In this case, $n = 2^6 - 1 = 63, r_1 = 7$, $r_2 = 9, a = 3$ and $b = 2$. So

$$\Gamma = \{(0,0),(1,0),(2,0),(3,0),(4,0),(5,0),(0,1),(1,1),(2,1),(3,1),$$
$$(4,1),(5,1),(0,2),(1,2),(2,2),(3,2),(4,2),(5,2),(0,3),(1,3),(2,3)\}.$$

Then, we obtain

$$T^{-1}(\Gamma) = \{0,1,2,9,10,11,18,19,21,28,29,30,36,37,38,45,46,47,54,56,57\},$$

so we have that $\{0,\alpha^i \mid i \in T^{-1}(\Gamma)\}$ is an information set for $R(2,6)$ and $\{\alpha^i \mid i \notin T^{-1}(\Gamma)\}$ is an information set for $R(3,6)$.

Finally, we see the case $m = 8$. We consider the decompositions of $n = 2^8 - 1 = 255$ defined by $(r_1 = 3, r_2 = 85)$. Then,

$$T^{-1}(\Gamma) = \{0,1,2,3,4,5,6,7,8,9,12,15,85,86,87,88,89,90,91,92,93,94,96,99,$$
$$170,171,172,173,174,175,176,177,178,179,180,183\}.$$

So we have that $\{0,\alpha^i \mid i \in T^{-1}(\Gamma)\}$ is an information set for $R(2,8)$ and $\{\alpha^i \mid i \notin T^{-1}(\Gamma)\}$ is an information set for $R(5,8)$.

8   Conclusions

We have described explicitly how to construct information sets for first and second order Reed-Muller codes only in terms of their defining sets as affine-invariant codes. To get these results we have used those given in [3]. The step from first to second order involves much more complicated technical results, so it is an open problem to study the generalization to order greater than 2. On the other hand, in [4] we show how to use the special structure of this kind of information sets to apply the permutation decoding algorithm, so it is expected that from the new information sets introduced in this paper we can study the applicability of that decoding procedure in a different way from that showed in [8,9] or [11].

References

1. Huffman, W.C.: Codes and groups. In: Pless, V.S., Huffman, W.C., Brualdi, R.A. (eds.) Handbook of Coding Theory, vol II. North-Holland, Amsterdam (1998)
2. Assmus Jr., E.F., Key, J.D.: Polynomial codes and finite geometries. In: Pless, V.S., Huffman, W.C., Brualdi, R.A. (eds.) Handbook of Coding Theory, vol II. North-Holland, Amsterdam (1998)
3. Bernal, J.J., Simón, J.J.: Information sets from defining sets in abelian codes. IEEE Trans. Inform. Theory **57**(12), 7990–7999 (2011)
4. Bernal, J.J.: Códigos de grupo. Conjuntos de información. Decodificación por permutación. Ph.D. thesis (2011)
5. Blokhuis, A., Moorhouse, G.E.: Some p-ranks related to orthogonal spaces. J. Algebraic Combin. **4**, 295–316 (1995)
6. Charpin, P.: Codes cycliques etendus invariants sous le group affine. These de Doctorat d'Etat, Universite Parés VII (1987)

7. Imai, H.: A theory of two-dimensional cyclic codes. Inform. Control **34**, 1–21 (1977)
8. Key, J.D., McDonough, T.P., Mavron, V.C.: Partial permutation decoding for codes from finite planes. Eur. J. Combin. **26**, 665–682 (2005)
9. Key, J.D., McDonough, T.P., Mavron, V.C.: Information sets and partial permutation decoding for codes from finite geometries. Finite Fields Appl. **12**, 232–247 (2006)
10. Moorhouse, G.E.: Bruck nets, codes and characters of loops. Des. Codes Cryptogr. **1**, 7–29 (1991)
11. Seneviratne, P.: Permutation decoding for hte first-order Reed-Muller codes. Discrete Math. **309**, 1967–1970 (2009)

Distance Properties of Short LDPC Codes and Their Impact on the BP, ML and Near-ML Decoding Performance

Irina E. Bocharova[1,2], Boris D. Kudryashov[1], Vitaly Skachek[2(✉)], and Yauhen Yakimenka[2]

[1] St. Petersburg University of Information Technologies, Mechanics and Optics, St. Petersburg 197101, Russia
kudryashov_boris@bk.ru
[2] University of Tartu, Tartu 50409, Estonia
{irinaboc,vitaly,yauhen}@ut.ee

Abstract. Parameters of LDPC codes, such as minimum distance, stopping distance, stopping redundancy, girth of the Tanner graph, and their influence on the frame error rate performance of the BP, ML and near-ML decoding over a BEC and an AWGN channel are studied. Both random and structured LDPC codes are considered. In particular, the BP decoding is applied to the code parity-check matrices with an increasing number of redundant rows, and the convergence of the performance to that of the ML decoding is analyzed. A comparison of the simulated BP, ML, and near-ML performance with the improved theoretical bounds on the error probability based on the exact weight spectrum coefficients and the exact stopping size spectrum coefficients is presented. It is observed that decoding performance very close to the ML decoding performance can be achieved with a relatively small number of redundant rows for some codes, for both the BEC and the AWGN channels.

Keywords: LDPC code · Minimum distance · Stopping distance · Stopping redundancy · BP decoding · ML decoding

1 Introduction

It is well-known that typically binary LDPC codes have minimum distances which are smaller than those of the best known linear codes of the same rate and length. It is not surprising, since minimum distance does not play an important role in iterative (belief propagation (BP)) decoding. On the other hand, a significant gap in the frame error rate (FER) performance of BP and maximum-likelihood (ML) decoding motivates developing near-ML decoding algorithms for LDPC codes.

This work is supported in part by the Norwegian-Estonian Research Cooperation Programme under the grant EMP133 and by the Estonian Research Council under the grant PUT405.

© Springer International Publishing AG 2017
A.I. Barbero et al. (Eds.): ICMCTA 2017, LNCS 10495, pp. 48–61, 2017.
DOI: 10.1007/978-3-319-66278-7_5

There are two main approaches to improving the BP decoding performance. First one is based on post-processing in case of BP decoder failure. Different post-processing techniques for decoding of binary LDPC codes over additive white Gaussian noise (AWGN) channels are studied in [7,11,24,31]. A similar approach to decoding of nonbinary LDPC codes over extensions of the binary Galois field is considered in [2]. Near-ML decoding algorithms for LDPC codes over binary erasure channel (BEC) can be found in [16,21,23].

The second approach is based on identifying and destroying specific structural configurations such as trapping and stopping sets of the Tanner graph of the code. In particular, this can be done by adding redundant rows to the code parity-check matrix (see, for example, [14,17,19]).

Suboptimality of the above modifications of BP decoding rises the following question: which properties of LDPC codes and to what extent influence their decoding FER performance? In this paper, we are trying to partially answer this question by studying short LDPC codes. We consider both binary LDPC codes and binary images of nonbinary random LDPC codes over extensions of the binary field, as well as quasi-cyclic (QC) LDPC codes constructed by using an optimization technique in [4]. Parameters such as minimum distance, stopping distance, girth of the Tanner graph and estimates on the stopping redundancy are tabulated. Near-ML decoding based on adding redundant rows to the code parity-check matrix is analyzed. Simulated over the BEC and the AWGN channel, the FER performance of the BP, ML and near-ML decoding of these classes of LDPC codes is presented and compared to the improved upper bounds on the performance of ML decoding of regular LDPC codes [5] over the corresponding channels. The presented error probability bounds rely on precise average enumerators for given ensembles which makes these bounds tighter than known bounds (see e.g. [28]). By using an approach similar to that in [5], an improved upper bound on the performance of the BP decoding of binary images of nonbinary regular LDPC codes over BEC is presented.

The paper is organized as follows. In Sect. 2, all necessary notations and definitions are given. In Sect. 3, a near-ML decoding method, which is based on adding redundant rows to the code parity-check matrix, is revisited. In Sect. 4, a recurrent procedure for computing the exact coefficients of the weight and stopping set size spectra is described. The improved upper bound on the ensemble average performance of the BP decoding over BEC is derived. Tables of the computed code parameters along with the simulation results for the BP, ML and near-ML decoding are presented in Sect. 5. A comparison with the theoretical bounds is done and conclusions are drawn in Sect. 6.

2   Preliminaries

2.1   Ensembles of Binary and Binary Images of Nonbinary Regular LDPC Codes

For a binary linear $[n, k]$ code $\mathcal{C}$ of rate $R = k/n$ denote by $r = n - k$ its redundancy. We use a notation $\{A_{n,w}\}_{0 \leq w \leq n}$ for a set of code weight enumerators, where $A_{n,w}$ is a number of codewords of weight w. Let $\boldsymbol{H}$ be an $r \times n$ parity-check matrix which defines $\mathcal{C}$.

By viewing $\boldsymbol{H}$ as a biadjacency matrix [1], we obtain a corresponding bipartite Tanner graph. The girth g is the length of the shortest cycle in the Tanner graph.

When decoded over a BEC, the FER performance of the BP decoding is determined by the size of the smallest stopping set called stopping distance d_{stop} (see, for example, [9]). In turn, a stopping set is defined as a subset of indices of columns in a parity-check matrix, such that a matrix constructed from these columns does not have a row of weight one. The asymptotic behavior of a stopping set distribution for ensembles of binary LDPC codes is studied in [22]. In this paper, we study both the average performance of the ensembles of random LDPC codes and of QC LDPC codes widely used in practical schemes.

Two ensembles of random regular LDPC codes are studied below. First we study the Gallager ensemble [13] of (J, K)-regular LDPC codes, where J and K denote the number of ones in each column and in each row of the code parity-check matrix, respectively. Codes of this ensemble are determined by random parity-check matrices $\boldsymbol{H}$, which consist of the strips $\boldsymbol{H}_i$ of width $M = r/J$ rows each, $i = 1, 2, \ldots, J$. All strips are random column permutations of the strip where the jth row contains K ones in positions $(j-1)K+1, (j-1)K+2, \ldots, jK$, for $j = 1, 2, \ldots, n/K$.

Next, we study the ensemble of binary (J, K)-regular LDPC codes, which is a special case of the ensemble described in [26, Definition 3.15]. We refer to this ensemble as the Richardson-Urbanke (RU) ensemble of (J, K)-regular LDPC codes.

For $a \in \{1, 2, \ldots\}$ denote by a^m a sequence $(a, a, \ldots, a)$ of m identical symbols a. In order to construct an $r \times n$ parity-check matrix $\boldsymbol{H}$ of an LDPC code from the RU ensemble, one does the following:

- construct the sequence $\boldsymbol{a} = (1^J, 2^J, \ldots, n^J)$;
- apply a random permutation $\boldsymbol{b} = \pi(\boldsymbol{a})$ to obtain a sequence $\boldsymbol{b} = (b_1, \ldots, b_N)$, where $N = Kr = Jn$;
- set to one the entries in the first row of $\boldsymbol{H}$ in columns $b_1, \ldots, b_K$, the entries in the second row of $\boldsymbol{H}$ in columns $b_{K+1}, \ldots, b_{2K}$, etc. The remaining entries of $\boldsymbol{H}$ are zeros.

In fact, an LDPC code from the RU ensemble is (J, K)-regular if for a given permutation π all elements of subsequences $(b_{iK-K+1}, \ldots, b_{iK})$ are different for all $i = 1, \ldots, r$. It is shown in [20] that the fraction of regular codes among the RU LDPC codes is roughly

$$\exp\left\{-\frac{1}{2}(K-1)(J-1)\right\},$$

which means that most of the RU codes are irregular. In what follows, we ignore this fact and interpret the RU LDPC codes as the (J, K)-regular codes, and call them "almost regular".

Generally, the design rate $R = 1 - J/K$ is a lower bound on the actual code rate since the rank of randomly constructed parity-check matrix can be smaller than the number of its rows. However, in our study the best generated almost regular RU codes always have the rate equal to the design rate. For this reason, we do not distinguish between the design rate and the actual rate.

In order to construct random binary images of nonbinary (J, K)-regular LDPC codes, we use the standard two-stage procedure. It consists of labeling a proper binary base parity-check matrix by random nonzero elements of the extension of the binary Galois field. In our work, we select a parity-check matrix of a binary LDPC code from the Gallager or the RU ensembles as the base matrix.

In what follows, the Gallager ensembles of binary regular LDPC codes and binary images of nonbinary regular LDPC codes are used only for the theoretical analysis, while for the simulations we use almost regular LDPC codes from the RU ensemble. The reason for this choice is that in the simulations, the RU LDPC ensembles outperform the Gallager LDPC codes with the same parameters.

2.2   QC LDPC Codes

The QC LDPC codes represent a class of LDPC codes which is very intensively used in communication standards. Rate $R = b/c$ QC LDPC codes are determined by a $(c - b) \times c$ polynomial parity-check matrix of their parent convolutional code [18]

$$
\boldsymbol{H}(D) = \begin{pmatrix}
h_{11}(D) & h_{12}(D) & \cdots & h_{1c}(D) \\
h_{21}(D) & h_{22}(D) & \cdots & h_{2c}(D) \\
\vdots & \vdots & \ddots & \vdots \\
h_{(c-b)1}(D) & h_{(c-b)2}(D) & \cdots & h_{(c-b)c}(D)
\end{pmatrix},
\tag{1}
$$

where $h_{ij}(D)$ is either zero or a monomial entry, that is, $h_{ij}(D) \in \{0, D^{w_{ij}}\}$ with w_{ij} being a nonnegative integer, $w_{ij} \leq \mu$, and $\mu = \max_{i,j}\{w_{ij}\}$ is the syndrome memory.

The polynomial matrix (1) determines an $[Mc, Mb]$ QC LDPC block code using a set of polynomials modulo $D^M - 1$. By tailbiting the parent convolutional code to length $M > \mu$, we obtain the binary parity-check matrix

$$
\boldsymbol{H}_{\text{TB}} = \begin{pmatrix}
\boldsymbol{H}_0 & \boldsymbol{H}_1 & \cdots & \boldsymbol{H}_{\mu-1} & \boldsymbol{H}_\mu & 0 & \cdots & 0 \\
0 & \boldsymbol{H}_0 & \boldsymbol{H}_1 & \cdots & \boldsymbol{H}_{\mu-1} & \boldsymbol{H}_\mu & \cdots & 0 \\
\vdots & \ddots & \vdots & \vdots & \vdots & \vdots & \ddots & \\
\boldsymbol{H}_\mu & 0 & \cdots & 0 & \boldsymbol{H}_0 & \boldsymbol{H}_1 & \cdots & \boldsymbol{H}_{\mu-1} \\
\vdots & \ddots & \vdots & \vdots & \vdots & \vdots & \vdots & \vdots \\
\boldsymbol{H}_1 & \cdots & \boldsymbol{H}_\mu & 0 & \cdots & 0 & \cdots & \boldsymbol{H}_0
\end{pmatrix}
\tag{2}
$$

of an equivalent (in the sense of column permutation) TB code (see [18, Chap. 2]), where $\boldsymbol{H}_i$, $i = 0, 1, \ldots, \mu$, are binary $(c - b) \times c$ matrices in the series expansion

$$
\boldsymbol{H}(D) = \boldsymbol{H}_0 + \boldsymbol{H}_1 D + \cdots + \boldsymbol{H}_\mu D^\mu,
$$

and $\mathbf{0}$ is the all-zero matrix of size $(c - b) \times c$. If each column of $\mathbf{H}(D)$ contains J nonzero elements, and each row contains K nonzero elements, the QC LDPC block code is (J, K)-regular. It is irregular otherwise.

Another form of the equivalent $[Mc, Mb]$ binary QC LDPC block code can be obtained by replacing the nonzero monomial elements of $\mathbf{H}(D)$ in (1) by the powers of the circulant $M \times M$ permutation matrix $\mathbf{P}$, whose rows are cyclic shifts by one position to the right of the rows of the identity matrix.

The polynomial parity-check matrix $\mathbf{H}(D)$ (1) can be interpreted as a $(c - b) \times c$ binary base matrix $\mathbf{B}$ labeled by monomials, where the entry in $\mathbf{B}$ is one if and only if the corresponding entry of $\mathbf{H}(D)$ is nonzero, i.e.

$$\mathbf{B} = \mathbf{H}(D)|_{D=1}.$$

All three matrices $\mathbf{B}$, $\mathbf{H}(D)$, and $\mathbf{H}$ can be interpreted as bi-adjacency matrices of the corresponding Tanner graphs.

3   Stopping Redundancy and Convergence to the ML Decoding Performance

The idea to improve the performance of iterative decoding of linear codes over a BEC by using redundant parity checks was studied, for example, in [27, 34]. This approach was further explored in [29] (for BEC) and in [35] (for BSC and AWGN). The idea of using redundant parity checks was also studied in the context of linear-programming decoding [12], the reader can refer, for example, to [32].

A straightforward method to extend a parity-check matrix of an LDPC code is based on appending a predetermined number of dual codewords to the parity-check matrix. In this approach, the BP decoder uses the redundant matrix instead of the original parity-check matrix. One of the strategies used to extend the parity-check matrix consists of appending dual codewords in the order of their increasing weights starting with the minimum weight d_{dual}. A problem of searching for low-weight dual codewords has high computational complexity in general, yet for short LDPC codes it is feasible. We apply this approach in the sequel, and study the convergence of the FER of BP decoding of LDPC codes determined by their extended parity-check matrices to the FER of the ML decoding (for both BEC and AWGN channels).

The stopping redundancy is defined as the minimum number of rows in a parity-check matrix required to ensure that the stopping distance of the code d_{stop} is equal to the code minimum distance d_{min}. For a set of the selected LDPC codes, we compute estimates on the minimum number of the rows required in order to ensure removal of stopping sets of a certain size. Next, we describe this approach in more detail. By ℓ-th stopping redundancy, ρ_ℓ, we denote the minimum number of rows in any parity-check matrix of the code, such that all ML-decodable stopping sets of size less than or equal to ℓ are removed. In particular, ρ_r is the minimum number of rows in any parity-check matrix of the code, such that there are no ML-decodable stopping sets of size up to r (incl.), i.e. no stopping sets which, if erased, still can be decoded by the ML decoder. Our definition of ℓ-th stopping redundancy is analogous to its counterpart in [15].

However, we stress the difference between the updated definition of the ℓ-th stopping redundancy for $\ell \geq d$ and its counterpart in [15]. In fact, the stopping sets of size $\ell \geq d$ that are *not* ML-decodable, are exactly the supports of the codewords.[1]

In order to calculate the upper bounds on the ℓ-th stopping redundancy with a method based on [33], we first estimate by sampling u_i, the number of ML-decodable stopping sets of size i in a particular parity-check matrix. Then, we use the estimates on u_i $(i = 1, 2, \ldots, r)$ with the method similar to [33, Theorem 1,2] in order to obtain the approximate upper bounds on the *stopping redundancy hierarchy*, i.e. the stopping redundancies $\rho_1, \rho_2, \ldots, \rho_r$.

In Table 1, we present estimates on ρ_ℓ, $\ell = d_{\min}, d_{\min} + 1, d_{\min} + 2$, and $\ell = r$, along with $d_{\min}$, d_{stop}, d_{dual} and g, for a set of selected LDPC codes. In Sect. 5, we also present the simulated FER performance of the BP and ML decoding over the BEC for this set of codes with varying number of redundant rows. The same set of LDPC codes with varying number of redundant rows in their parity-check matrices is also simulated over the AWGN channel.

The LDPC codes from the following four families were selected:

- Random regular LDPC codes from the RU ensemble (rows 2 and 3 in Table 1)
- QC LDPC codes (row 4)
- Binary images of nonbinary regular LDPC codes (row 5)
- Linear codes represented in a "sparse form" (row 1)

Two random RU codes were selected by an exhaustive search among 100000 code candidates. As a search criteria, we used the minimum distance and the first spectrum coefficient $A_{d_{\min},n}$. The QC LDPC code was obtained by optimization of lifting degrees for a constructed base matrix in order to guarantee the best possible minimum distance under a given restriction on the girth value of the code Tanner graph. For comparison, we simulated the best linear code with the same length and dimension determined by a parity-check matrix with the lowest possible correlation between its rows. Next, we refer to this form of the

Table 1. Parameters of studied $[48, 24]$ codes

Code	$d_{\min}$	$A_{d_{\min},n}$	d_{stop}	d_{dual}	g	J,K	$\rho_{d_{\min}}, \rho_{d_{\min}+1}, \rho_{d_{\min}+2}$	ρ_r	Type
1	12	17296	4	12	4	6,12	6240,12151,23468	13 761 585	'L'
2	8	13	4	6	4	6,12	261,581,1254	13 683 513	'RU'
3	7	1	5	5	4	4,8	83,175,380	12 549 204	'RU'
4	7	8	7	5	6	3,6	58,130,274	9 876 964	'QC'
5	8	7	4	7	4	3,6	355,751,1551	13 819 276	'NB'

---

[1] We recall that a support of a codeword is a stopping set.

parity-check matrix as a "sparse form". Parameters of the selected codes are presented in Table 1. Here we use the notations 'RU' for random LDPC codes, 'L' for the best linear code with parity-check matrix in 'sparse form', 'NB' for the binary image of nonbinary regular LDPC code and 'QC' for QC LDPC code, respectively.

4   Upper Bounds on ML and BP Decoding Error Probability for Ensembles of LDPC Codes

In this section, we analyze the Gallager ensembles of binary and binary images of nonbinary (J, K)-regular LDPC codes. By following the approach in [5] we derive estimates on the decoding error probability of the ML and BP decoding by using precise coefficients of the average weight spectrum and average stopping set size spectrum, respectively. Additionally to the bounds on the performance of the ML decoding obtained in [5], in this paper we derive the improved bounds on the performance of BP decoding for both binary LDPC codes and binary images of nonbinary regular LDPC codes.

The main idea behind the approach in [5] is computing the average spectra coefficients recurrently with complexity linear in n (see also [8]). The resulting coefficients are substituted into the union-type upper bound on the error probability of the ML decoding over a BEC [3]

$$P_e \leq \sum_{i=d}^{n} \min \left\{ \binom{n}{i}, \sum_{w=d}^{i} S_w \binom{n-w}{i-w} \right\} \varepsilon^i (1-\varepsilon)^{n-i}, \qquad (3)$$

where S_w is the w-th weight (stopping set size) spectrum coefficient, ε is the erasure probability and d denotes the minimum distance (stopping distance). In order to upper-bound the error probability of the ML decoding over an AWGN channel, the average weight spectrum coefficients are substituted into the tangential-sphere bound [25].

Consider the Gallager ensemble of q-ary LDPC codes, where $q = 2^m$, $m \geq 1$ is an integer. The weight generating function of q-ary sequences of length n satisfying the nonzero part of one q-ary parity-check equation is given in [13] as

$$g(s) = \frac{(1 + (q-1)s)^K + (q-1)(1-s)^K}{q}. \qquad (4)$$

It is easy to derive the weight generating function of q-ary sequences of length K and q-ary weight not equal to 1:

$$g_{\text{stop}}(s) = \sum_{w=0,2,3,\ldots,K} \binom{K}{w} (q-1)^w s^w = (1 + (q-1)s)^K - K(q-1)s. \qquad (5)$$

Each q-ary symbol can be represented as a binary sequence (image) of length m. It is easy to see that different representations of a finite field of characteristic two will lead to different generating functions of binary images for the same ensemble of nonbinary LDPC codes. Following the techniques in [10], we study an average binary weight spectrum for the ensemble of m-dimensional binary images. By assuming uniform distribution on the m-dimensional binary images of the non-zero q-ary symbols, we obtain the generating function of the average binary weights of a q-ary symbol in the form

$$\phi(s) = \frac{1}{q-1} \sum_{w=1}^{m} \binom{m}{w} s^w = \frac{(1+s)^m - 1}{q-1}. \tag{6}$$

The average binary weight generating function for one strip is given by

$$G(s) = \left(g(\phi(s))\right)^M = \sum_{w=0}^{nm} N_{nm,w} s^w,$$

where $N_{nm,w}$ denotes the average number of binary sequences β of weight w and of length nm satisfying $\beta \mathcal{B}_i^{T} = \mathbf{0}$. Here, $\mathcal{B}_i$ denotes the average binary image of $\boldsymbol{H}_i$. We obtain the average binary weight enumerator of nonbinary regular LDPC code as

$$\mathrm{E}\{A_{nm,w}\} = \binom{nm}{w} \left(p(w)\right)^J = \binom{nm}{w}^{1-J} N_{nm,w}^J, \tag{7}$$

where $p(w) = \binom{nm}{w}^{-1} N_{nm,w}$. By substituting (6) into (5), similarly to (7), we obtain the average binary stopping set size spectrum coefficient.

It is known that if the generating function is represented as a degree of another generating function it can be easily computed by applying a recurrent procedure. Details of the recurrent procedure for computing coefficients of the average weight spectra can be found in [5]. We proceed by computing $N_{nm,w}$ recursively.

5   Simulation Results

We simulate the BP and ML decoding over the BEC and AWGN channel for the five LDPC codes whose parameters are presented in Table 1. In Fig. 1, the FER performance of the BP and ML decoding over the BEC and the AWGN channel is compared. It is easy to see that the best BP decoding performance both over the BEC and over the AWGN channel (and at the same time the worse ML decoding performance) is shown by the QC LDPC code with the most sparse parity-check matrix and the largest girth value of its Tanner graph. We remark that the best linear [48,24,12] code determined by a parity-check matrix in a "sparse form", as expected, has the best ML decoding performance over the both channels. Its BP decoding performance is worse than that of the selected LDPC codes except for the binary image of nonbinary LDPC code.

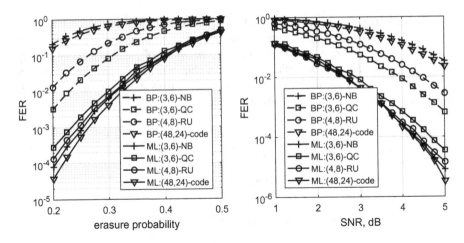

Fig. 1. Comparison of the FER performance of BP and ML decoding over the BEC and the AWGN channel for LDPC codes of length $n = 48$ and rate $R = 1/2$

Figure 2 shows the BP decoding performance over the BEC and AWGN channel of the codes 'QC' and 'L' from Table 1, when their parity-check matrices are extended. We call the corresponding decoding technique "redundant parity check" (RPC) decoding. The number next to "RPC" in Fig. 2 indicates the number of redundant rows that was added. The best convergence of the FER performance of the BP decoding over the BEC to that of the ML decoding is demonstrated by the QC LDPC code, while the best linear code has the slowest convergence of its BP performance to the ML decoding performance. We observe that the obtained simulation results are consistent with the estimates on the stopping redundancy hierarchy given in Table 1. Surprisingly, similar behavior can also be observed for the FER performance of RPC decoding over the AWGN channel.

6   Discussion

In this section, we compare the simulated FER performance of the BP, ML and near-ML (RPC) decoding over the BEC and the AWGN channel with improved bounds on the ML and BP decoding performance. In Fig. 3, the FER performance over the BEC for the binary image of nonbinary $(3, 6)$-regular LDPC code over $GF(2^4)$ ('NB' code in Table 1) and the corresponding bounds are shown.

As it is shown in the presented plots, the ML performance of the 'NB' code is rather close to the ML performance of the 'L' code, but the convergence of the FER performance of the RPC decoding to the performance of the ML decoding for the 'NB' code is much faster than for the 'L' code.

In Fig. 4, the FER performance of the BP, ML and RPC decoding over the BEC and the AWGN channel is compared to the corresponding upper and lower bounds on the performance of the ML decoding. In particular, for comparison

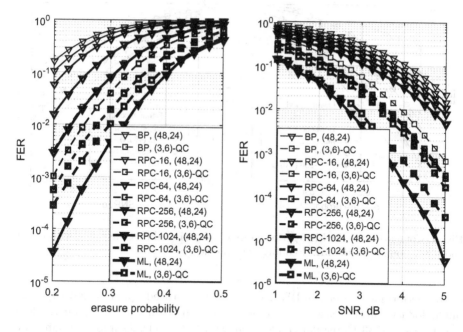

Fig. 2. FER performance of RPC decoding over the BEC and the AWGN channel for 'L' and 'QC' codes.

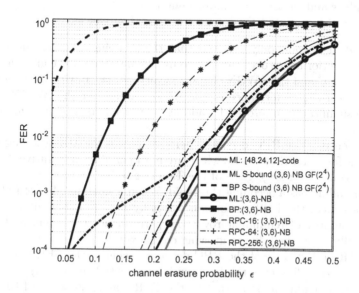

Fig. 3. Comparison of the FER performance of BP and RPC decoding over the BEC with improved union-type bounds (3) on the ML and BP decoding performance.

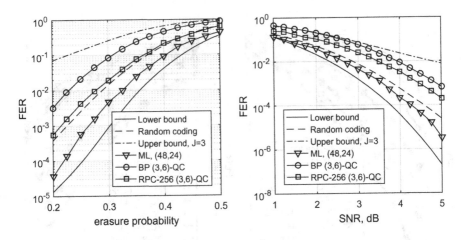

Fig. 4. Comparison of the FER performance of BP, ML and RPC decoding with upper and lower bounds on the ML decoding performance.

of the performance over the BEC, we use the improved upper bound (3) computed for the precise ensemble average spectrum coefficients for both random linear code and $(3, 6)$-regular random binary LDPC code. As a lower bound, we consider the tighten sphere-packing bound in [6]. For comparison of the performance over the AWGN channel, we show the tangential-sphere upper bound [25] computed with the precise ensemble average spectrum coefficients for the same two ensembles and the Shannon lower bound [30].

Based on the presented results, we conclude the following:

- Although it is commonly believed that the stopping sets influence the BP decoding performance over the BEC only, the behavior of the analyzed codes over the BEC and the AWGN channel is very similar. In particular, for short codes, the FER performance of the BP decoding over the AWGN channel can be significantly improved by adding redundant rows to the parity-check matrix.
- Convergence of the RPC decoding performance to the ML decoding performance is faster for those codes which are most suitable for iterative decoding, that is, codes with large girth of the Tanner graph.
- RPC decoding has a decoding threshold. When a small number of redundant rows is added, the FER performance rapidly improves, but after adding a certain number of redundant rows, the performance improvement becomes practically unjustified due to growing complexity.
- The FER performance of the RPC decoding achieves the FER performance of the ML decoding over the BEC with exponential (in length) complexity. However, a significant reduction in the FER compared to the FER of BP decoding can be achieved with a significantly lower complexity than that of the ML decoding.

– Binary images of nonbinary LDPC codes with RPC decoding demonstrate good FER performance over the BEC. In order to apply RPC decoding to these codes over the AWGN channel it is required to add q-ary parity-checks to their parity-check matrices. This method looks promising and is subject of our future research.

References

1. Asratian, A.S.: Bipartite Graphs and Their Applications, vol. 131. Cambridge University Press, Cambridge (1998)
2. Baldi, M., Chiaraluce, F., Maturo, N., Liva, G., Paolini, E.: A hybrid decoding scheme for short non-binary LDPC codes. IEEE Commun. Lett. **18**(12), 2093–2096 (2014)
3. Berlekamp, E.R.: The technology of error-correcting codes. Proc. IEEE **68**(5), 564–593 (1980)
4. Bocharova, I.E., Kudryashov, B., Johannesson, R.: Searching for binary and nonbinary block and convolutional LDPC codes. IEEE Trans. Inform. Theory **62**(1), 163–183 (2016)
5. Bocharova, I.E., Kudryashov, B.D., Skachek, V.: Performance of ML decoding for ensembles of binary and nonbinary regular LDPC codes of finite lengths. In: Proceeding of IEEE International Symposium on Information Theory (ISIT), 2017, pp. 794–798 (2017)
6. Bocharova, I.E., Kudryashov, B.D., Skachek, V., Rosnes, E., Ytrehus, Ø.: ML and Near-ML Decoding Performance of LDPC Codes over BEC: Bounds and Decoding Algorithms (preprint) (August 2017). http://kodu.ut.ee/~vitaly/Papers/BEC-bounds/becbounds.pdf
7. Bocharova, I.E., Kudryashov, B.D., Skachek, V., Yakimenka, Y.: Low complexity algorithm approaching the ML decoding of binary LDPC codes. In: Proceeding of IEEE International Symposium on Information Theory (ISIT), 2016, pp. 2704–2708 (2016)
8. Bocharova, I.E., Kudryashov, B.D., Skachek, V., Yakimenka, Y.: Average spectra for ensembles of LDPC codes and their applications. In: Proceeding of IEEE International Symposium on Information (ISIT), 2017, pp. 361–365 (2017)
9. Di, C., Proietti, D., Telatar, I.E., Richardson, T.J., Urbanke, R.L.: Finite-length analysis of low-density parity-check codes on the binary erasure channel. IEEE Trans. Inform. Theory **48**(6), 1570–1579 (2002)
10. El-Khamy, M., McEliece, R.J.: Bounds on the average binary minimum distance and the maximum likelihood performance of Reed Solomon codes. In: Proceeding of 42nd Allerton Conference on Communication, Control and Computing (2004)
11. Fang, Y., Zhang, J., Wang, L., Lau, L.: BP-Maxwell decoding algorithm for LDPC codes over AWGN channels. In: 6th International Conference on Wireless Communications, Networking and Mobile Computing (WiCOM), 2010, pp. 1–4 (2010)
12. Feldman, J., Wainwright, M.J., Karger, D.R.: Using linear programming to decode binary linear codes. IEEE Trans. Inform. Theory **51**(3), 954–972 (2005)
13. Gallager, R.G.: Low-Density Parity-Check Codes. M.I.T. Press, Cambridge (1963)
14. Hehn, T., Huber, J.B., Laendner, S.: Improved iterative decoding of LDPC codes from the IEEE WiMAX standard. In: Proceeding International ITG Conference on Source and Channel Coding (SCC), 2010, pp. 1–6 (2010)

15. Hehn, T., Milenkovic, O., Laendner, S., Huber, J.B.: Permutation decoding and the stopping redundancy hierarchy of cyclic and extended cyclic codes. IEEE Trans. Inform. Theory **54**(12), 5308–5331 (2008)

16. Hosoya, G., Matsushima, T., Hirasawa, S.: A decoding method of low-density parity-check codes over the binary erasure channel. In: Proceeding of 27th Symposium on Information Theory and its Applications (SITA), 2004, pp. 263–266 (2004)

17. Jianjun, M., Xiaopeng, J., Jianguang, L., Rong, S.: Parity-check matrix extension to lower the error floors of irregular LDPC codes. IEICE Trans. Commun. **94**(6), 1725–1727 (2011)

18. Johannesson, R., Zigangirov, K.S.: Fundamentals of Convolutional Coding, vol. 2015. Wiley (2015)

19. Laendner, S., Hehn, T., Milenkovic, O., Huber, J.B.: When does one redundant parity-check equation matter? In: Global Telecommunications Conference (Globecom), 2006, pp. 1–6 (2006)

20. Litsyn, S., Shevelev, V.: On ensembles of low-density parity-check codes: asymptotic distance distributions. IEEE Trans. Inform. Theory **48**(4), 887–908 (2002)

21. Olmos, P.M., Murillo-Fuentes, J.J., Pérez-Cruz, F.: Tree-structure expectation propagation for decoding LDPC codes over binary erasure channels. In: Proceeding of IEEE International Symposium on Information Theory (ISIT), 2010, pp. 799–803 (2010)

22. Orlitsky, A., Viswanathan, K., Zhang, J.: Stopping set distribution of LDPC code ensembles. IEEE Trans. Inform. Theory **51**(3), 929–953 (2005)

23. Pishro-Nik, H., Fekri, F.: On decoding of low-density parity-check codes over the binary erasure channel. IEEE Trans. Inform. Theory **50**(3), 439–454 (2004)

24. Pishro-Nik, H., Fekri, F.: Results on punctured low-density parity-check codes and improved iterative decoding techniques. IEEE Trans. Inform. Theory **53**(2), 599–614 (2007)

25. Poltyrev, G.: Bounds on the decoding error probability of binary linear codes via their spectra. IEEE Trans. Inform. Theory **40**(4), 1284–1292 (1994)

26. Richardson, T., Urbanke, R.: Modern Coding Theory. Cambridge University Press, Cambridge (2008)

27. Santhi, N., Vardy, A.: On the effect of parity-check weights in iterative decoding. In: Proceeding of International Symposium on Information Theory (ISIT), Chicago, IL, p. 322 (2004)

28. Sason, I., Shamai, S.: Improved upper bounds on the ensemble performance of ML decoded low density parity check codes. IEEE Commun. Lett. **4**(3), 89–91 (2000)

29. Schwartz, M., Vardy, A.: On the stopping distance and the stopping redundancy of codes. IEEE Trans. Inform. Theory **52**(3), 922–932 (2006)

30. Shannon, C.E.: Probability of error for optimal codes in a Gaussian channel. Bell Syst. Tech. J. **38**(3), 611–656 (1959)

31. Varnica, N., Fossorier, M.P., Kavcic, A.: Augmented belief propagation decoding of low-density parity check codes. IEEE Trans. Commun. **55**(7), 1308–1317 (2007)

32. Vontobel, P.O., Kötter, R.: Graph-cover decoding and finite-length analysis of message-passing iterative decoding of LDPC codes, arXiv preprint cs/0512078 (2005)

33. Yakimenka, Y., Skachek, V.: Refined upper bounds on stopping redundancy of binary linear codes. In: Proceeding of IEEE Information Theory Workshop (ITW), 2015, pp. 1–5 (2015)

34. Yedidia, J.S., Chen, J., Fossorier, M.: Generating code representations suitable for belief propagation decoding. In: Proceeding of 40-th Allerton Conference Communication, Control, and Computing, Monticello, IL (2002)
35. Zumbrägel, J., Flanagan, M.F., Skachek, V.: On the pseudocodeword redundancy of binary linear codes. IEEE Trans. Inform. Theory $58(7)$, 4848–4861 (2012)

Decoding a Perturbed Sequence Generated by an LFSR

Sara D. Cardell[1], Joan-Josep Climent[2(✉)], and Alicia Roca[3]

[1] Instituto de Matemática, Estatística e Computação Científica,
Universidade Estadual de Campinas (UNICAMP), R. Sérgio Buarque de Holanda,
651, Cidade Universitária, Campinas, SP 13083-859, Brazil
sdcardell@ime.unicamp.br
[2] Departament de Matemàtiques, Universitat d'Alacant,
Ap. Correus 99, 03080 Alacant, Spain
jcliment@ua.es
[3] Departamento de Matemática Aplicada, IMM,
Universitat Politècnica de València, Camí de Vera, s/n, 46022 València, Spain
aroca@mat.upv.es

Abstract. Given a sequence of bits produced by a linear feedback shift register (LFSR), the Berlekamp-Massey algorithm finds a register of minimal length able to generate the sequence. The situation is different when the sequence is perturbed; for instance, when it is sent through a transmission channel. LFSRs can be described as autonomous systems. A perturbed sequence of bits generated by an LFSR can be interpreted as a codeword in the binary linear code generated by the corresponding observability matrix. The problem of finding the original sequence can then be stated as the decoding problem, "given the received codeword, find the information transmitted". We propose two decoding algorithms, one based on a brute force attack and the other one based on the representation technique of the syndromes introduced by Becker, Joux, May, and Meurer (2012).

Keywords: LFSR · Correlation attack · Keystream sequence · Companion matrix · Autonomous system · Syndrome decoding · Decoding representation technique

1 Introduction

Binary linear feedback shift registers (LFSR) are used to generate pseudorandom sequences with desirable properties for keystream (see, for example, [7,11]). However, given a sequence of bits produced by an LFSR, the Berlekamp-Massey algorithm [13] finds a register of minimal length able to generate the sequence, without any further information. Linearity must be destroyed in order to produce a sequence useful as a keystream.

A common method aiming at destroying the predictability of the output of LFSRs is to use the outputs of several LFSRs as inputs of a suitably designed

© Springer International Publishing AG 2017
A.I. Barbero et al. (Eds.): ICMCTA 2017, LNCS 10495, pp. 62–71, 2017.
DOI: 10.1007/978-3-319-66278-7_6

nonlinear Boolean function in order to produce a keystream. Very frequently, in these cases, a correlation appears between the keystream of the nonlinear generator and the output of one of its LFSR components (like the Geffe generator [5] or the A5/1 algorithm [6]). Correlation attacks are a class of plaintext attacks that exploit the statistical weakness that arises from a poor choice of the Boolean function.

The first correlation attack was devised by Siegenthaler [18] and is based on a model where the keystream is viewed as a noisy version of the output of some of the constituent LFSRs. He assumed that noise was additive and independent of the underlying LFSR sequence. Later, Meier and Staffelbach [15] presented two different algorithms for fast correlation attacks using a correlation between the keystream and the output stream of an LFSR. Other correlations attacks have been developed since then; see [1,3,4,8,12,14,17,19], among others.

In this paper the LFSR sequence is understood as a codeword of a certain linear block code, and this work is devoted to decode it.

The paper is organized as follows. In Sect. 2, the correlation problem is expressed as a coding theory problem. In Sect. 3, we propose two different algorithms to recover the initial state of the LFSR given the keystream are proposed.

2   A Decoding Problem

Assume that the characteristic polynomial of an LFSR is

$$f(x) = c_0 + c_1 x + c_2 x^2 + \cdots + c_{k-1} x^{k-1} + x^k \in \mathbb{F}_2[x].$$

Let us consider the **companion matrix**

$$A = \begin{bmatrix} 0 & 0 & 0 & \cdots & 0 & c_0 \\ 1 & 0 & 0 & \cdots & 0 & c_1 \\ 0 & 1 & 0 & \cdots & 0 & c_2 \\ \vdots & \vdots & \vdots & & \vdots & \vdots \\ 0 & 0 & 0 & \cdots & 0 & c_{k-2} \\ 0 & 0 & 0 & \cdots & 1 & c_{k-1} \end{bmatrix} \in \mathbb{F}_2^{k \times k}$$

of $f(x)$, and the column matrix

$$C = \begin{bmatrix} 1 & 0 & 0 & \cdots & 0 & 0 \end{bmatrix}^T \in \mathbb{F}_2^{k \times 1}.$$

The LFSR defined by $f(x)$ can be described as the autonomous system

$$\left. \begin{array}{c} x_{t+1} = x_t A \\ y_t = x_t C \end{array} \right\} \qquad t = 0, 1, 2 \ldots \qquad (1)$$

where x_t is the state of the system at instant t. We denote by x_0 the **initial state** and we say that system (1) is in **canonical observability form** (see [9]). If we consider an initial state $x_0 = (y_0, y_1, \ldots, y_{k-1})$, then the t-th stream bit y_t, for $t \geq k$, can be computed using expression (1). Moreover, assuming that $f(\mathsf{x})$ is a primitive polynomial, then we know that the output sequence $y_0, y_1, \ldots, y_{k-1}, y_k, y_{k+1}, \ldots$ has the maximum period $2^k - 1$.

Let n be a positive integer such that $k < n < 2^k - 1$. We can compute any output sequence $\boldsymbol{y} = (y_0, y_1, y_2, \ldots, y_{k-1}, y_k, y_{k+1}, \ldots, y_{n-1})$ as $\boldsymbol{y} = \boldsymbol{x}_0 G$ where

$$G = \begin{bmatrix} C & AC & A^2C & A^3C & \cdots & A^{n-2}C & A^{n-1}C \end{bmatrix}$$

is the **observability matrix** of the system given by expression (1).

Assume that we know neither the sequence $\boldsymbol{y}$ nor the initial state $\boldsymbol{x}_0$. Assume also that we know the sequence

$$\boldsymbol{z} = (z_0, z_1, z_2, \ldots, z_{k-1}, z_k, z_{k+1}, \ldots, z_{n-1})$$

which is correlated to the sequence $\boldsymbol{y}$ with correlation probability $1 - \varepsilon$ (usually $0.25 \le \varepsilon \le 0.5$). The idea of the correlation attack is to view the sequence $\boldsymbol{z}$ as a perturbation of the sequence $\boldsymbol{y}$ by a binary symmetric memoryless noise channel (see Fig. 1) with $Pr(z_t = y_t) = 1 - \varepsilon$ (see [15, 16]).

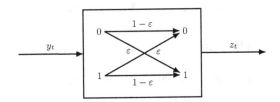

Fig. 1. Transmission over a binary symmetric channel

Thus, the LFSR sequence $\boldsymbol{y}$ is interpreted as a codeword in the $[n, k]$ binary linear code $\mathcal{C}$ generated by matrix G, and the keystream sequence $\boldsymbol{z}$ as the received channel output. Therefore, the correlation attack can now be reformulated as: *Given a received word $\boldsymbol{z} \in \mathbb{F}_2^n$, find the transmitted codeword $\boldsymbol{y} \in \mathcal{C}$.* This means that $\boldsymbol{z} = \boldsymbol{y} + \boldsymbol{e}$, where $\boldsymbol{e} \in \mathbb{F}_2^n$ is a vector with average Hamming weight $\mathrm{wt}(\boldsymbol{e}) = \lfloor \varepsilon n \rfloor$.

Assume now that $n = mk$ for some positive integer m, and that $k < n < 2^k - 1$. It follows that $1 < m < \frac{2^k - 1}{k}$. As the system (1) is in canonical observability form, it follows that

$$\begin{bmatrix} C & AC & A^2C & A^3C & \cdots & A^{k-1}C \end{bmatrix} = I_k$$

where I_k denotes the $k \times k$ identity matrix. Consequently

$$\begin{bmatrix} A^kC & A^{k+1}C & A^{k+2}C & A^{k+3}C & \cdots & A^{2k-1}C \end{bmatrix}$$
$$= A^k \begin{bmatrix} C & AC & A^2 & A^3 & \cdots & A^{k-1}C \end{bmatrix} = A^k.$$

So, we can write the generator matrix G of $\mathcal{C}$ as

$$G = \begin{bmatrix} I_k & A^k & A^{2k} & A^{3k} & \cdots & A^{(m-2)k} & A^{(m-1)k} \end{bmatrix}$$

which is in systematic form. Consequently,

$$
H = \begin{bmatrix}
(A^k)^T & I_k & & & & \\
(A^{2k})^T & & I_k & & & \\
(A^{3k})^T & & & I_k & & \\
\vdots & & & & \ddots & \\
(A^{(m-2)k})^T & & & & & I_k & \\
(A^{(m-1)k})^T & & & & & & I_k
\end{bmatrix}
\tag{2}
$$

is a parity-check matrix of $\mathcal{C}$.

3   The Decoding Algorithm

From now on, we will assume that the output sequence of the LFSR defined by $f(\mathsf{x})$, and the received sequence are

$$
\boldsymbol{y} = (\boldsymbol{y}_0, \boldsymbol{y}_1, \boldsymbol{y}_2, \ldots, \boldsymbol{y}_{m-2}, \boldsymbol{y}_{m-1}) \quad \text{and} \quad \boldsymbol{z} = (\boldsymbol{z}_0, \boldsymbol{z}_1, \boldsymbol{z}_2, \ldots, \boldsymbol{z}_{m-2}, \boldsymbol{z}_{m-1})
$$

respectively, where

$$
\boldsymbol{y}_i = (y_{ik}, y_{ik+1}, y_{ik+2}, \ldots, y_{(i+1)k-2}, y_{(i+1)k-1})
$$
$$
\boldsymbol{z}_i = (z_{ik}, z_{ik+1}, z_{ik+2}, \ldots, z_{(i+1)k-2}, z_{(i+1)k-1})
$$

for $i = 0, 1, 2, \ldots, m-2, m-1$.

Since $\boldsymbol{y} \in \mathcal{C}$, it follows that $\boldsymbol{y} H^T = \boldsymbol{0}$, i.e.,

$$
\boldsymbol{y}_0 \begin{bmatrix} A^k & A^{2k} & A^{3k} & \cdots & A^{(m-2)k} & A^{(m-1)k} \end{bmatrix} + \begin{bmatrix} \boldsymbol{y}_1 & \boldsymbol{y}_2 & \boldsymbol{y}_3 & \cdots & \boldsymbol{y}_{m-2} & \boldsymbol{y}_{m-1} \end{bmatrix}
$$
$$
= \begin{bmatrix} \boldsymbol{0} & \boldsymbol{0} & \boldsymbol{0} & \cdots & \boldsymbol{0} & \boldsymbol{0} \end{bmatrix}
$$

and therefore

$$
\boldsymbol{y}_0 A^{ik} + \boldsymbol{y}_i = \boldsymbol{0}, \quad \text{for} \quad i = 1, 2, \ldots, m-1.
$$

This means that $(\boldsymbol{y}_0, \boldsymbol{y}_i)$ is a codeword of the $[2k, k]$ binary code $\mathcal{C}_i$ whose parity-check matrix is $H_i = \begin{bmatrix} (A^{ik})^T & I_k \end{bmatrix}$, for $i = 1, 2, \ldots, m-1$.

On the other hand, let $\boldsymbol{s} \in \mathbb{F}_2^{(m-2)k}$ be the syndrome of $\boldsymbol{z}$, i.e., $\boldsymbol{z} H^T = \boldsymbol{s}$, i.e.,

$$
\boldsymbol{z}_0 \begin{bmatrix} A^k & A^{2k} & A^{3k} & \cdots & A^{(m-2)k} & A^{(m-1)k} \end{bmatrix} + \begin{bmatrix} \boldsymbol{z}_1 & \boldsymbol{z}_2 & \boldsymbol{z}_3 & \cdots & \boldsymbol{z}_{m-2} & \boldsymbol{z}_{m-1} \end{bmatrix}
$$
$$
= \begin{bmatrix} \boldsymbol{s}_0 & \boldsymbol{s}_1 & \boldsymbol{s}_2 & \cdots & \boldsymbol{s}_{m-3} & \boldsymbol{s}_{m-2} \end{bmatrix}
$$

or equivalently,

$$
\boldsymbol{z}_0 A^{(i+1)k} + \boldsymbol{z}_{i+1} = \boldsymbol{s}_i, \quad \text{for} \quad i = 0, 1, 2, \ldots, m-2,
$$

where

$$
\boldsymbol{s}_i = (s_{ik}, s_{ik+1}, s_{ik+2}, \ldots, s_{(i+1)k-2}, s_{(i+1)k-1}), \quad \text{for} \quad i = 0, 1, 2, \ldots, m-2.
$$

Since $z = y + e$; i.e., $z_i = y_i + e_i$, for $i = 0, 1, 2, \ldots, m - 1$, it follows that

$$e_0 A^{(i+1)k} + e_{i+1} = s_i, \quad \text{for} \quad i = 0, 1, 2, \ldots, m - 2,$$

or equivalently,

$$e_0 B^T + \tilde{e} = s, \tag{3}$$

where

$$B^T = \begin{bmatrix} A^k \ A^{2k} \ A^{3k} \cdots A^{(m-2)k} \ A^{(m-1)k} \end{bmatrix},$$
$$\tilde{e} = \begin{bmatrix} e_1 \ e_2 \ e_3 \cdots e_{m-2} \ e_{m-1} \end{bmatrix}.$$

Remember that we are looking for a vector $e \in \mathbb{F}_2^n$ such that $eH^T = s$ and $\text{wt}(e) = w$. If we assume that the ω nonzero components of e are uniformly distributed, we can assume that $\text{wt}(e_0) = \lfloor \varepsilon n/m \rfloor = \lfloor \varepsilon k \rfloor$.

Now, expression (3) suggests the brute force attack given by Algorithm 1 below.

Algorithm 1.

---

input : H, s, and ω
output: $e \in \mathbb{F}_2^n$ such that $eH^T = s$ and $\text{wt}(e) = \omega$

1 **for** $e_0 \in \mathbb{F}_2^k$ *such that* $\text{wt}(e_0) = \lfloor \varepsilon k \rfloor$ **do**
2 $\quad$ compute $\tilde{e}$ as $\tilde{e} = e_0 B^T + s$ $\qquad$ % see expression (3)
3 $\quad$ let $e = (e_0, \tilde{e})$
4 $\quad$ **if** $\text{wt}(e) = \omega$ **then**
5 $\quad\quad$ | STOP
6 $\quad$ **end**
7 **end**

The second algorithm is based on a modification of the representation technique of the syndromes proposed by Becker, Joux, May, and Meurer [2]. Recall that we are looking for a vector $e \in \mathbb{F}_2^n$ such that $eH^T = s$ and $\text{wt}(e) = w$. To improve calculations, we introduce additional hypotheses on the weight distribution of the vector e. We search for vectors e, which can be decomposed into $e = (\tilde{e}_0, \tilde{e}_1) \in \mathbb{F}_2^{2k} \times \mathbb{F}_2^{n-2k}$ where $\text{wt}(\tilde{e}_0) = p$, for some p such that $1 \le p \le 2k$, $\text{wt}(\tilde{e}_1) = w - p$, and $eH^T = s$. As we are assuming that the errors are uniformly distributed, in average, $p = wt(\tilde{e}_0) = \lfloor 2k\varepsilon \rfloor$. Then, as a consequence of expression (2), after obtaining the vector $\tilde{e}_0$ from

$$s_0 = \tilde{e}_0 H_1^T \tag{4}$$

we can find the vector $\tilde{e}_1$ as being

$$\tilde{e}_1 = e_0 B_1^T + \tilde{s}_1 \tag{5}$$

where $\tilde{e}_0 = (e_0, e_1)$, $\tilde{s}_1 = (s_1, s_2, \ldots, s_{m-2})$ and

$$B_1^T = \begin{bmatrix} A^{2k} & A^{3k} & \cdots & A^{(m-2)k} & A^{(m-1)k} \end{bmatrix}.$$

Therefore, we must focus on efficiently computing $\tilde{e}_0$. In paper [2], the authors propose an algorithm to decode random linear codes. The algorithm contains two main parts. The first one consists of transforming the parity check matrix H into a systematic form through elementary operations, and the second constitutes the proper decoding algorithm.

From now on, we use the following notation: if $u = (u_1, u_2, \ldots, u_\ell) \in \mathbb{F}_2^\ell$ is a given vector and $1 \le i \le j \le \ell$; then we denote by $u[i : j]$ the vector $(u_i, u_{i+1}, \ldots, u_j) \in \mathbb{F}_2^{j-i+1}$.

Because of the structure of matrix H (see (2)), our decoding problem is already stated in systematic form. Therefore, we can apply directly the second part of the algorithm of [2] on it. We describe next the main ideas on which the algorithm is based.

- Essentially the method approaches the value of $\tilde{e}_0$ in three steps. In each step a set of intermediate error vectors $u \in \mathbb{F}_2^{2k}$ of certain weight is obtained such that for a given number r, the last r components of uH_1^T match a chosen target vector in $\mathbb{F}_2^r$. The process is gradual in such a way that in the last step the vectors $u \in \mathbb{F}_2^{2k}$ selected have weight p and uH_1^T coincides with s_0. Therefore satisfying (4), as desired.
- In each step the procedure is analogous, and uses the **Merge-Join algorithm** of [2], which in turn is based on a classical matching algorithm by Knuth [10]. For given ρ, ν where $1 \le \rho \le 2k$, $1 \le \nu \le k$ and $t \in \mathbb{F}_2^\nu$, and two sets of vectors $\mathcal{U}$ and $\mathcal{V}$ in $\mathbb{F}_2^{2k}$ of weight approximately $\rho/2$, the Merge-Join procedure produces a new set of vectors $\mathcal{W}$ of weight ρ satisfying a matching condition as follows:

$$\mathcal{W} = \Big\{ u + v \in \mathbb{F}_2^{2k} \mid u \in \mathcal{U}, v \in \mathcal{V},$$

$$\mathrm{wt}(u + v) = \rho, \ \big((u + v) H_1^T\big) [k - \nu + 1 : k] = t \Big\}.$$

- The Merge-Join algorithm is run four times in step 2, two times in step 1, and one time in the final step 0. The input sets to each step of the procedure come from the outputs of previous steps in the same procedure. To start the procedure, four pairs of sets of $p_2/2$-weight vectors are prepared by brute search, in a way that the nonzero components of the vectors in each pair set appear in disjoint positions with respect to the other one.
- Another idea to be highlighted. As mentioned previously, in each step sets of p_i-weight vectors are obtained adding vectors coming from pairs of sets of p_{i+1}-weight vectors, where $p_{i+1} = \frac{p_i}{2} + \varepsilon_{i+1}$, for $i = 0, 1, 2$, and $p_0 = p$ and $\varepsilon_3 = 0$. This means that the added vectors must match in ε_{i+1} nonzero components. Notice that it allows the existence of $\binom{k}{p_{i+1}}$ possible vectors

instead of $\binom{k}{p_i/2}$ and as $\binom{k}{p_{i+1}} > \binom{k}{p_i/2}$, whenever $p_{i+1} \leq k/2$. It follows that we have more possible vectors to analyze.

The algorithm is described next (see Algorithm 2 below). Given the matrix H_1, the vector s_0, and the integer p with $1 \leq p \leq 2k$, the algorithm allows us to obtain a set $\mathcal{L}$ of vectors in $\mathbb{F}_2^{2k}$ whose elements are candidates for the solution of (4).

First we analyze the calculations performed in the different layers. The remarks below are intended to be an explanation of the behavior of the algorithm.

Remark 1. Assume that vectors $v_1 \in \mathbb{F}_2^{r_1}$ and $u_1, u_3 \in \mathbb{F}_2^{r_2}$ are chosen (see line 3 of Algorithm 2), and the vectors $v_2 \in \mathbb{F}_2^{r_1}$ and $u_2, u_4 \in \mathbb{F}_2^{r_2}$ defined (see line 4 of Algorithm 2). Then, for $i = 1, 2, 3, 4$, the basic lists $\left(\mathcal{L}_{2i-1}^{(3)}, \mathcal{L}_{2i}^{(3)} \right)$ are created.

Next, for $i = 1, 2, 3, 4$, if $a_1^{(i)} \in \mathcal{L}_{2i-1}^{(3)}$ and $a_2^{(i)} \in \mathcal{L}_{2i}^{(3)}$, then

$$a_1^{(i)} + a_2^{(i)} \in \mathcal{L}_i^{(2)} \quad \text{and} \quad \left(\left(a_1^{(i)} + a_2^{(i)} \right) H_1^T \right) [k - r_2 + 1 : k] = u_i,$$

and therefore

$$\left(\left(a_1^{(1)} + a_2^{(1)} + a_1^{(2)} + a_2^{(2)} + a_1^{(3)} + a_2^{(3)} + a_1^{(4)} + a_2^{(4)} \right) H_1^T \right) [k - r_2 + 1 : k]$$
$$= u_1 + u_2 + u_3 + u_4 = (v_1 + v_2) [k - r_2 + 1 : k] = s_0 [k - r_2 + 1 : k].$$

As the vectors of the final list will be selected as sums of vectors in $\mathcal{L}_i^{(2)}$, this property guarantees that the final selected vectors will match the last r_2 components of the syndrome block s_0.

Similarly, for $j = 1, 2$, if $b_1^{(j)} \in \mathcal{L}_{2j-1}^{(2)}$ and $b_2^{(j)} \in \mathcal{L}_{2j}^{(2)}$, then

$$b_1^{(j)} + b_2^{(j)} \in \mathcal{L}_j^{(1)} \quad \text{and} \quad \left(\left(b_1^{(j)} + b_2^{(j)} \right) H_1^T \right) [k - r_1 + 1 : k] = v_j$$

and therefore

$$\left(\left(b_1^{(1)} + b_2^{(1)} + b_1^{(2)} + b_2^{(2)} \right) H_1^T \right) [k - r_1 + 1 : k] = v_1 + v_2 = s_0 [k - r_1 + 1 : k].$$

In the same way, we see that with the selection of vectors at this step, the last r_1 components of the syndrome block s_0 are guaranteed, for $r_1 \geq r_2$.

Finally, if $c_1 \in \mathcal{L}_1^{(1)}$ and $c_2 \in \mathcal{L}_2^{(1)}$, then

$$c_1 + c_2 \in \mathcal{L} \quad \text{and} \quad (c_1 + c_2) H_1^T = s_0,$$

which was our target.

Once the set $\mathcal{L}$ has been determined, we use Algorithm 3 to obtain a vector e such that $\text{wt}(e) = w$ y $eH^T = s$.

Algorithm 2.

---

input : H_1, s_0, and p with $1 \leq p \leq 2k$

output: $\mathcal{L} = \left\{ \tilde{e}_0 \in \mathbb{F}_2^{2k} \mid \text{wt}(\tilde{e}_0) = p, \tilde{e}_0 H_1^T = s_0 \right\}$ % see expression (4)

1 Define $p_1 = p/2 + \varepsilon_1$ and $p_2 = p_1/2 + \varepsilon_2$ such that p_1 and p_2 are even numbers;

2 Choose random integers r_1, r_2 such that $1 \leq r_2 \leq r_1 \leq k$;

3 Choose random vectors $v_1 \in \mathbb{F}_2^{r_1}$, and $u_1, u_3 \in \mathbb{F}_2^{r_2}$;

4 Define vectors $\quad v_2 = s_0[k - r_1 + 1 : k] + v_1$,
$u_2 = v_1[r_1 - r_2 + 1 : r_1] + u_1, \quad u_4 = v_2[r_1 - r_2 + 1 : r_1] + u_3$;

5 **for** $i = 1, 2, 3, 4$ **do**

6 $\quad$ choose random partitions $\mathbb{P}_i = (\mathcal{P}_{2i-1}, \mathcal{P}_{2i})$ of $\{1, 2, \ldots, 2k\}$ such that $|\mathcal{P}_{2i-1}| = |\mathcal{P}_{2i}|$;

7 $\quad$ obtain the pairs of basic sets $\left(\mathcal{L}_{2i-1}^{(3)}, \mathcal{L}_{2i}^{(3)} \right)$ such that

$$\mathcal{L}_{2i-1}^{(3)} = \left\{ a = (a_1, a_2, \ldots, a_{2k}) \in \mathbb{F}_2^{2k} \mid \text{wt}(a) = p_2/2 \right.$$
$$\left. a_l = 0 \text{ for all } l \in \mathcal{P}_{2i} \right\},$$

$$\mathcal{L}_{2i}^{(3)} = \left\{ a = (a_1, a_2, \ldots, a_{2k}) \in \mathbb{F}_2^{2k} \mid \text{wt}(a) = p_2/2 \right.$$
$$\left. a_l = 0 \text{ for all } l \in \mathcal{P}_{2i-1} \right\},$$

8 **end**

9 **for** $i = 1, 2, 3, 4$ **do**

10 $\quad$ use the basic sets $\left(\mathcal{L}_{2i-1}^{(3)}, \mathcal{L}_{2i}^{(3)} \right)$, the vector u_i, the integer p_2, and the Merge-Join algorithm to obtain the set

$$\mathcal{L}_i^{(2)} = \left\{ a_1^{(i)} + a_2^{(i)} \in \mathbb{F}_2^{2k} \mid a_1^{(i)} \in \mathcal{L}_{2i-1}^{(3)}, a_2^{(i)} \in \mathcal{L}_{2i}^{(3)}, \right.$$
$$\left. \text{wt}\left(a_1^{(i)} + a_2^{(i)} \right) = p_2 \text{ and } \left((a_1^{(i)} + a_2^{(i)}) H_1^T \right)[k - r_2 + 1 : k] = u_i \right\}$$

11 **end**

12 **for** $j = 1, 2$ **do**

13 $\quad$ use the sets $\left(\mathcal{L}_{2j-1}^{(2)}, \mathcal{L}_{2j}^{(2)} \right)$, the vector v_j, the integer p_1, and the Merge-Join algorithm to obtain the set

$$\mathcal{L}_j^{(1)} = \left\{ b_1^{(j)} + b_2^{(j)} \in \mathbb{F}_2^{2k} \mid b_1^{(j)} \in \mathcal{L}_{2j-1}^{(2)}, b_2^{(j)} \in \mathcal{L}_{2j}^{(2)}, \right.$$
$$\left. \text{wt}\left(b_1^{(j)} + b_2^{(j)} \right) = p_1 \text{ and } \left((b_1^{(j)} + b_2^{(j)}) H_1^T \right)[k - r_1 + 1 : k] = v_j \right\}$$

14 **end**

15 Use the sets $\left(\mathcal{L}_1^{(1)}, \mathcal{L}_2^{(1)} \right)$, the vector s_0, the integer p, and the Merge-Join algorithm to obtain the set

$$\mathcal{L} = \left\{ c_1 + c_2 \in \mathbb{F}_2^{2k} \mid c_1 \in \mathcal{L}_1^{(1)}, c_2 \in \mathcal{L}_2^{(1)}, \right.$$
$$\left. \text{wt}(c_1 + c_2) = p \text{ and } (c_1 + c_2) H_1^T = s_0 \right\}$$

Algorithm 3.

input : $H, s, \mathcal{L}, \omega$
output: $e \in \mathbb{F}_2^n$ such that $eH^T = s$ and $\mathrm{wt}(e) = \omega$

1 **for** $\tilde{e}_0 \in \mathcal{L}$ **do**   % $\mathrm{wt}(\tilde{e}_0) = p$, $\tilde{e}_0 = (e_0, e_1)$
2 | Compute $\tilde{e}_1 \in \mathbb{F}_2^{n-2k}$ as $\tilde{e}_1 = e_0 B_1^T + \tilde{s}_1$ % see expression (5)
3 | **if** $\mathrm{wt}(\tilde{e}_1) = \omega - p$ **then**
4 | | define $e = (\tilde{e}_0, \tilde{e}_1)$
5 | **end**
6 **end**

Remark 2. Preliminary calculations performed for registers of length k in the range $8 \leq k \leq 16$ show that in 95 % of the cases we obtain the desired solution. Algorithm 1 presents better performance in terms of running time for small values of k, whilst Algorithms 2 and 3 seem to be more efficient when k increases.

In the case of Algorithm 2, we have taken $p = \mathrm{wt}(\tilde{e}_0) = \lfloor 2k\varepsilon \rfloor$, and for the rest of parameters: $\varepsilon_1, \varepsilon_2, r_1$ and r_2, we assign the values recommended in [2]; i.e.,

- $0 < \varepsilon_1 < 2k - p$,
- $0 < \varepsilon_2 < 2k - p_1$,
- $r_1 \approx \log_2 R_1$, where $R_1 = \binom{p}{p/2}\binom{2k-p}{\varepsilon_1}$,
- $r_2 \approx \log_2 R_2$, where $R_2 = \binom{p_1}{p_1/2}\binom{2k-p_1}{\varepsilon_2}$.

As future work, we aim to run the algorithms for higher values of k (in the range of $16 < k \leq 64$). Strategies to decrease the amount of computation in order to prepare the lists are being explored.

Acknowledgements. The first author was supported by FAPESP with number of process 2015/07246-0. The second author was partially supported by grants MIMECO MTM2015-68805-REDT and MTM2015-69138-REDT. The third author was partially supported by grants MINECO MTM2013-40960-P and MTM2015-68805-REDT.

References

1. Ågren, M., Löndahl, C., Hell, M., Johansson, T.: A survey on fast correlation attacks. Crypt. Commun. 4(3–4), 173–202 (2012)
2. Becker, A., Joux, A., May, A., Meurer, A.: Decoding random binary linear codes in $2^{n/20}$: how $1 + 1 = 0$ improves information set decoding. In: Pointcheval, D., Johansson, T. (eds.) EUROCRYPT 2012. LNCS, vol. 7237, pp. 520–536. Springer, Heidelberg (2012). doi:10.1007/978-3-642-29011-4_31
3. Canteaut, A., Naya-Plasencia, M.: Correlation attacks on combination generators. Crypt. Commun. 4(3–4), 147–171 (2012)
4. Chepyzhov, V.V., Johansson, T., Smeets, B.: A simple algorithm for fast correlation attacks on stream ciphers. In: Goos, G., Hartmanis, J., Leeuwen, J., Schneier, B. (eds.) FSE 2000. LNCS, vol. 1978, pp. 181–195. Springer, Heidelberg (2001). doi:10.1007/3-540-44706-7_13

5. Geffe, P.: How to protect data with ciphers that are really hard to break. Electronics **46**(1), 99–101 (1973)
6. Golić, J.D.: Cryptanalysis of alleged A5 stream cipher. In: Fumy, W. (ed.) EURO-CRYPT 1997. LNCS, vol. 1233, pp. 239–255. Springer, Heidelberg (1997). doi:10. 1007/3-540-69053-0_17
7. Golomb, S.W.: Shift Register-Sequences. Aegean Park Press, Laguna Hill (1982)
8. Johansson, T., Jönsson, F.: Theoretical analysis of a correlation attack based on convolutional codes. IEEE Trans. Inf. Theory **48**(8), 2173–2181 (2002)
9. Kailath, T.: Linear Systems. Prentice-Hall, Upper Saddle River (1980)
10. Knuth, D.E.: The Art of Computer Programming. Sorting and Searching. Addison-Wesley, Boston (1998)
11. Lidl, R., Niederreiter, H.: Introduction to Finite Fields and Their Applications. Cambridge University Press, New York (1986)
12. Lu, P., Huang, L.: A new correlation attack on LFSR sequences with high error tolerance. Prog. Comput. Sci. Appl. Logic **23**, 67–83 (2004)
13. Massey, J.L.: Shift-register synthesis and BCH decoding. IEEE Trans. Inf. Theory **15**(1), 122–127 (1969)
14. Meier, W.: Fast correlation attacks: methods and countermeasures. In: Joux, A. (ed.) FSE 2011. LNCS, vol. 6733, pp. 55–67. Springer, Heidelberg (2011). doi:10. 1007/978-3-642-21702-9_4
15. Meier, W., Staffelbach, O.: Fast correlation attacks on stream ciphers. In: Barstow, D., Brauer, W., Brinch Hansen, P., Gries, D., Luckham, D., Moler, C., Pnueli, A., Seegmüller, G., Stoer, J., Wirth, N., Günther, C.G. (eds.) EUROCRYPT 1988. LNCS, vol. 330, pp. 301–314. Springer, Heidelberg (1988). doi:10.1007/3-540-45961-8_28
16. Meier, W., Staffelbach, O.: Fast correlation attacks on certain stream ciphers. J. Cryptology **1**(3), 159–176 (1989)
17. Molland, H., Mathiassen, J.E., Helleseth, T.: Improved fast correlation attack using low rate codes. In: Paterson, K.G. (ed.) Cryptography and Coding 2003. LNCS, vol. 2898, pp. 67–81. Springer, Heidelberg (2003). doi:10.1007/978-3-540-40974-8_7
18. Siegenthaler, T.: Decrypting a class of stream ciphers using ciphertext only. IEEE Trans. Comput. **34**(1), 81–85 (1985)
19. Zhang, B., Wu, H., Feng, D., Bao, F.: A fast correlation attack on the shrinking generator. In: Menezes, A. (ed.) CT-RSA 2005. LNCS, vol. 3376, pp. 72–86. Springer, Heidelberg (2005). doi:10.1007/978-3-540-30574-3_7

A Construction of Orbit Codes

Joan-Josep Climent[1]([⊠]), Verónica Requena[2], and Xaro Soler-Escrivà[1]

[1] Departament de Matemàtiques, Universitat d'Alacant,
Ap. Correus 99, 03080 Alacant, Spain
{jcliment,xaro.soler}@ua.es
[2] Departamento de Estadística, Matemáticas e Informática,
Universidad Miguel Hernández de Elche,
Avda. Universidad, s/n, Elche, 03202 Alicante, Spain
vrequena@umh.es

Abstract. Given a finite field $\mathbb{F}_q$, a constant dimension code is a set of k-dimensional subspaces of $\mathbb{F}_q^n$. Orbit codes are constant dimension codes which are defined as orbits when the action of a subgroup of the general linear group on the set of all subspaces of $\mathbb{F}_q^n$ is considered. In this paper we present a construction of an Abelian non-cyclic orbit code whose minimum subspace distance is maximal.

Keywords: Random linear network coding · Subspace codes · Grassmannian · Group action · General linear group

1 Introduction

Random linear network coding is a powerful tool for disseminating information in networks [10]. Kötter and Kschischang [11] proposed a mathematical description of network communications in which the transmitted messages are vector subspaces, rather than vectors, of a given vector space over a finite field. This approach is known as subspace codes and their study has led to many papers in recent years (see, for example, [2–5,7–9,12,13,15–17] to mention only the most recent). Most of these authors focus on the study of constant dimension codes; that is, codes in which all subspaces have the same dimension.

We focus our attention on constant dimension codes which are obtained as orbits of certain groups. This concept traces back to Slepian [14], where Euclidean spaces were considered. In the linear network coding setting, Trautmann et al. [17] defined such codes as orbit codes by the action of subgroups of the general linear group. Since then, several papers have been written about the structure of orbit codes (see for instance [1,6,7,15,16]).

When we consider cyclic subgroups of the general linear group, we talk about cyclic orbit codes. These codes were introduced by Trautmann et al. in [16], where the authors gave a characterization of a particular subclass of them.

In this paper we are interested on *Abelian orbit codes*; that is, orbit codes which are generated by Abelian subgroups of the general linear group. Since Abelian subgroups are direct product of cyclic subgroups, it seems a natural

© Springer International Publishing AG 2017
A.I. Barbero et al. (Eds.): ICMCTA 2017, LNCS 10495, pp. 72–83, 2017.
DOI: 10.1007/978-3-319-66278-7_7

extension of the research on cyclic orbit codes. Specifically, we present a construction of an Abelian non-cyclic orbit code for which we calculate its cardinality and its minimum subspace distance. Moreover, the minimum subspace distance of our code is maximal.

The rest of the paper is structured as follows: In Sect. 2 we give some preliminaries, first about random linear network coding and orbit codes and then a result about upper triangular matrices. The main body of the paper is Sect. 3, where the whole construction of our Abelian non-cyclic orbit code is given. We finish the paper with some open questions in Sect. 4.

2 Preliminaries

Let $\mathbb{F}_q$ be the finite field of q elements, where q is a prime power. For any integer $n \geq 1$, the set of all vector subspaces of $\mathbb{F}_q^n$, denoted by $\mathcal{P}_q(n)$, forms a metric space with respect to the subspace distance defined by (see [11])

$$d(\mathcal{U}, \mathcal{V}) = \dim(\mathcal{U} + \mathcal{V}) - \dim(\mathcal{U} \cap \mathcal{V}), \quad \text{for all } \mathcal{U}, \mathcal{V} \in \mathcal{P}_q(n). \tag{1}$$

For any integer k, where $0 \leq k \leq n$, the set of all k-dimensional vector subspaces of $\mathbb{F}_q^n$ is called **Grassmann variety** (or simply **Grassmannian**) and is denoted by $\mathcal{G}_q(k, n)$. Obviously, $\mathcal{P}_q(n) = \cup_{k=0}^n \mathcal{G}_q(k, n)$.

A **subspace code** of length n is a nonempty subset $\mathfrak{C}$ of $\mathcal{P}_q(n)$; a codeword of such a code is a vector subspace of $\mathfrak{C}$. We call $\mathfrak{C}$ a **constant dimension code** if all codewords of $\mathfrak{C}$ have the same dimension, i.e., if $\mathfrak{C} \subseteq \mathcal{G}_q(k, n)$ for some k. The **minimum distance** [11] of a subspace code $\mathfrak{C} \subseteq \mathcal{P}_q(n)$ is defined as

$$d(\mathfrak{C}) = \min \left\{ d(\mathcal{U}, \mathcal{V}) \mid \mathcal{U}, \mathcal{V} \in \mathfrak{C}, \ \mathcal{U} \neq \mathcal{V} \right\}.$$

The **complementary code** (or **dual code**) corresponding to a subspace code $\mathfrak{C}$ is the subspace code

$$\mathfrak{C}^\perp = \{ \mathcal{U}^\perp \subseteq \mathcal{P}_q(n) \mid \mathcal{U} \in \mathfrak{C} \}$$

obtained from the orthogonal subspaces of the codewords of $\mathfrak{C}$. It is easy to see that $d(\mathcal{U}^\perp, \mathcal{V}^\perp) = d(\mathcal{U}, \mathcal{V})$, and therefore, $d(\mathfrak{C}^\perp) = d(\mathfrak{C})$ (see [11]).

In this paper we consider $\mathcal{G}_q(k, n)$, and we will assume, without loss of generality, that $k \leq n/2$. Otherwise, we can take $\mathfrak{C}^\perp$ which has the same length, cardinality and minimum distance.

In order to represent a k-dimensional subspace $\mathcal{U} \in \mathcal{G}_q(k, n)$, we use a **generator matrix** $U \in \mathbb{F}_q^{k \times n}$ whose rows form a basis of $\mathcal{U}$, that is

$$\mathcal{U} = \text{rowspace}\,(U) = \left\{ \boldsymbol{x} U \mid \boldsymbol{x} \in \mathbb{F}_q^k \right\}.$$

Note that in $\mathcal{G}_q(k, n)$ the subspace distance (1) is given by

$$d(\mathcal{U}, \mathcal{V}) = 2(k - \dim(\mathcal{U} \cap \mathcal{V})) = 2 \ \text{rank} \begin{bmatrix} U \\ V \end{bmatrix} - 2k \tag{2}$$

for any $\mathcal{U}, \mathcal{V} \in \mathcal{G}_q(k, n)$ and some respective matrix representations $U, V \in \mathbb{F}_q^{k \times n}$.

We focus on constant dimension codes arising from group actions, which are simply called **orbit codes** and which were introduced in [17] (see also [16]).

The general linear group of degree n, denoted by GL_n, is the set of all invertible $n \times n$ matrices with entries in $\mathbb{F}_q$. Given a full-rank matrix $U \in \mathbb{F}_q^{k \times n}$, $\mathcal{U} = \mathrm{rowspace}\,(U) \in \mathcal{G}_q(k, n)$ its rowspace, and $A \in \mathrm{GL}_n$, we define

$$\mathcal{U} \cdot A = \mathrm{rowspace}\,(UA).$$

Provided that this operation is independent from the representation of $\mathcal{U}$ (see [17]) and since any invertible matrix maps vector subspaces to vector subspaces of the same dimension, we obtain a group action (from the right) on the Grassmann variety:

$$\begin{aligned} \mathcal{G}_q(k, n) \times \mathrm{GL}_n &\longrightarrow \mathcal{G}_q(k, n) \\ (\mathcal{U}, A) &\longmapsto \mathcal{U} \cdot A \end{aligned}$$

It is well-known that two k-dimensional subspaces can be mapped onto each other by an invertible matrix. Therefore, the orbit of any k-dimensional subspace $\mathcal{U}$ under the action of GL_n is the whole set $\mathcal{G}_q(k, n)$. That is, GL_n acts transitively on $\mathcal{G}_q(k, n)$.

Now, let $\mathbf{G}$ be a subgroup of GL_n and consider the action of $\mathbf{G}$ on $\mathcal{G}_q(k, n)$. Then the orbit of a subspace $\mathcal{U} \in \mathcal{G}_q(k, n)$ under the action of $\mathbf{G}$ is

$$\mathfrak{C} = \{\mathcal{U} \cdot A \mid A \in \mathbf{G}\},$$

which is called the **orbit code** generated by the action of $\mathbf{G}$ on $\mathcal{U}$ (see [17]). The stabilizer of $\mathcal{U}$ under this action is

$$\mathrm{stab}_{\mathbf{G}}\,(\mathcal{U}) = \{A \in \mathbf{G} \mid \mathcal{U} \cdot A = \mathcal{U}\} = \mathbf{G} \cap \mathrm{stab}_{\mathrm{GL}_n}\,(\mathcal{U})$$

and the size of the orbit code $\mathfrak{C}$ is

$$|\mathfrak{C}| = \frac{|\mathbf{G}|}{|\mathrm{stab}_{\mathbf{G}}\,(\mathcal{U})|}.$$

On the other hand (see [17, Proposition 8]), the subspace distance is GL_n-invariant, that is, $d(\mathcal{U}, \mathcal{V}) = d(\mathcal{U} \cdot A, \mathcal{V} \cdot A)$, for all $A \in \mathrm{GL}_n$; this property allow us to compute the minimum distance of orbit codes in a simple manner:

$$d(\mathfrak{C}) = \min \{d(\mathcal{U}, \mathcal{U} \cdot A) \mid A \in \mathbf{G} \backslash \mathrm{stab}_{\mathbf{G}}\,(\mathcal{U})\}$$

The following lemma on block upper triangular matrices will be useful for the rest of the paper. The proof is straightforward and we omit it.

Lemma 1. *Assume that $A \in \mathbb{F}^{r \times r}$, $B \in \mathbb{F}^{s \times s}$, and $P \in \mathbb{F}^{r \times s}$ and consider the block upper triangular matrix*

$$M = \begin{bmatrix} A & P \\ O & B \end{bmatrix}$$

where O denotes the zero matrix of the appropriate size. If h is a nonnegative integer, then

$$M^h = \begin{bmatrix} A^h & \sigma_h(A,B,P) \\ O & B^h \end{bmatrix}$$

where

$$\sigma_h(A,B,P) = \begin{cases} O, & \text{if } h = 0, \\ \sum_{i=1}^{h} A^{h-i} P B^{i-1}, & \text{if } h \geq 1. \end{cases}$$

Our aim is to construct an orbit code $\mathfrak{C}$ by considering the orbit of $\mathcal{U}_k$, the standard k-dimensional vector subspace, i.e., the subspace generated by the first k unit vectors of $\mathbb{F}_q^n$, under the action of a concrete subgroup $\mathbf{H}$ of GL_n. If I_k denotes the identity matrix of size $k \times k$, then $\mathcal{U}_k = \mathrm{rowspace}\left(\begin{bmatrix} I_k & O \end{bmatrix}\right)$. In order to choose our group $\mathbf{H}$, we have taken into account different factors. First, Lemma 1 which provides an easy tool for working with block upper triangular matrices and for this reason we take $\mathbf{H}$ as a subgroup of upper triangular matrices. Secondly, since the stabilizer of $\mathcal{U}_k$ in GL_n is composed of block lower triangular matrices, it turns out that if $\mathbf{H}$ consists of upper triangular matrices, the stabilizer of $\mathcal{U}_k$ in $\mathbf{H}$ is readily described (see expression (6) below). Thirdly, our group $\mathbf{H}$ must be a product of cyclic groups to obtain an Abelian group. Moreover, we have chosen the generators of $\mathbf{H}$ in such a way that $\mathbf{H}$ is non-cyclic.

Assume that $\mathbf{H}$ is a subgroup of upper triangular matrices of GL_n and consider the following block partition of any matrix

$$H = \begin{bmatrix} H_{11} & H_{12} \\ O & H_{22} \end{bmatrix} \in \mathbf{H}, \tag{3}$$

where $H_{11} \in \mathrm{GL}_k$, $H_{12} \in \mathbb{F}_q^{k \times (n-k)}$, $H_{22} \in \mathrm{GL}_{n-k}$, and O denotes the zero matrix of the appropriate size. Then, the orbit code $\mathfrak{C}$ can be written as

$$\mathfrak{C} = \{\mathcal{U}_k \cdot H \mid H \in \mathbf{H}\} = \{\mathrm{rowspace}\left(\begin{bmatrix} H_{11} & H_{12} \end{bmatrix}\right) \mid H \in \mathbf{H}\}. \tag{4}$$

Moreover, according to (2) and (3), for any $H \in \mathbf{H}$ we have

$$d(\mathcal{U}_k, \mathcal{U}_k \cdot H) = 2\,\mathrm{rank}\begin{bmatrix} I_k & O \\ H_{11} & H_{12} \end{bmatrix} - 2k = 2\,\mathrm{rank}\,(H_{12}) \leq 2k \tag{5}$$

since $k \leq n - k$. Therefore $d(\mathfrak{C}) \leq 2k$.

On the other hand, it can be readily verifies that the stabilizer of $\mathcal{U}_k$ in GL_n is

$$\mathrm{stab}_{\mathrm{GL}_n}\,(\mathcal{U}_k) = \left\{ \begin{bmatrix} A & O \\ B & C \end{bmatrix} \mid A \in \mathrm{GL}_k,\ B \in \mathbb{F}_q^{(n-k) \times k} \text{ and } C \in \mathrm{GL}_{n-k} \right\}.$$

Since $\mathrm{stab}_{\mathbf{H}}\,(\mathcal{U}_k) = \mathbf{H} \cap \mathrm{stab}_{\mathrm{GL}_n}\,(\mathcal{U}_k)$, and using the previous partition of matrices of $\mathbf{H}$ given by (3), the stabilizer of $\mathcal{U}_k$ in $\mathbf{H}$ can be written as

$$\mathrm{stab}_{\mathbf{H}}\,(\mathcal{U}_k) = \left\{ \begin{bmatrix} H_{11} & H_{12} \\ O & H_{22} \end{bmatrix} \in \mathbf{H} \mid H_{12} = O \right\}. \tag{6}$$

3   Our Construction

From now on, we will assume that q is a prime number such that $q \geq n - k$, and let λ be a generator of the multiplicative cyclic group $\mathbb{F}_q^*$, i.e., λ is a primitive element of $\mathbb{F}_q$ (in particular $o(\lambda) = q - 1$ is the multiplicative order of λ).

Let us denote by J_ℓ the upper unitriangular matrix of size $\ell \times \ell$, with 1's on the second upper diagonal and 0's elsewhere, that is, J_ℓ is a Jordan block of size $\ell \times \ell$ associated to 1. The following result gives the multiplicative order of J_ℓ.

Lemma 2. *If $q^{t-1} < \ell \leq q^t$ for some nonnegative integer t, then $o(J_\ell) = q^t$.*

Proof. Let $J_\ell = I_\ell + N_\ell$ where

$$N_\ell = \begin{bmatrix} 0 & 1 & 0 & \cdots & 0 \\ 0 & 0 & 1 & \cdots & 0 \\ \vdots & \vdots & \vdots & & \vdots \\ 0 & 0 & 0 & \cdots & 1 \\ 0 & 0 & 0 & \cdots & 0 \end{bmatrix} \in \mathbb{F}_q^{\ell \times \ell}$$

is the canonical nilpotent matrix with index ℓ. Then $J_\ell^{q^t} = I_\ell^{q^t} + N_\ell^{q^t} = I_\ell$. Now, since $q^{t-1} < \ell \leq q^t$ it follows that $o(J_\ell) = q^t$. $\square$

In order to construct our subgroup **H** of GL_n , we consider the following two matrices of GL_n :

$$S = \begin{bmatrix} \lambda I_{n-k} & X \\ O & I_k \end{bmatrix}, \qquad T = \begin{bmatrix} J_{n-k} & Y \\ O & \lambda I_k \end{bmatrix} \tag{7}$$

where $X, Y \in \mathbb{F}_q^{(n-k) \times k}$ are arbitrary matrices.

Lemma 3. *For the matrices S and T given by (7), it follows that*

(a) $o(S) = o(\lambda) = q - 1$.
(b) $o(T) = o(J_{n-k}) \, o(\lambda) = q(q - 1)$.

Proof. The proof follows from the choice of λ, the definition of S and T, and Lemma 2. $\square$

As a consequence of the previous result the subgroups of GL_n generated by the matrices S and T are cyclic subgroups of order $q - 1$ and $q(q - 1)$ respectively, that is:

$$\langle S \rangle = \{ S^a \mid 0 \leq a < q - 1 \} \cong C_{q-1}, \quad \langle T \rangle = \{ T^b \mid 0 \leq b < q(q - 1) \} \cong C_{q(q-1)}.$$

Lemma 4. *The intersection of the groups $\langle S \rangle$ and $\langle T \rangle$ is trivial.*

Proof. If $S^a = T^b$ for some a and b where $0 \leq a < q - 1$ and $0 \leq b < q(q - 1)$, then it follows that $a = b = 0$, by the definition of S and T, and Lemma 3. $\square$

The following lemma is crucial for our construction. The proof is straightforward and we omit it.

Lemma 5. *If* $X \neq O$ *and* $Y = \frac{1}{1-\lambda}\left(\lambda I_{n-k} - J_{n-k}\right)X$ *in* (7), *then* $ST = TS$.

Now we are ready to present our group. From now on, we assume that matrices X and Y are defined as in Lemma 5. Since $ST = TS$ and $\langle S \rangle \cap \langle T \rangle = \{I_n\}$ we obtain

$$
\begin{aligned}
\mathbf{H} = \langle S, T \rangle &= \langle S \rangle \times \langle T \rangle \\
&= \left\{ S^a T^b \mid 0 \leq a < q-1,\ 0 \leq b < q(q-1) \right\} \cong C_{q-1} \times C_{q(q-1)},
\end{aligned}
\tag{8}
$$

that is, $\mathbf{H}$ is an Abelian non-cyclic group of order $q(q-1)^2$ and the elements of $\mathbf{H}$ present the following form

$$
\begin{aligned}
S^a T^b &= \begin{bmatrix} \lambda^a I_{n-k} & \sigma_a(\lambda I_{n-k}, I_k, X) \\ O & I_k \end{bmatrix} \begin{bmatrix} J_{n-k}^b & \sigma_b(J_{n-k}, \lambda I_k, Y) \\ O & \lambda^b I_k \end{bmatrix} \\
&= \begin{bmatrix} \lambda^a J_{n-k}^b & \lambda^a \sigma_b(J_{n-k}, \lambda I_k, Y) + \lambda^b \sigma_a(\lambda I_{n-k}, I_k, X) \\ O & \lambda^b I_k \end{bmatrix}.
\end{aligned}
\tag{9}
$$

The following results will help us to understand the structure of $S^a T^b$. The first one is well known and, therefore, we omit the proof.

Lemma 6. *For any nonnegative integer* b *it follows that*

$$
J_{n-k}^b = \begin{bmatrix} 1 & \binom{b}{1} & \binom{b}{2} & \cdots & \binom{b}{n-k-1} \\ 0 & 1 & \binom{b}{1} & \cdots & \binom{b}{n-k-2} \\ \vdots & \vdots & \vdots & & \vdots \\ 0 & 0 & 0 & \cdots & 1 \end{bmatrix}.
$$

Lemma 7. *For matrices* X *and* Y *as in Lemma 5, then*

$$
\begin{aligned}
&\lambda^a \sigma_b(J_{n-k}, \lambda I_k, Y) + \lambda^b \sigma_a(\lambda I_{n-k}, I_k, X) \\
&= \begin{bmatrix} c & e_1 & e_2 & \cdots & e_{n-k-2} & e_{n-k-1} \\ 0 & c & e_1 & \cdots & e_{n-k-3} & e_{n-k-2} \\ 0 & 0 & c & \cdots & e_{n-k-4} & e_{n-k-3} \\ \vdots & \vdots & \vdots & & \vdots & \vdots \\ 0 & 0 & 0 & \cdots & c & e_1 \\ 0 & 0 & 0 & \cdots & 0 & c \end{bmatrix} X,
\end{aligned}
$$

where $c = \frac{\lambda^a - \lambda^b}{\lambda - 1}$ *and* $e_i = \frac{\lambda^a}{\lambda - 1}\binom{b}{i}$, *for* $i = 1, 2, \ldots, n-k-1$.

Proof. According to Lemma 1 we have that

$$
\lambda^a \sigma_b(J_{n-k}, \lambda I_k, Y) + \lambda^b \sigma_a(\lambda I_{n-k}, I_k, X)
$$
$$
= \lambda^a \left(J_{n-k}^{b-1} + \lambda J_{n-k}^{b-2} + \cdots + \lambda^{b-2} J_{n-k} + \lambda^{b-1} I_{n-k} \right) Y + \lambda^b d_a X
$$
$$
= \left(\frac{\lambda^a}{\lambda - 1} \underbrace{\left(J_{n-k}^{b-1} + \lambda J_{n-k}^{b-2} + \cdots + \lambda^{b-2} J_{n-k} + \lambda^{b-1} I_{n-k} \right)}_{Z} (\lambda I_{n-k} - J_{n-k}) \right.
$$
$$
\left. + \lambda^b d_a I_{n-k} \right) X.
$$

Notice that in the last expression, we have an upper triangular matrix whose diagonal elements are all equal to c, multiplying X. Moreover,

$$
c = \frac{\lambda^a}{1 - \lambda} d_b (\lambda - 1) + \lambda^b d_a = \lambda^b d_a - \lambda^a d_b
$$
$$
= \lambda^b \left(\frac{\lambda^a - 1}{\lambda - 1} \right) - \lambda^a \left(\frac{\lambda^b - 1}{\lambda - 1} \right) = \frac{\lambda^a - \lambda^b}{\lambda - 1}.
$$

Now we describe the elements e_i, for $i = 1, 2, \ldots, n-k-1$. First, if we denote by z_{ij} the (i,j)-entry of Z, then

$$
Z(\lambda I_{n-k} - J_{n-k})
$$
$$
= \begin{bmatrix}
(\lambda - 1)z_{11} & -z_{11} + (\lambda - 1)z_{12} & \cdots & -z_{1(n-k-1)} + (\lambda - 1)z_{1(n-k)} \\
0 & (\lambda - 1)z_{11} & \cdots & -z_{1(n-k-2)} + (\lambda - 1)z_{1(n-k-1)} \\
\vdots & \vdots & & \vdots \\
0 & 0 & \cdots & -z_{11} + (\lambda - 1)z_{12} \\
0 & 0 & \cdots & (\lambda - 1)z_{11}
\end{bmatrix}.
$$

Therefore, for $i = 1, 2, \ldots, n - k - 1$, it follows that

$$
e_i = \frac{\lambda^a}{1 - \lambda} \left(-z_{1i} + (\lambda - 1)z_{1(i+1)} \right). \tag{10}
$$

Now, notice that for $i = 2, 3, \ldots, n - k$, we have

$$
z_{1i} = \sum_{j=0}^{b-1} \lambda^j \binom{b-1-j}{i-1}.
$$

Therefore, using that $\binom{n+1}{i} - \binom{n}{i} = \binom{n}{i-1}$, we obtain

$$
(\lambda - 1)z_{1(i+1)} = -\binom{b-1}{i} + \lambda \left[\binom{b-1}{i} - \binom{b-2}{i} \right] + \lambda^2 \left[\binom{b-2}{i} - \binom{b-3}{i} \right]
$$
$$
+ \cdots + \lambda^{b-(i+1)} \left[\binom{i+1}{i} - \binom{i}{i} \right] + \lambda^{b-1} \binom{i}{i}
$$
$$
= -\binom{b}{i} + z_{1i}.
$$

and by means of (10) it follows that

$$e_i = \frac{\lambda^a}{\lambda - 1} \binom{b}{i},$$

for $i = 1, 2, \ldots, n - k - 1$.

$\square$

The proof of the following result is straightforward and we omit it.

Lemma 8. *With the notation of Lemma 7, we have that $c = 0$ if, and only if, $a \equiv b \pmod{(q - 1)}$.*

Note that, according to (4) and (5), to define the code $\mathfrak{C}$, we lay aside the block partition of $S^a T^b$ given by (9); instead, we focus on the first k rows of $S^a T^b$; and more specifically, on the first k rows and the last $n - k$ columns of $S^a T^b$. So, we consider the following partition

$$S^a T^b = \begin{bmatrix} (S^a T^b)_{11} & (S^a T^b)_{12} & (S^a T^b)_{13} \\ O & (S^a T^b)_{22} & (S^a T^b)_{23} \\ O & O & (S^a T^b)_{33} \end{bmatrix} \begin{matrix} \} k \\ \} n-2k \\ \\ \end{matrix}$$

where, according to (9),

$$\lambda^a J_{n-k}^b = \begin{bmatrix} (S^a T^b)_{11} & (S^a T^b)_{12} \\ O & (S^a T^b)_{22} \end{bmatrix},$$

$$\lambda^a \sigma_b(J_{n-k}, \lambda I_k, Y) + \lambda^b \sigma_a(\lambda I_{n-k}, I_k, X) = \begin{bmatrix} (S^a T^b)_{13} \\ (S^a T^b)_{23} \end{bmatrix},$$

$$(S^a T^b)_{11} = \lambda^a J_k^b, \quad (S^a T^b)_{22} = \lambda^a J_{n-2k}^b, \quad (S^a T^b)_{33} = \lambda^b I_k,$$

and, $(S^a T^b)_{12}$ is the submatrix of $\lambda^a J_{n-k}^b$ formed by the first k rows and the last $n - 2k$ columns; therefore, as a consequence of Lemma 6,

$$(S^a T^b)_{12} = \lambda^a \begin{bmatrix} \binom{b}{k} & \binom{b}{k+1} & \cdots & \binom{b}{n-k-1} \\ \binom{b}{k-1} & \binom{b}{k} & \cdots & \binom{b}{n-k-2} \\ \vdots & \vdots & & \vdots \\ \binom{b}{1} & \binom{b}{2} & \cdots & \binom{b}{n-2k} \end{bmatrix}. \tag{12}$$

Moreover, if we assume that

$$X = \begin{bmatrix} X_1 \\ X_2 \end{bmatrix} \quad \text{where } X_1 \in \mathbb{F}_q^{k \times k}, \; X_2 \in \mathbb{F}_q^{(n-2k) \times k},$$

then, from Lemma 7, it follows that

$$(S^aT^b)_{13} = \begin{bmatrix} c & e_1 & e_2 & \cdots & e_{k-2} & e_{k-1} \\ 0 & c & e_1 & \cdots & e_{k-3} & e_{k-2} \\ 0 & 0 & c & \cdots & e_{k-4} & e_{k-3} \\ \vdots & \vdots & \vdots & & \vdots & \vdots \\ 0 & 0 & 0 & \cdots & c & e_1 \\ 0 & 0 & 0 & \cdots & 0 & c \end{bmatrix} X_1$$

$$+ \begin{bmatrix} e_k & e_{k+1} & e_{k+2} & \cdots & e_{n-k-2} & e_{n-k-1} \\ e_{k-1} & e_k & e_{k+1} & \cdots & e_{n-k-3} & e_{n-k-2} \\ e_{k-2} & e_{k-1} & e_k & \cdots & e_{n-k-4} & e_{n-k-3} \\ \vdots & \vdots & \vdots & & \vdots & \vdots \\ e_2 & e_3 & e_4 & \cdots & e_{n-2k-2} & e_{n-2k-1} \\ e_1 & e_2 & e_3 & \cdots & e_{n-2k-3} & e_{n-2k} \end{bmatrix} X_2. \qquad (13)$$

According to (4), the orbit code $\mathfrak{C}$ generated by the action of $\mathbf{H}$ on $\mathcal{U}_k$ is

$$\mathfrak{C} = \{\mathcal{U}_k \cdot S^aT^b \mid 0 \le a < q-1,\ 0 \le b < q(q-1)\}$$
$$= \{\text{rowspace}\left(\left[(S^aT^b)_{11}\ (S^aT^b)_{12}\ (S^aT^b)_{13}\right]\right)$$
$$\mid 0 \le a < q-1,\ 0 \le b < q(q-1)\}.$$

Notice that, according to (5),

$$d(\mathcal{U}_k, \mathcal{U}_k \cdot S^aT^b) = 2\ \text{rank}\left[(S^aT^b)_{12}\ (S^aT^b)_{13}\right] \le 2k.$$

Our next goal will be to choose X in such a way that

$$\text{rank}\left[(S^aT^b)_{12}\ (S^aT^b)_{13}\right] = k,$$

for all $S^aT^b \in \mathbf{H}\backslash \text{stab}_{\mathbf{H}}(\mathcal{U}_k)$. Consequently, $d(\mathfrak{C}) = 2k$ and the minimum distance of our code will be maximal for a concrete choice of X.

Theorem 1. *If $X_1 \ne O$, then the stabilizer in $\mathbf{H}$ of $\mathcal{U}_k$ is the cyclic subgroup of $\mathbf{H}$ generated by ST^q; i.e., $\text{stab}_{\mathbf{H}}(\mathcal{U}_k) = \langle ST^q \rangle$.*

Proof. Recall that by means of (6) we obtain

$$\text{stab}_{\mathbf{H}}(\mathcal{U}_k) = \{S^aT^b \in \mathbf{H} \mid (S^aT^b)_{12} = O \text{ and } (S^aT^b)_{13} = O\}.$$

Moreover, $\langle ST^q \rangle = \{S^aT^b \in \mathbf{H} \mid b = qa,\ 0 \le a < q-1\}$ which is a cyclic subgroup of order $q-1$.

First, we prove that $\langle ST^q \rangle$ is included into $\text{stab}_{\mathbf{H}}(\mathcal{U}_k)$. Consider $S^aT^b \in \langle ST^q \rangle$, and let us see that $(S^aT^b)_{12} = O$ and $(S^aT^b)_{13} = O$. Since b is a multiple of q, it follows that $(S^aT^b)_{12} = O$, by (12), and that

$$\lambda^a \sigma_b(J_{n-k}, \lambda I_k, Y) + \lambda^b \sigma_a(\lambda I_{n-k}, I_k, X) = cX$$

by Lemma 7. Moreover, since $b = qa$, then $c = 0$ by Lemma 8. In particular $(S^a T^b)_{13} = O$ and we obtain $\langle ST^q \rangle \subseteq \mathrm{stab}_{\mathbf{H}}(\mathcal{U}_k)$.

On the other hand, if $S^a T^b \in \mathrm{stab}_{\mathbf{H}}(\mathcal{U}_k)$, it follows that $(S^a T^b)_{12} = O$ and $(S^a T^b)_{13} = O$. Since $(S^a T^b)_{12} = O$, its $(k, 1)$-entry will be zero; this entry is $\lambda^a \binom{b}{1} = \lambda^a b$. Then $b \equiv 0 \pmod{q}$ and then $b = qu$ where $0 \le u < q - 1$. This means in particular that $b \equiv u \pmod{(q-1)}$. Moreover, by (13) we have

$$O = (S^a T^b)_{13} = c X_1.$$

Now, since $X_1 \neq O$, we obtain $c = 0$. Then by Lemma 8, $a \equiv b \pmod{(q-1)}$. Therefore $a \equiv u \pmod{(q-1)}$ and then $a = u$, since $0 \le a, u < q - 1$. Thus $b = qa$ and $S^a T^b \in \langle ST^q \rangle$. $\square$

Now, we are going to specify matrix X such that the minimum distance of the code $\mathfrak{C}$ is maximal. So we consider

$$X_1 = I_k \quad \text{and} \quad X_2 = \begin{bmatrix} 1 & 0 & \cdots & 0 \\ 0 & 0 & \cdots & 0 \\ \vdots & \vdots & & \vdots \\ 0 & 0 & \cdots & 0 \end{bmatrix}. \tag{14}$$

Theorem 2. *For X_1 and X_2 in* (14), *it follows that*

$$\mathrm{rank}\left[(S^a T^b)_{12}\, (S^a T^b)_{13} \right] = k, \quad \text{for all } S^a T^b \in \mathbf{H} \backslash \mathrm{stab}_{\mathbf{H}}(\mathcal{U}_k).$$

Proof. Let $S^a T^b \in \mathbf{H} \backslash \mathrm{stab}_{\mathbf{H}}(\mathcal{U}_k)$; that is, $0 \le a < q - 1$, $0 \le b < q(q-1)$ and $b \neq qa$ by Theorem 1. In particular, this means that $e_1 \neq 0$ by Lemma 7.

Now, by (13) and (14), it follows that

$$(S^a T^b)_{13} = \begin{bmatrix} c + e_k & e_1 & e_2 & e_3 & \cdots & e_{k-1} \\ e_{k-1} & c & e_1 & e_2 & \cdots & e_{k-2} \\ e_{k-2} & 0 & c & e_1 & \cdots & e_{k-3} \\ e_{k-3} & 0 & 0 & c & \cdots & e_{k-4} \\ \vdots & \vdots & \vdots & \vdots & & \vdots \\ e_1 & 0 & 0 & 0 & \cdots & c \end{bmatrix}.$$

If $c = 0$ then $\det(S^a T^b)_{13} = (-1)^k e_1^k \neq 0$ and we obtain the desired result.

Thus, we may assume that $c \neq 0$. In this case, it follows that

$$k - 1 \le \mathrm{rank}(S^a T^b)_{13} \le k.$$

If $\mathrm{rank}(S^a T^b)_{13} = k$, the result holds. On the other hand, if $\mathrm{rank}(S^a T^b)_{13} = k-1$, then

$$0 = \det(S^a T^b)_{13} = \det \begin{bmatrix} c + e_k & \boldsymbol{u} \\ \boldsymbol{v}^T & U \end{bmatrix} \tag{15}$$

where $\boldsymbol{u} = \begin{bmatrix} e_1 & e_2 & e_3 & \cdots & e_{k-2} \end{bmatrix}$ and $\boldsymbol{v} = \begin{bmatrix} e_{k-1} & e_{k-2} & e_{k-3} & \cdots & e_1 \end{bmatrix}$.

Arguing by contradiction, assume that rank $\left[(S^a T^b)_{12} \, (S^a T^b)_{13} \right] = k - 1$. Therefore, by (11), if we consider the first column of $\frac{1}{\lambda^a}(S^a T^b)_{12}$ and the last $k - 1$ columns of $(S^a T^b)_{13}$, we have that

$$\text{rank} \begin{bmatrix} -\binom{b}{k} & e_1 & e_2 & e_3 & \cdots & e_{k-1} \\ \binom{b}{k-1} & c & e_1 & e_2 & \cdots & e_{k-2} \\ \binom{b}{k-2} & 0 & c & e_1 & \cdots & e_{k-3} \\ \binom{b}{k-3} & 0 & 0 & c & \cdots & e_{k-4} \\ \vdots & \vdots & \vdots & \vdots & & \vdots \\ \binom{b}{1} & 0 & 0 & 0 & \cdots & c \end{bmatrix} = \text{rank} \begin{bmatrix} e_k & u \\ v^T & U \end{bmatrix} = k - 1 \qquad (16)$$

and by means of (15) and (16),

$$0 = \det \begin{bmatrix} c + e_k & u \\ v^T & U \end{bmatrix} = \det \begin{bmatrix} c & u \\ 0^T & U \end{bmatrix} + \det \begin{bmatrix} e_k & u \\ v^T & U \end{bmatrix} = c^k + 0,$$

which is a contradiction since $c \neq 0$. It follows that rank $\left[(S^a T^b)_{12} \, (S^a T^b)_{13} \right] = k$, for all $S^a T^b \in \mathbf{H} \setminus \text{stab}_{\mathbf{H}}(\mathcal{U}_k)$ and the result holds. $\qquad \square$

Finally, as a consequence of the above results, we obtain our main theorem.

Theorem 3. *Assume that q is a prime number such that $q \geq n - k$ where k and n are nonnegative integers where $k \leq n/2$. Assume also that $\lambda \in \mathbb{F}_q$ is a primitive element. Consider the upper triangular matrices S and T defined in (7), and $X = \begin{bmatrix} X_1 \\ X_2 \end{bmatrix}$ where X_1 and X_2 are defined in (14). If $\mathbf{H} = \langle S, T \rangle$ is the Abelian non-cyclic subgroup of GL_n given in (8), and $\mathfrak{C}$ is the orbit code defined in (4), then $|\mathfrak{C}| = q(q - 1)$ and $d(\mathfrak{C}) = 2k$.*

Proof. $|\mathfrak{C}| = q(q-1)$ as a consequence of (8) and Theorem 1. Moreover, $d(\mathfrak{C}) = 2k$ results from (4), (5) and Theorem 2. $\qquad \square$

4   Open Questions

We would like to know if it is possible to extend our construction to any finite field $\mathbb{F}_q$, where q is a prime power. Moreover, we wonder if it is possible to obtain longer, constant dimension codes starting from our Abelian orbit code without compromising the distance.

Acknowledgements. The first author was supported by grants MIMECO MTM2015-68805-REDT and MTM2015-69138-REDT.

References

1. Bardestani, F., Iranmanesh, A.: Cyclic orbit codes with the normalizer of a Singer subgroup. J. Sci. Islamic Republic of Iran **26**(1), 49–55 (2015)
2. Bartoli, D., Pavese, F.: A note on equidistant subspace codes. Discrete Appl. Math. **198**, 291–296 (2016)
3. Ben-Sasson, E., Etzion, T., Gabizon, A., Raviv, N.: Subspace polynomials and cyclic subspace codes. IEEE Trans. Inf. Theory **62**(3), 1157–1165 (2016)
4. Cossidente, A., Pavese, F.: On subspace codes. Des. Codes Crypt. **78**(2), 527–531 (2016)
5. Etzion, T., Vardy, A.: Error-correcting codes in projective space. IEEE Trans. Inf. Theory **57**(2), 1165–1173 (2011)
6. Ghatak, A.: Construction of Singer subgroup orbit codes based on cyclic difference sets. In: Proceedings of the Twentieth National Conference on Communications (NCC 2014), pp. 1–4. IEEE, Kanpur, India, February 2014
7. Gluesing-Luerssen, H., Morrison, K., Troha, C.: Cyclic orbit codes and stabilizer subfields. Adv. Math. Commun. **9**(2), 177–197 (2015)
8. Gluesing-Luerssen, H., Troha, C.: Construction of subspace codes through linkage. Adv. Math. Commun. **10**(3), 525–540 (2016)
9. Gorla, E.G., Ravagnani, A.: Equidistant subspace codes. Linear Algebra Appl. **490**, 48–65 (2016)
10. Ho, T., Koetter, R., Médard, M., Karger, D.R., Effros, M.: The benefits of coding over routing in a randomized setting. In: Proceedings of the 2003 IEEE International Symposium on Information Theory (ISIT 2003), p. 442. IEEE, Yokohama, Japan, June/July 2003
11. Koetter, R., Kschischang, F.R.: Coding for errors and erasures in random network coding. IEEE Trans. Inf. Theory **54**(8), 3579–3591 (2008)
12. Rosenthal, J., Trautmann, A.L.: A complete characterization of irreducible cyclic orbit codes and their Plücker embedding. Des. Codes Crypt. **66**, 275–289 (2013)
13. Silberstein, N., Trautmann, A.L.: New lower bounds for constant dimension codes. In: Proceedings of the 2013 IEEE International Symposium on Information Theory (ISIT 2013), pp. 514–518. IEEE, Istanbul, July 2013
14. Slepian, D.: Group codes for the Gaussian channel. Bell Syst. Tech. J. **47**(4), 575–602 (1968)
15. Trautmann, A.L.: Isometry and automorphisms of constant dimension codes. Adv. Math. Commun. **7**(2), 147–160 (2013)
16. Trautmann, A.L., Manganiello, F., Braun, M., Rosenthal, J.: Cyclic orbit codes. IEEE Trans. Inf. Theory **59**(11), 7386–7404 (2013)
17. Trautmann, A.L., Manganiello, F., Rosenthal, J.: Orbit codes - a new concept in the area of network coding. In: Proceedings of the 2010 IEEE Information Theory Workshop (ITW 2010). IEEE, Dublin, Ireland, August 2010

Analysis of Two Tracing Traitor Schemes via Coding Theory

Elena Egorova and Grigory Kabatiansky[(✉)]

Skolkovo Institute of Science and Technology (Skoltech),
Moscow Region 143025, Russia
elena.egorova@skolkovotech.ru, G.Kabatyansky@skoltech.ru

Abstract. We compare two popular tracing traitor schemes (1) using non-binary *codes* with identifiable parent property (IPP-codes) and (2) using *family of sets* with identifiable parent property. We establish a natural basis for comparing and show that the second approach is stronger than IPP-codes. We also establish a new lower bound on the cardinality of the family of sets with identifiable parent property.

1 Introduction

In [1] Chor, Fiat and Naor introduced traitor tracing schemes in the context of broadcast encryption. Their main idea was to encrypt data in such a way that it prevents illegal redistribution of digital content. Namely, when a malicious coalition of users ("traitors") of a limited size creates a "device" for an unauthorized user then the distributor is able to identify at least one traitor. To this end, the distributor encrypts a data block with the corresponding session key and gives the authorized users personal keys to decrypt them. In order to create unauthorized decryption keys (decoders), some coalition of traitors creates a forged key/decoder based on their common knowledge of keys/decoders. Assuming that the cardinality of the coalition is not greater than t, once a forged key is observed, the distributor should be able to trace back at least one traitor from a malicious coalition. Such schemes could be "open" and "secret" in the notations of [1], where a "secret" scheme is in fact a family of "open" schemes but the particular choice of one of them is unknown to the traitors. Another difference is that for a secret scheme it is allowed to trace traitor with the probability of error enough small (tending to zero with the "length" of a scheme) but an open scheme should provide traitor traicing with zero-error probability. One simple class of open schemes was proposed in [1] and developed further under the name of codes with the identifiable parent property (IPP-codes), and its particular case called t-traceability codes, in [2]. These codes were extensively studied, see e.g. [3–5], also a detailed overview can be found in [6].

The original idea of [1] was to establish a traitor tracing (TT) schemes (like t-IPP codes) based on perfect secret sharing schemes (SSS for short). Perfect SSS were discovered in [7,8]. From this point of view t-IPP codes can be considered as TT schemes based on the simplest (n, n)-threshold SSS. Extension of this idea

© Springer International Publishing AG 2017
Á.I. Barbero et al. (Eds.): ICMCTA 2017, LNCS 10495, pp. 84–92, 2017.
DOI: 10.1007/978-3-319-66278-7_8

to arbitrary (w, n)-threshold SSS was proposed in [9] under the name of family of IPP-sets, and further developed in [10,11]. Surprisingly no relationship as well as comparison of these two types of TT schemes were given before. We do it in this paper, by returning, in some sense, to the paper [1]. Another result is a new lower bound on the cardinality of the best family of IPP-sets.

2   Open Tracing Traitor Schemes - How Do They Work?

Consider a broadcasting scenario where a dealer distributes some digital content to M users. In order to prevent illegal redistribution, the dealer sends the content in an encrypted form obtained by using of some secret key k, which serves as a session key and should be changed for distributing another portion of digital content. For distribution the key k to users, the dealer transmits some blocks of information $e_1, \ldots, e_N$, which allow any legal user to recover k. In order to trace member(s) of a malicious coalition the dealer forms e_j as encrypted version of some secret information s_j. The i-th user receives the corresponding set of keys $\{f_{1i}, \ldots, f_{ni}\}$ during the initialization phase. Let us cite [1]:

"We devise t-resilient traceability scheme with the following properties: 1. Either the cleartext information itself is continuously transmitted to the enemy by a traitor, or 2. Any captured pirate decoder will correctly identify a traitor and will protect the innocent even if up to t traitors combine and collude....

Definition 1. *An n user open TT scheme is called t resilient if for every coalition of at most t traitors the following holds: suppose the coalition uses the information its members got in the initialization phase to construct a pirate decoder. If this decoder is capable of applying the decryption scheme, then the tracing traitors algorithm will correctly identify one of the coalition members."*

Below we make this definition more precise for two particular and most popular tracing traitors schemes. Let us start from the scheme proposed in [2]. We assume that the session key k belongs to some q-ary alphabet. Below it will be the finite field $GF(q)$. The corresponding secret information "symbols" $s_1, \ldots, s_n$ are random uniformly distributed variables with values from $GF(q)$ with the property that

$$s_1 + \ldots + s_n = k \tag{1}$$

where summation is taken in the field $GF(q)$. In other words, $s_1, \ldots, s_n \in GF(q)$, where $s_1, \ldots, s_{n-1}$ are independent random uniformly distributed elements of $GF(q)$ and $s_n = k - \sum_1^{n-1} s_i$. The values s_j are called *shares* of k (see below for Secret Sharing Scheme) and every share s_j is encrypted on q different subkeys from the set $F^{(j)} = \{f_{j1}, \ldots, f_{jq}\}$, resulting in the set of q encrypted shares $\{e_{j1}, \ldots, e_{jq}\}$. The set of nq corresponding encrypted shares $\{e_{11}, \ldots, e_{1q}, \ldots, e_{n1}, \ldots, e_{nq}\}$ is transmitted by the dealer along with the encrypted portion of digital content.

The i-th user receives during initialization phase the set of keys $\{f_{1i_1}, \ldots, f_{ni_n}\}$, where $f_{ji_j} \in F^{(j)}$, i.e., each user receives exactly one key from every set $F^{(j)}$.

Since $|F^{(j)}| = q$ for all $j = 1, 2, \ldots, n$ one can consider the set $\{f_{1i_1}, \ldots, f_{ni_n}\}$ as a q-ary codevector $c_i \in GF(q)^n$ assigned to the i-th user, and the set of all such vectors is called a *fingerprinting code* $C \subset GF(q)^n$. If a coalition of malicious users (traitors) $U \subset \{1, \ldots, M\}$ wants to create a "device" ("decoder") which will be able to decrypt every transmitted encrypted portion of digital content then the coalition have to create a new set $Y = \{y_1, \ldots, y_n\}$ of subkeys with the property that $y_j \in F^{(j)}$ for all $j \in \{1, 2, \ldots, n\}$. It is very important to note that $y_j \in \{f_{ju_j} : u \in U\}$, i.e. that the coalition can choose subkeys only from the set of the coalition subkeys. Hence the resulting problem can be formulated in the language of codes as it was done in [2]. Namely, define for any set $U \subset GF(q)^n$ and any coordinate i its the i-th projection $P_i(U)$ of the set V as

$$P_i(U) = \bigcup_{u \in U} u_i.$$

For a fingerprinting code C we shall denote by U a coalition of traitors as well as the corresponding to them set of codevectors. Then we denote by $<U>$ the set of all false fingerprints (also called descendants [2]) that the coalition U can create, namely,

$$<U> = \{\mathbf{x} = (x_1, \ldots, x_n) \in F_q^n : \forall i \; x_i \in P_i(U)\} \tag{2}$$

Let

$$E_t(C) = \cup_{U \subset C: \, |U| \leq t} <U>$$

denote the set of all false fingerprints which can be created by all coalitions U of size at most t.

Definition 1 ([2]). *A code C has the identifiable parent property of order t, or C is t-IPP code for short, if for all $\mathbf{z} \in E_t(C)$*

$$C_t(\mathbf{z}) := \bigcap_{U: \, \mathbf{z} \in <\varphi(U)>, \, |U| \leq t} U \neq \emptyset \tag{3}$$

Hence, if the fingerprints form a code possessing the *Identifiable Parent Property*, then from any false fingerprint $\mathbf{z}$ created by a coalition U, at least one user from U will be identified without any doubt.

It is easy to see that q-ary t-IPP codes do not exist for $t \geq q$. On the other hand, for $t < q$ there exist families of t-IPP codes with non-vanishing rate, i.e. with a number of codewords growing exponential in n, see [2–4]. Note that for the *binary* case there are no t-IPP codes even for coalitions of size 2.

Nevertheless there is a natural way to represent q-ary IPP-codes as binary codes but with q times larger length. Namely, consider one of the simplest concatenated codes, proposed in [15], when j-th q-ary symbols is replaced by the corresponding binary vector of length q, which all coordinates are zeroes except of j-th coordinate which is 1. Denote by $X_q^N \subset \{0,1\}^N$ the set of all binary vectors $\mathbf{x} = (x_1, \ldots, x_N)$ of length $N = nq$ such that the Hamming weight of every block

$x^j = (x_{(j-1)q+1}, \ldots, x_{jq})$ equals to 1, $j = 1, \ldots, n$. We shall say that a binary vector $\mathbf{x} = (x_1, \ldots, x_N)$ is covered by a binary vector $\mathbf{y} = (y_1, \ldots, y_N)$ and denote it as $\mathbf{y} \succ \mathbf{x}$ if $y_i \geq x_i$ for all i. For every coalition $U = \{\mathbf{u}^1, \ldots, \mathbf{u}^t\} \subset X^N$ the corresponding set of descendants $<U>$ consists of all vectors $\mathbf{z} \in X^N$ which are covered by the vector $\mathbf{U} = \mathbf{u}^1 \vee \ldots \vee \mathbf{u}^t$. Then the corresponding set $<U>_{bin}$ of all false *binary* fingerprints (*binary* descendants) that the coalition U can create equals $<U>_{bin} = \{\mathbf{x} \in X^N : \mathbf{U} \succ \mathbf{x}\}$.

Now the Identifiable Parent Property can be reformulated for a binary fingerprinting code C_{bin} of length $N = nq$ in the following way: for every $\mathbf{z} \in X^N$ either all coalitions which can create $\mathbf{z}$ have at least one common member, i.e.,

$$\bigcap_{U:\ \mathbf{z} \in <U>_{bin},\ |U| \leq t} U \neq \emptyset, \qquad (4)$$

or no coalition of size at most t can create a given $\mathbf{z}$.

The IPP-codes can be explained by the following toy example. Imagine that the dealer should allow access to a treasury chest for M users, but some of them are dishonest. The dealer locks the treasury chest by n different locks and each lock has q different key holes and correspondingly q different keys. Therefore the dealer provides to every user a unique set of n keys - by one key for every lock. Then a malicious coalition U can create a fraud set of keys by giving one key for every lock from the keys at their disposal. The aim of the dealer is to form the users' sets of keys in such a way that from any sufficient set of keys created by a coalition, the dealer can for sure reveal at least one member of the coalition.

In fact, the basic property of the above described TT scheme is that it is based on *perfect secret sharing scheme*, and as it was noted in [1] that any such scheme can be used for constructing a TT scheme. This was demonstrated in [1] by the so-called two-level scheme. But there is a more natural and simple generalization of IPP TT-scheme, namely, by using of w-out-of-n SSS. Note that IPP-codes correspond to the case of n-out-of-n SSS. Let us recall the well known Shamir's polynomial w-out-of-n SSS.

Let there be N participants, k be a common secret key, $GF(q)$ be the field of q elements with $q > N$ and let $\alpha_1, \ldots, \alpha_N$ be N different nonzero elements of $GF(q)$. The dealer chooses a random polynomial

$$f(x) = k + f_1 x + \ldots + f_{w-1} x^{w-1} \qquad (5)$$

with coefficients $f_1, \ldots, f_{w-1} \in GF(q)$ which are independent random variables, uniformly distributed on $GF(q)$. The share assigned to the i-th participant is $s_i = f(\alpha_i)$. Then any w (or more) participants can recover the secret key k since the degree of polynomial f is $w - 1$, and any set of less than w participants get no information about k. This is the famous w-out-of-n threshold secret sharing scheme [7].

Based on this secret sharing scheme, the following TT scheme, called the t-IPP family of sets, was proposed in [9]. Namely, the dealer generates a random polynomial $f(x)$ of degree $w - 1$, chooses for the i-th user its personal subset $A_i \subset \{1, \ldots, N\}$, where $|A_i| = w$, and *secretly* distributes to the i-th user the values of the polynomial $f(x)$ at points $\alpha_j : j \in A_i$. In order to distribute shares to users in a secret way, the dealer encrypts each share $s_j = f(\alpha_j)$ by its key E_j, distributes during the initialization step to the i-th user its encryption keys $E_j : j \in A_i$, and finally sends all N encrypted shares via a broadcast channel. Any authorized user can decrypt w shares, then recover from them the secret key k, and hence can decrypt the transmitted digital content. A malicious coalition U can create a fraud "decoder" by arranging together at least w different key E_j which belong to members of U. Denote the corresponding set of keys by

$$\hat{E} = \bigcup_{j \in \hat{A}} E_j,$$

where $\hat{A} \subset \bigcup_{i \in U} A_i$ and its cardinality $|\hat{A}|$ should be at least w. Thus the set of descendants of the coalition U equals to

$$<U>_{set} = \{B \subset \{1, ..., N\} : B \subset \bigcup_{i \in U} A_i, |B| \geq w\} \tag{6}$$

Now the Identifiable Parent Property can be reformulated for a family of sets in the following way.

Definition 2 ([9]). *We shall say that a family $\mathbf{F}$ of w-subsets of a N-set $\{1, ..., N\}$ has the identifiable parent property of order t, or $\mathbf{F}$ is t-IPP family of sets for short, if for any $\hat{A} \subset \{1, ..., N\}$ such that $|\hat{A}| \geq w$ either*

$$\bigcap_{U:\ \hat{A} \in <U>_{set},\ |U| \leq t} U \neq \emptyset, \tag{7}$$

or there is no U such that $|U| \leq t$ and $\hat{A} \in <U>_{set}$

In words, a family $\mathbf{F}$ of w-subsets of a N-set $\{1, ..., N\}$ is a t-IPP family of sets if for any w-subset which belongs to the union of some t sets of $\mathbf{F}$ at least on of these sets can be uniquely determined. In particular, it means that no one set of $\mathbf{F}$ belongs to the union of t other sets of $\mathbf{F}$. Such families of sets are very popular, thanks to [13, 14]. They appeared first in coding theory under the name of superimposed codes, see [15–17].

3   How to Compare t-IPP Codes and t-IPP Family of Sets?

In order to compare t-IPP codes and t-IPP family of sets we use descriptions of both schemes in the language of binary vectors. Recall that a q-ary t-IPP code of length n can be represented as a binary code of length $N = nq$, in which vectors have block structure where every block of length q consists of all zeroes except for a single 1, and the code satisfies the property (4).

As for t-IPP family of sets, it can be reformulated in the language of binary vectors almost in the similar way. To construct a binary t-IPPS code C of length

N and weight w, one should replace every set of a t-IPP family of sets with its characteristic vector. We call the code C a binary t-IPPS code. A collusion attack of a malicious coalition $U \subset C$ in binary-vectors language proceeds as follows. The coalition U creates a fraud vector $\mathbf{z} = (z_1, \ldots, z_N)$ of the Hamming weight at least w under the following rule which is an analog of *Marking Assumption* for digital fingerprinting codes, see [12]:

if all users of U have 0 in j-th coordinate then $z_j = 0$,
and z_j could be 0 or 1 if at least one user has 1 in this coordinate.

This description reveals the main difference between IPPS codes and IPP codes. In the binary representation of q-ary IPP codes, if all vectors of a coalition in the j-th block have 1 at the same position, then this position in a fraud vector must also be 1. But in IPPS code, both 0 an 1 can be placed in a fraud vector in this case, nevertheless the resulting fraud vector should have weight at least t. Hence IPPS codes resists a more general attack than IPP codes. On the other hand, IPPS codes have less restrictions as codes, namely, code-vectors should have weight $w = n$ without the need for having each of the n blocks has weight exactly one. Therefore our concatenated construction of IPPS codes is of particular interest since the corresponding codes have the structure of IPP codes (every of n blocks has weight exactly one) but resists a wider attacks. To compare both types of codes, it is natural to put $w := n$ for IPPS codes and it is very interesting to construct IPPS codes with cardinality larger than of known IPP codes.

Another reason to compare these two types of TT schemes on the basis of equal parameter N is that N is proportional to the overhead which the dealer should transmit over a broadcast channel. Indeed, in the case of IPP codes, the dealer should send n shares and each of them q times differently encrypted. In the case of the IPP family of sets the dealer should send encrypted values of the polynomial $f(x)$ in all N points. Note that for the case of IPP codes it is inefficient to choose a large value of q. Let $M_t(n, q)$ denotes the maximal possible cardinality of t-IPP code, and $R_t(n, q) = N^{-1} \log_2 M_t(n, q)$ denotes the corresponding *binary* code rate. Then

$$R_t(n, q) \le (nq)^{-1} n \log_2 q = \frac{log_2 q}{q}$$

and it tends to zero with growing q. For example, the best known lower bound on the rate of 2-IPP codes [2] achieves its maximum for $q = 3$ (the minimal possible q).

4   New Lower Bound on the Size of IPPS Codes

Let $\omega := w/N$ be the relative weight, and $M(N, \omega)$ be the maximum possible size of t-IPPS code of length N and relative weight ω. Denote by

$$R_t(n, \omega) := \frac{\log_2 M(N, \omega)}{N}$$

the maximum possible rate and by $R_t(N)$ denote the maximum of $R_t(N, \omega)$ over $\omega \in [0, 1]$.

Definition 3. *A family $F = \{S_1, ..., S_M\}$ of w-subsets of $\{1, ..., N\}$ is called a* **t-traceability set system (t-TSS)** *if for any coalition $U = \{i_1, ..., i_k\} \subset [M]$, $k \leq t$ and any $S \in <U>_{set}$, it holds*

$$|S \cap S_j| < |S \cap S_{i_l}| \text{ for any } l=1,...,k \text{ and any } j \notin U$$

This definition means that for a t-traceability set system, search for malicious user(s) reduces to the search of "closest" sets. Moreover, any t-TSS is also t-IPP family of sets, so a lower bound on the size of t-TSS is also a lower bound for the cardinality of t-IPPS codes.

The notion of t-traceability was coined in [9] and further studied in [10,11]. These papers provide results about upper bounds as well as about lower bounds on the size of traceability set systems. We are interested in finding a new lower bound for t-TSS since it automatically gives us a new lower bound on the size of t-IPPS codes.

4.1   The Previous Results

In paper [1] the authors provided the following bound: $R_t \geq \frac{1}{8t^4}$ with $\omega = \frac{1}{2t^2}$. In recent paper [11] the authors proved that

$$|M(n, \omega)| \geq \frac{\binom{n}{\lceil w/t^2 \rceil}}{\binom{w}{\lceil w/t^2 \rceil}},$$

which asymptotically means $R_t(\omega) \geq H(\frac{\omega}{t^2}) - 2\omega H(\frac{1}{t^2})$.

Below we improve the aforementioned bounds. The following simple lemma establishes a natural condition on a set system to be t-TSS (very similar to original approach of [1]):

Lemma 1. *If $|S_i \cap S_j| < w/t^2$ for any $S_i, S_j \in F, i \neq j$, then F is a t-traceability set system.*

Proof. Consider any coalition $U = \{i_1, ..., i_k\}$, $k \leq t$ and any $S \in <U>$. Then, $\max_{l \in \{1,...,k\}} |S_{i_l} \cap S| \geq w/t$ since $|S| \geq w$. On the other hand, for any $j \notin U$,

$$|S_j \cap S| < |S_j \cap (S_{i_1} \cup ... \cup S_{i_k})| < |S_j \cap S_{i_1}| + ... + |S_j \cap S_{i_k}| < t \cdot \frac{w}{t^2} = \frac{w}{t}.$$

4.2   Constant Weight Codes with Large Distance and Traceability Set System

Obviously, from the set systems paradigm we can move to a language of binary vectors. This will imply the change of "distance". Indeed, if we consider the set systems we appeal to the cardinality of intersections, but if we consider the code consisting of binary vectors we will appeal to the Hamming distance.

More formally, consider the binary code C_F, corresponding to family F, consisting of the characteristic vectors of sets $S_i \in F$. Then, C_F is a constant weight code of weight $w = \omega n$ with the minimal Hamming distance $d \geq 2\tau n$ where $\tau = \omega(1 - 1/t^2)$ and vice versa. Application of VG-bound for the constant weight codes, see [18], leads to the following new lower bound.

Theorem 1. *There exists a t-IPPS code with rate*

$$R_t(\omega) \geq H(\omega) - \omega H\left(\frac{\tau}{\omega}\right) - (1 - \omega) H\left(\frac{\tau}{1 - \omega}\right)$$

where H is the binary entropy function and $\tau = \omega(1 - 1/t^2)$.

If we compare the received bound with previous ones, we can note that the new one is better for arbitrary values of parameter t. Indeed, straight calculations show that **for $t = 2$:** Theorem 1 implies that $R_2 = \max_\omega R_t(\omega) \geq 0.0181$ and the best known bound is $R_{2,M} = 0.0159$ [11]. **For $t = 3$:** the new bound is $R_3 = 0.00316$ and the best known bound is $R_{3,M} = 0.0027$ [11]. **For** arbitrary t it is also true since it can be seen from the following inequality

$$H(\omega) + H(t^{-2}) \geq (1 - \omega)H\left(t^{-2}\omega/(1 - \omega)\right) + H(t^{-2}\omega)$$

that the new bound is always better than previously known ones.

References

1. Chor, B., Fiat, A., Naor, M.: Tracing traitors. In: Desmedt, Y.G. (ed.) CRYPTO 1994. LNCS, vol. 839, pp. 257–270. Springer, Heidelberg (1994). doi:10.1007/3-540-48658-5_25
2. Hollmann, H.D., van Lint, J.H., Linnartz, J.P., Tolhuizen, L.M.: On codes with the identifiable parent property. J. Comb. Theor. Ser. A **82**(2), 121–133 (1998)
3. Barg, A., Cohen, G., Encheva, S., Kabatiansky, G., Zémor, G.: A hypergraph approach to the identifying parent property: the case of multiple parents. SIAM J. Discrete Math. **14**(3), 423–431 (2001)
4. Alon, N., Cohen, G., Krivelevich, M., Litsyn, S.: Generalized hashing and parent-identifying codes. J. Comb. Theor. Ser. A **10**(1), 207–215 (2003)
5. Staddon, J.N., Stinson, D.R., Wei, R.: Combinatorial properties of frameproof and traceability codes. IEEE Trans. Inf. Theor. **47**, 1042–1049 (2001)
6. Blackburn, S.R.: Combinatorial schemes for protecting digital content. Surv. Comb. **307**, 43–78 (2003)
7. Shamir, A.: How to share a secret. Commun. ACM **22**(11), 612–613 (1979)
8. Blakley, G.R.: Safeguarding cryptographic keys. Proc. Natl. Comput. Conf. **48**, 313–317 (1979)
9. Stinson, D.R., Wei, R.: Combinatorial properties and constructions of traceability schemes and frameproof codes. SIAM J. Discrete Math. **11**(1), 41–53 (1998)
10. Collins, M.J.: Upper bounds for parent-identifying set systems. Des. Codes Cryptogr. **51**(2), 167–173 (2009)
11. Gu, Y., Miao, Y.: Bounds on traceability schemes. arXiv preprint arXiv:1609.08336 (2016)

12. Boneh, D., Shaw, J.: Collusion-secure fingerprinting for digital data. IEEE Trans. Inf. Theor. **44**, 1897–1905 (1998)
13. Erdos, P., Frankl, P., Furedi, Z.: Families of finite sets in which no set is covered by the union of two others. J. Comb. Theor. Ser. A **33**(2), 158–166 (1982)
14. Furedi, Z., Erdos, P., Frankl, P.: Families of finite sets in which no set is covered by the union ofr others. Isr. J. Math. **51**(1), 79–89 (1985)
15. Kautz, W., Singleton, R.: Nonrandom binary superimposed codes. IEEE Trans. Inf. Theor. **10**(4), 363–377 (1964)
16. Dyachkov, A.G., Rykov, V.V.: Bounds on the length of disjunctive codes. Probl. Inf. Transm. **18**(2), 166–171 (1982)
17. Quang, A.N., Zeisel, T.: Bounds on constant weight binary superimposed codes. Probl. Control Inf. Theor. **17**, 223–230 (1988)
18. Zinov'ev, V.A., Ericson, T.: On concatenated constant-weight codes beyond the Varshamov-Gilbert bound. Probl. Inf. Transm. **23**(1), 110–111 (1987)

Reliable Communication Across Parallel Asynchronous Channels with Glitches

Shlomo Engelberg[1] and Osnat Keren[2(✉)]

[1] Department of Electrical and Electronics Engineering,
School of Engineering and Computer Science,
Jerusalem College of Technology, Jerusalem, Israel
[2] Faculty of Engineering, Bar-Ilan University, Ramat Gan, Israel
osnat.keren@biu.ac.il

Abstract. Transmission across asynchronous communication channels is subject to laser injection attacks which cause glitches, pulses that are added to the transmitted signal at arbitrary times, and delays. We present self-synchronizing coding schemes with low latency at the receiver that require no acknowledgement and can decode transmissions subject to random delays and distorted by random glitches.

Keywords: Random delays · Glitch · Parallel asynchronous communications

1 Introduction

In this work, we consider a channel composed of multiple wires each of which suffers from random delays and glitches. Because the wires suffer from random delays, a receiver that sees their output must decide how to group the bits from the signals seen on the N wires to form N bit binary vectors. Such a channel is an example of an *asynchronous channel*. The error mechanisms and models in such channels differ significantly from those of synchronous channels.

Synchronization errors in synchronous channels may cause insertions, deletions or substitution of *symbols* [9] which change the block boundaries. For example, the channel input may be a sequence of, say, n symbols and the output a sequence of m symbols, m being a random variable depending on the number of insertions/deletions. The positions of insertions, deletions, and substitutions are random and unknown to both the transmitter and the receiver. Capacity bounds and coding schemes for such channels are described in the literature [5,6,8]. Synchronous channels may also suffer from delays that change the arrival time of symbols by a random (yet, bounded) number of time slots. Consider, for example, the parallel indistinguishable channels studied in [11] in which the information is conveyed via the number of "particles" transmitted in each time slot.

The research of the second author was supported by the ISRAEL SCIENCE FOUNDATION (grant No. 923/16). A preliminary version of part of this work was presented at TRUEDEVICE 2016, Barcelona, Spain.

© Springer International Publishing AG 2017
A.I. Barbero et al. (Eds.): ICMCTA 2017, LNCS 10495, pp. 93–106, 2017.
DOI: 10.1007/978-3-319-66278-7_9

In many synchronous binary channels, the receiver "sees" both zeros and ones. This is not the case in communication over asynchronous binary channels where the sender and receiver have different clock frequencies. A typical parallel asynchronous channel without feedback consists of N wires connecting two units. When transmitting a zero, the transmitter leaves a wire in its default state, and when transmitting a one, the transmitter sends a (positive) pulse down the wire. Transmissions on all wires take place simultaneously, however the arrival times may vary.

The receiver is composed of an asynchronous front-end followed by a buffer and a synchronous block. The asynchronous front-end consists of edge detectors and an $N \rightarrow \log_2(N)$ encoder. (See Fig. 1.) Upon detecting the *rising edge of a pulse* on a wire, the asynchronous front-end determines that it has seen a pulse on that wire and updates the buffer accordingly. The buffer maintains an ordered list of integers that represent the wires on which pulses have appeared and the order in which they appeared since the list was last read by the receiving block. Unless a properly crafted code is used, the receiver will not be able to distinguish between a single transmission over k wires and k (or fewer) transmissions over the same wires. In this work, all channels are taken to be asynchronous channels without feedback.

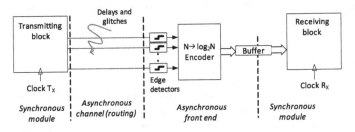

Fig. 1. Communication scheme over an asynchronous channels. The transmitting and receiving blocks have different (unsynchronized) clocks which may experience jitter and vary in time.

In some asynchronous channels, the propagation time over the parallel wires may cause a *skew*. That is, pulses from different transmissions may mix with one another. In a skew-less asynchronous channel, the pulses sent in any transmission can be received in any order but all the pulses from the ith transmission arrive at the receiver before any pulse from the $(i + 1)^{\text{th}}$ transmission arrives. In a skew-less, noise-free channel, unordered codes (codes for which no valid codeword is "contained" in another valid codeword [2]) are zero-error codes.

Some asynchronous channels suffer from glitches – informally, unwanted signals that are erroneously identified by the receiver as valid pulses. Our contribution is the description of techniques for designing zero-latency (instantaneously decodable) and low-latency codes for asynchronous skew-less channels that suffer from glitches and providing bounds on the rate such codes can achieve.

The techniques described in this paper allow one to ameliorate the effects of both *naturally occurring* and *intentionally* inserted glitches on asynchronous channels. As error injection can sometimes be used to obtain information about cryptographic keys [1], the techniques presented here can also play a role in securing certain communications systems against fault injection attacks.

The rest of this paper is organized as follows. The problem is formulated in the next section. Section 3 reviews related work, and in Sect. 4 we describe three techniques for reliable low-latency communication across asynchronous channels with glitches. Section 5 concludes the paper.

2   Problem Formulation

Let $p(t)$ be the pulse transmitted when a one is to be sent, let $\hat{p}(t)$ represent a typical (physical) glitch, and let T_k be the time between the $(k-1)^{\text{th}}$ and the kth transmission; T_k is *unknown* to the receiver. Let the waveform transmitted on the ith wire be

$$S_i(t) = \sum_{k=1}^{W} c_{i,k} p\left(t - \left(\sum_{l=1}^{k} T_l\right)\right)$$

where $\{c_{i,k}\}$ is the sequence of ones and zeros transmitted on the ith wire. We say that the vector

$$c_{:,k} = (c_{1,k}, c_{2,k}, \ldots, c_{N,k})$$

is the codeword transmitted over the channel at the kth time slot. The set of possible codewords that can be transmitted at the kth time slot forms the code $\mathcal{C}_k$ for this time slot. (That is, different codes can be used in different time slots.) In what follows, we describe coding schemes in which the set of possible codewords at time-slot k is a function of $k \mod W, W \geq 1$. We call the set of all sequences of W codewords $\{c_{i,k}\}_{i=1,k=1}^{N,W} \subseteq \mathbb{Z}_2^{N \times W}$ a *coding framework* and denote it by $\mathcal{F}$.

We are interested in reliable communication in the presence of physical-glitches. The received waveform is modeled as

$$R_i(t) = \sum_{k=1}^{W} c_{i,k} p\left(t - \left(\sum_{l=0}^{k} T_l\right) - \tau - \tau_{i,k}\right)$$
$$+ \sum_{k=1}^{W} e_{i,k} \hat{p}\left(t - \left(\sum_{l=0}^{k} T_l\right) - \tau - \theta_{i,k}\right)$$

where $\{e_{i,k}\}$ represents the sequence of glitches on the ith wire, τ is the average of the propagation delay over all the wires, where $\tau_{i,k}$ and $\theta_{i,k}$ are the delays of the pulses and the glitches, respectively, and where $0 \leq \tau_{i,k}, \theta_{i,k} < T_{i+1}$ do not cause pulses on the same line to interfere with one another.

We assume that the physical-glitches are uncorrelated with the transmitted pulses and that the delays between the wires are uncorrelated; that is, for all $i \neq j, m, k$, the delays $\tau_{i,k}$ and $\tau_{j,m}$ are *statistically independent*. (In practice, this may not be the case. As the glitches and delays may be being caused by a laser beam, they may be correlated. Any correlation could be used to enable reliable communication with less redundancy.) We also assume that there are N wires in use and that the probability of a glitch occurring on a wire in a single time-slot is ϵ. Thus, the expected number of glitches on N wires is ϵN.

A simple example of a possible set of transmitted signals in which a single glitch has been inserted in several time-slots is given in Fig. 2. The coding framework used $\mathcal{F} \subset \mathbb{Z}_2^{4 \times 1}$ has three words – $c_1 = 1010, c_2 = 1100, c_3 = 0011$ where the leftmost bit indicates the value transmitted on the first wire, the next bit indicates the value transmitted on the second wire, etc. For clarity, the transmitted pulses are depicted with solid lines and the physical-glitches with dotted lines. In this example, the transmitter sends the word c_1 twice. The receiver sees the sequence $1, 3, 2, 1, 3, 4 \ldots$ where each number indicates the wire on which a pulse was received and the order of the number indicate the order in which the pulses were received. The receiver cannot distinguish between the pulses and the physical-glitches; hence it may decide that the sequence c_1, c_2, c_3 was sent.

In our analysis of the communication system, it is necessary to distinguish between a physical glitch and a logical glitch; a physical glitch is an unwanted pulse – $\hat{p}\left(t - \left(\sum_{l=0}^{k} T_l\right) - \tau - \theta_{i,k}\right)$. When it arrives at the receiver, if it appears on a wire where no pulse is supposed to arrive ($a_{i,k+1} = 0$) then the physical-glitch is also a logical-glitch. However, when it appears on a wire where a pulse is expected to arrive ($a_{i,k+1} = 1$), then if the glitch precedes the pulse, the glitch is treated as the pulse, and the true pulse will (temporarily) be referred to as a logical-glitch. In Fig. 3, for example, the physical glitch on the second wire is not a logical glitch as there is supposed to be a one in that location. The second "true" pulse becomes a logical-glitch as after the glitch is identified as a one, the "true" pulse "becomes" an unwanted pulse – becomes a logical glitch. In fact, at any given time a logical glitch can be renamed as a pulse if it takes the place of an expected pulse. We show both that it is possible to remove logical glitches and to correctly decode the transmission (in the presence of physical-glitches).

In general, physical glitches may:

1. Cause a decoding error. In Fig. 2, the second c_1 was decoded as c_2.
2. Create a new word. In Fig. 2, the word c_3 was inserted after c_2.
3. Aggregate. Namely, up to a certain point the correct sequences are identified, and the "leftovers" are pushed forward.

Figure 3 shows a scenario in which three codewords are used: $c_1 = 110, c_2 = 101$ and $c_3 = 011$. The sequence c_1, c_3 is transmitted, and the received sequence, to which glitches have been added, is correctly decoded, yet, the decoder cannot get rid of the additional pulses, and they are aggregated and then interpreted as a codeword.

Since glitches can aggregate, a conventional unordered code with a predefined minimum distance cannot provide reliable transmission over this channel. The question addressed in this paper is: **What is the maximal code rate of a coding framework that enables near-zero-latency and reliable communication?** That is, we are looking to find coding frameworks $\mathcal{F} \subset \mathbb{Z}_2^{N \times W}$

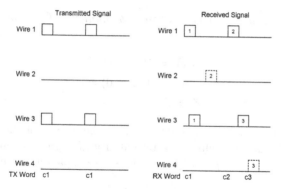

Fig. 2. The transmitted pulses (left). There are four wires and two transmission periods. The received signals (right). The numbers inside the pulses and glitches indicate the transmission with which the receiver associates the pulse or glitch.

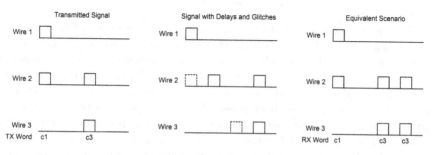

Fig. 3. The transmitted pulses with no more than a single glitch per time-slot. There are three wires and two transmission periods. The transmitted signals are shown on the left. The signals with delays and glitches is shown in the middle. An equivalent scenario is shown on the right hand side.

that maximize

$$R = \lim_{N,W \to \infty} \frac{\log_2(|\mathcal{F}|)}{NW}.$$

3   Related Work

The problem of delay-insensitive codes is considered in [16]. Various ways of dealing with parallel asynchronous channels that suffer from a relatively *small number* of skews and of dealing with a single channel ($N = 1$) that suffers from bit shifts have been considered in the past [2,3,13]. A recent work on communication via parallel channels that suffer from skews of *arbitrary* size [7] presented a zero-latency, zero-error code that achieves a rate of $\log_2((\sqrt{5}+1)/2)$ asymptotically as the bus width, N, tends to infinity.

Skews and glitches as asymmetric errors (Z-channels): Consider a Z-channel – a channel described by:

$$P(Y = 1|X = 1) = 1$$
$$P(Y = 1|X = 0) = \epsilon$$
$$P(Y = 0|X = 0) = 1 - \epsilon$$

where X is the input to the channel and Y is the output of the channel. That is, consider a channel for which the probability of a glitch occurring is ϵ. Let $H_b(p) = -p\log_b(p) - (1-p)\log_b(1-p)$. Then the capacity of such a channel is

$$C_Z(\epsilon) = H_2\left(\frac{1}{1 + \exp(H_e(\epsilon)/(1-\epsilon))}\right)$$
$$-\frac{1}{1-\epsilon}\frac{1}{1 + \exp(H_e(\epsilon)/(1-\epsilon))}H_2(\epsilon). \tag{1}$$

(See, for example, [10,14].) As long as the probability of error is not too large, the capacity of a Z-channel is greater than that of the binary symmetric channel [15].

One way to decode transmissions across asynchronous channels subject to glitches is to wait until the transmitted codeword (and additional ones) is "seen" at the decoder. That is, the decoder only starts decoding the current transmission when enough ones have been received since the previous transmission was decoded that it can be (almost) certain that every one from the current transmission has been received. Of course, glitches, pieces of the previous transmission or pieces of the next transmission may have arrived too causing additional ones to be present at the decoder. Though our wires are not Z-channels, when we treat the wires in this way, the data seen on the N wires can be treated as N bits received at the output of a Z-channel. Because the decoder waits until it is (almost) certain that all the bits of the current transmission have arrived we know that (with very high probability) all errors are due to zeros that were seen as ones – all errors are asymmetric. In what follows we define coding schemes for which the decoder can interpret the sequence of pulses over the bus as a sequence of codewords transmitted over a Z-channel.

4   Coding Schemes

We now consider several codes to correct glitches in a skew-less asynchronous channel. In the following, we assume that the decoder has successfully decoded all previous transmissions. This serves as a kind of inductive hypothesis.

We face three principal challenges.

1. Decoding words despite not knowing precisely where one word ends and the next begins (and this challenge exists even in a glitch-free channel).
2. Correctly decoding a word despite the presence of a number of glitches.
3. Preventing glitches from accumulating to the point that they make transmission of information impossible or impractical.

Challenge 1 can be dealt with by using fixed weight codes. The code's weight (and the number of wires and the probability of a glitch occurring) serves to tell the decoder by which point the current codeword ends (with high probability) and allows the decoder to determine the point by which the next word may begin (with high probability). Challenge 3 can be dealt with by not allowing two ones to be transmitted back-to-back – by forcing ones to appear alone. That is, assuming that c_k was decoded correctly, when decoding the $(k+1)^{\text{th}}$ codeword we require that $\text{supp}(c_k) \cap \text{supp}(c_{k+1}) = \varnothing$. This property enables us to eliminate logical glitches and prevent aggregation (though, as we shall see, there are other ways of dealing with challenge 3). Challenge 2 can be addressed by methods that allow an asynchronous bus suffering from glitches to be treated like a Z-channel. A sketch of a proof that coding frameworks with these three properties can be used to provide practically error free transmission of information is given in Appendix A.1. (Appendix A.1 should be read after reading this section.) We develop several coding frameworks that make use of the various combinations of the methods described above. We will see that when looking to maximize the rate, the choice of coding framework depends upon the probability of a glitch appearing on a particular wire.

We start with by presenting a property and several definitions.

Property 1. One may choose $p \in [0,1]$ and use the characteristics of the channel and the ideas used in the proof of Shannon's noisy channel theorem in [4] to generate constant weight codes appropriate to the channel where the density of ones in the code can be made arbitrarily close to p.

Definition 1. *A Δ-self-synchronizing code is a code that allows one to correctly decode the kth transmission before signals from the $(k+1+\Delta)^{\text{th}}$ transmission arrive at the decoder without using external synchronization signals or feedback. An instantaneous (zero latency) code is one for which $\Delta = 0$, and a near instantaneous (single-slot-latency) code is one for which $\Delta = 1$.*

In what follows, all the techniques lead to near instantaneous (1-self-synchronizing) codes and, in this way, allow us to overcome challenge 1.

Definition 2. *An admissible sequence of length W is a sequence in which no more than a single one occurs between successive zeros (and this definition differs from that in [7]).*

Though it is not necessary to use admissible sequences in each transmission in order to meet challenge 3, we will find it necessary to arrange our transmissions in such a way that from time-to-time we can identify and delete ones that have aggregated.

Property 2. If one uses admissible sequences on each wire – as the rows of a coding framework, then the coding framework's rate is upper bounded by

$$R \le \lim_{W \to \infty} \left(\frac{\log_2(A(W))}{W} \right) = \log_2(\phi) \tag{2}$$

where $A(W)$ is the number of admissible sequences of length W and ϕ is the golden ratio: $\phi \equiv (1 + \sqrt{5})/2$.

The correctness of (2) follows from [12, p. 70] because $A(W)$ satisfies $A(W) = A(W-1) + A(W-2)$ and is equal to the $(W+1)^{\text{th}}$ Fibonacci number. Thus, asymptotically, the number of admissible sequences tends to ϕ^W.

In principle, if we knew when each new codeword began (and hence, when the previous codeword ended), if we had a genie who could give us this piece of information, we could accumulate all the ones related to a single transmission and all the glitches that arrived during that period, and then we would have a collection of Z-channels.

We now consider several techniques that use constant-rate codes and allow us to "convert" our channel into a channel whose rate is closely related to that of the Z-channel. By making use of the admissible sequence idea, we are able to formulate coding frameworks that allow for (near) error-free communication.

4.1 Single-Code Based Coding Frameworks ($W = 1$)

Single-code based frameworks make use of the same underlying code in each time-slot, and the error correcting capability in each time slot is fixed.

Transmitting on Half the Wires. Consider a system in which one transmits information on $N/2$ wires in each time-slot and one transmits zeros on the rest of the wires. In the next time-slot, one transmits information on the other $N/2$ channels while transmitting zeros on the channel on which information was transmitted in the preceding time-slot.

Let ψ_k be the probability of at least one glitch appearing on a wire between k transmission. Then $\psi_k \equiv 1 - (1 - \epsilon)^k$. Following our assumption that a (single) glitch may be injected on a wire in between transmissions, a glitch may occur before the transmission and a glitch that may be injected after the transmission may arrive before the last (wanted) symbol (due to delays). Thus, at most two glitches may appear on a single wire before decoding begins, and the probability of at least one glitch appearing on a wire is $\psi_2 = 1 - (1 - \epsilon)^2 = 2\epsilon - \epsilon^2$.

Let μ be a very small positive number, let

$$\nu \equiv N^{-1/2+\mu}, 0 < \mu << 1,$$

and let the minimal weight of any codeword be at least $(2\epsilon - \epsilon^2 + 2\nu)(N/2)$. At the receiver, at time slot i the decoder starts by waiting for $(2\epsilon - \epsilon^2 + \nu)(N/2)$ ones on the *set of wires on which information was not transmitted in the ith time slot*. When this condition is met (and as $N \to \infty$, it will be met with probability $\to 1$), the receiver knows that ones from the $(i+1)^{\text{th}}$ time-slot are arriving and all the ones from the ith time-slot have arrived. At this point, the set of wires used for information transmission at the ith time slot forms a Z-channel in which the probability of a zero becoming a one is $2\epsilon - \epsilon^2$. The capacity of the $N/2$ wires is at least $(N/2)C_Z(2\epsilon - \epsilon^2)$. After determining what the transmitted word is, the receiver knows that all other ones that had been aggregated on this set of wires until now are glitches (as no ones are being transmitted on these $N/2$ lines in this time-slot) and, therefore, erases them.

This technique uses admissible sequences, but it is relatively inefficient. In Sect. 4.2, we find that it can profitably be used as a part of another technique.

The Ones-Alone Code. In the next single-code based coding framework we consider, the encoder works by always placing m ones on the N wires and never placing ones in the same locations in which they were placed in the previous transmission. Consequently, the sequences transmitted over each wire are admissible sequences.

The value of m determines the code rate; at each stage, there are $C(N-m, m)$ possible ways to place the ones, and thus the maximal rate per column is $R_{\max} = \log_2(C(N-m, m))/N$.

Theorem 1. *For $N \to \infty$, the maximal rate $R_{\max}$ tends to $\log_2(\phi)$ for $m \to \delta N$ where*

$$\delta \equiv \frac{1}{\phi + 2}.$$

Proof. See Appendix A.2.

In this scheme, the receiver waits to see $(\delta + 2\epsilon - \epsilon^2 + \nu)N$ ones and then determines the transmitted codeword. It then searches for the codeword in the received pulses on the $(1 - \delta)N$ wires on which ones or zeros may appear and "declares" that the codeword was seen as soon as all the bits in the codeword appeared on those wires. At this point, it "discards" all the pulses seen before the last bit of the received word appeared – and this prevents aggregation of logical glitches.

As $N \to \infty$, the best rate for a code for which the density of ones is δ tends towards

$$(1 - \delta)I(X; Y)$$

where in the distribution associated with X the probability of a one is $\delta/(1-\delta)$, where the distribution associated with Y is the distribution that follows from the conditional probabilities that define the channel, and where $I(X; Y)$ is the mutual information between X and Y. (It follows from Property 1 that the codebook can be made into a constant rate code and that as $N \to \infty$ the weight tends towards δN.)

Let $p = \delta/(1 - \delta)$ and, consequently, $\delta(p) = p/(1 + p)$, and let ϵ be the probability of a glitch occurring. Then the probability of at least one glitch occurring on a given wire between two "receptions" is $\epsilon_{\mathrm{eff}} \equiv \psi_2$.

The information capacity per "active" wire of a our channel when a constant weight code with weight δN is used and when the glitch probability is ϵ is the mutual information between a binary random variable for which the probability of a one occuring is p and the output of a Z-channel with glitch probability ϵ with such input. When dealing with a constant weight one's alone code, there are always δN wires that carry no information. Making use of the properties of the Z-channel we arrive at Theorem 2.

Theorem 2. *Let $a = \epsilon_{\text{eff}}(1-p) + p$, let $b = \epsilon_{\text{eff}}(1-p)$, and let $c = 1-p$. Then the information capacity of our channel is greater than or equal to*

$$\max_p C_{\text{ones alone}}(p), \tag{3}$$

where

$$C_{\text{ones alone}}(p) = -\left(a\log_2(a) - b\log_2(b) + c\log_2(c)\right)(1 - \delta(p))$$

is the average information capacity per wire of a Z-channel for which the probability of a one occurring on a wire in which ones may occur is p.

4.2   A Multiple-Code Based Coding Framework ($W > 1$)

A second general technique for protecting transmissions that may suffer from (a fairly limited number of) glitches is to add error correction to each transmission and only to use a ones-alone transmission periodically to clear the receiver of accumulated glitches. In this section the coding framework is designed to use the different codes in each time-slot. Here, the error correcting capability increases in each time slot.

We make use of a transmission scheme in which two sets of $N/2$ wires are treated somewhat separately. In each period of W transmissions, each set is used to transmit data $W - 1$ times. For each set of $N/2$ wires, one of the W time-slots is used to transmit zeros – and this cannot be the time-slot used for the other $N/2$ wires. In this way, for each set of $N/2$ wires, we "wipe the slate clean" once every W transmissions. This leads to error probabilities of $\psi_2, \psi_3, \dots \psi_{W-1}$ on each set of $N/2$ channels (though not in the same time-slot). Asymptotically, the maximal rate for this coding scheme is

$$R = \frac{1}{W} \sum_{k=1}^{W-1} C_Z(\psi_{k+1}) \tag{4}$$

and when $W = 2$, we find that we revert to the technique of Sect. 4.1. In order to make the coding as efficient as possible, in time-slots in which all N wires are used to transmit data, a single, efficient code should be used to code the data for all N wires.

4.3   A Comparison of the Efficiency of the Techniques

In Fig. 4 we compare:

- the capacity of a Z-channel (given in (1)) for which the probability of a zero becoming a one is ϵ;

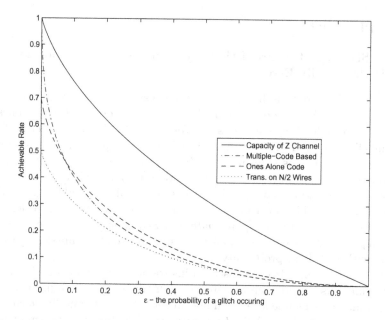

Fig. 4. The capacity of the Z-channel and of the three techniques we consider. For small probabilities of error, the multiple-code based coding framework is best. For higher probabilities of error, the ones alone – Z-channel is preferable.

- the best achievable rate for the multiple-code based coding framework where W is chosen so that (4) is maximized;
- the best achievable rate for the ones alone code as given by (3);
- and the best achievable rate when using $N/2$ channels – which is half the capacity of a Z-channel (given in (1)) for which the probability of a zero becoming a one is $2\epsilon - \epsilon^2$.

We find that the multiple-code based coding framework is most useful when there are not too many glitches and aggregation is not too much of a problem. The ones alone – Z channel is most useful for large probabilities of error as it prevents the aggregation of glitches. Though using only $N/2$ wires in any given transmission is not in and of itself an effective strategy, it is an integral part of the multiple-code based coding framework.

5   Summary

A globally asynchronous locally synchronous system is prone to delays and fault injection attacks which may cause glitches. This paper presents three simple to use, zero-latency, self-synchronizing coding frameworks. These coding frameworks enable us to treat an asynchronous channel as a Z-channel, prevent aggregation of glitches, and provide reliable transmission over the channel.

A   Appendices

A.1   A Sketch of the Proof that the Transmission of Information Is Practically Error Free

We consider a transmitter that makes use of constant-weight codes appropriate to a Z-channel and a decoder that looks for a valid codeword whose ones are contained in a buffer that maintains an ordered list of the pulses seen so far on the channel's wires and that were not identified as true pulses or glitches during the decoding of previous transmissions. (In our channel, the ones of the actual transmission are always present. Glitches add additional ones.)

When the ones-alone code of Sect. 4.1 is used, all "old" glitches are disregarded. The probability of a glitch affecting any given wire will be less than or equal to ψ_2. Using a code appropriate to a Z-channel in which zeros become ones with probability ψ_2, one can make certain that each transmission is decoded properly with very high probability.

If one uses the multiple-code based coding framework of Sect. 4.2, then at each stage one must use a code appropriate to a Z-channel with glitch probability ψ_n – where n increases by one after each not-all-zeros transmission. For such a code and large enough N the probability of correctly decoding the transmission can be made to approach one arbitrarily closely. Every $W - 1$ transmissions, zeros are transmitted on half the wires, and this reveal the glitches on those wires. We find that the probability of correctly decoding each transmission tends to one as N tends to infinity.

We assume that from time to time we resynchronize so that if there was a mistake in the reception of a transmission, the mistake does not propagate for too long.

A.2   The Proof that When $R_{\max}$ Is Maximized, the Density of Ones Tends to $\frac{1}{\phi+2}$

In order to determine the optimal m, we first determine the density of ones among admissible sequences, δ. We then calculate $\log_2(C(N - \lceil \delta \cdot N \rceil, \lceil \delta \cdot N \rceil))/N$ and show that as $N \to \infty$ this limit tends to $\log_2(\phi)$.

We know that $A(W) = A(w-1)+A(w-2)$. Let $O(W)$ be the number of ones in all the admissible sequences of length W. Then $O(W) = O(W-1)+O(W-2)+ A(W - 2)$ as one builds the sequences of length W by taking all the sequences of length $W - 1$ and appending zeros to them and by taking all sequences of length $W - 2$ and appending a zero and a one to them. Asymptotically, the number of solutions satisfies $A(W) \to \alpha\phi^W$. It is easy to see that when calculating the density of ones only this term need be considered. As ϕ satisfies the homogeneous version of this recurrence relation, it is not hard to show that

$$\frac{1}{\phi + 2}\alpha W \phi^W$$

a particular solution of $O(W) = O(W-1) + O(W-2) + \alpha\phi^W$. As the total number of sequences of length W tends to $\alpha\phi^W$ and the number of elements in each sequence is W, the average number of ones per element is $\delta = \frac{1}{\phi+2}$.

Estimating $C(N-m, m)$ while keeping in mind that we intend to take this function's logarithm and divide by N, we find that

$$C(N-m, m) \sim \left(\frac{(N/m)-1}{(N/m)-2}\right)^{N-m} (N/m - 2)^m.$$

Assuming that as $N \to \infty$ we choose m to satisfy $m/N \to \delta$, we find that the rate of the code tends towards

$$\left(1 - \frac{1}{\phi+2}\right) \times \log_2\left(\frac{\phi+1}{\phi}\right) + \frac{1}{\phi+2}\log_2\phi$$

$$= \frac{\phi+1}{\phi+2}\log_2(\phi+1) - \frac{\phi}{\phi+2}\log_2(\phi).$$

Note, however, that $\phi^2 = \phi+1$ (as ϕ solves the homogeneous recurrence relation), so that the above term is equal to

$$2\frac{\phi+1}{\phi+2}\log_2(\phi) - \frac{\phi}{\phi+2}\log_2(\phi) = \log_2\phi.$$

References

1. Barenghi, A., Breveglieri, L., Koren, I., Naccache, D.: Fault injection attacks on cryptographic devices: theory, practice and countermeasures. Proc. IEEE **100**(11), 3056–3076 (2012)
2. Blaum, M., Bruck, J.: Unordered error-correcting codes and their applications. In: FTSC-22, pp. 486–493 (1992)
3. Blaum, M., Bruck, J.: Coding for tolerance and detection of skew in parallel asynchronous communications. IEEE Trans. Inf. Theor. **46**(7), 2329–2335 (2000)
4. Cover, T.M., Thomas, J.A.: Elements of Information Theory. Wiley, New York (1991)
5. Davey, M.C., MacKay, D.J.C.: Reliable communication over channels with insertions, deletions, and substitutions. IEEE Trans. Inf. Theor. **47**, 687–698 (2001)
6. Dobrushin, R.L.: Shannon's theorems for channels with synchronization errors. Prob. Inf. Transm. **3**(4), 11–26 (1967)
7. Engelberg, S., Keren, O.: Reliable communications across parallel asynchronous channels with arbitrary skews. IEEE Trans. Inf. Theor. **63**(1), 1120–1129 (2017)
8. Fertonani, D., Duman, T.M., Erden, M.F.: Bounds on the capacity of channels with insertions, deletions and substitutions. IEEE Trans. Commun. **59**(1), 2–6 (2011)
9. Gallager, R.: Sequential decoding for binary channels with noise and synchronization errors. Lincoln Group Report, 2502 (1961)
10. Golomb, S.W.: The limiting behavior of the Z-channel. IEEE. Trans. Inf. Theor. **26**(3), 372 (1980)
11. Kovacevic, M., Popovski, P.: Zero-error capacity of a class of timing channels. IEEE. Trans. Inf. Theor. **60**(11), 6796–6800 (2014)

12. Marcus, B., Roth, R., Siegel, P.: Introduction to coding for constrained systems. http://www.math.ubc.ca/~marcus/Handbook/index.html
13. Shamai, S., Zehavi, E.: Bounds on the capacity of a channel with bit shift. IEEE Trans. Inf. Theor. **37**(3), 863–872 (1991)
14. Tallini, L.G., Al-Bassam, S., Bose, B.: On the capacity and codes for the Z-channel. ISIT 2002, Lausanne, Switzerland, 30 June–5 July, p. 422 (2002)
15. Tallini, L.G., Al-Bassam, S., Bose, B.: Feedback codes achieving the capacity of the Z-channel. IEEE Trans. Inf. Theor **54**(3), 1357–1362 (2008)
16. Verhoeff, T.: Delay-Insensitive codes-an overview. Distrib. Comput. **3**(1), 1–8 (1988)

On the Kernel of $\mathbb{Z}_{2^s}$-Linear Hadamard Codes

Cristina Fernández-Córdoba, Carlos Vela$^{(\boxtimes)}$, and Mercè Villanueva

Department of Information and Communications Engineering,
Universitat Autònoma de Barcelona, Edificio Q, 08193 Cerdanyola del Vallès, Spain
{cristina.fernandez,carlos.vela,merce.villanueva}@uab.cat
http://ccsg.uab.cat

Abstract. The $\mathbb{Z}_{2^s}$-additive codes are subgroups of $\mathbb{Z}_{2^s}^n$, and can be seen as a generalization of linear codes over $\mathbb{Z}_2$ and $\mathbb{Z}_4$. A $\mathbb{Z}_{2^s}$-linear Hadamard code is a binary Hadamard code which is the Gray map image of a $\mathbb{Z}_{2^s}$-additive code. For $s = 2$, the kernel of $\mathbb{Z}_4$-linear Hadamard codes can be used for their complete classification. In this paper, the kernel of $\mathbb{Z}_{2^s}$-linear Hadamard codes is given for $s > 2$. However, unlike for $s = 2$, we show that the dimension of the kernel just allows a partial classification of these Hadamard codes.

Keywords: Kernel · Hadamard code · $\mathbb{Z}_{2^s}$-linear code · $\mathbb{Z}_{2^s}$-additive code · Gray map · Classification

1 Introduction

Let $\mathbb{Z}_{2^s}$ be the ring of integers modulo 2^s with $s \geq 1$. The set of n-tuples over $\mathbb{Z}_{2^s}$ is denoted by $\mathbb{Z}_{2^s}^n$. In this paper, the elements of $\mathbb{Z}_{2^s}^n$ will also be called vectors over $\mathbb{Z}_{2^s}$ of length n. A binary code of length n is a nonempty subset of $\mathbb{Z}_2^n$, and it is linear if it is a subspace of $\mathbb{Z}_2^n$. Equivalently, a nonempty subset of $\mathbb{Z}_{2^s}^n$ is a $\mathbb{Z}_{2^s}$-additive code if it is a subgroup of $\mathbb{Z}_{2^s}^n$. Note that, when $s = 1$, a $\mathbb{Z}_{2^s}$-additive code is a binary linear code and, when $s = 2$, it is a quaternary linear code or a linear code over $\mathbb{Z}_4$.

Two binary codes C_1 and C_2 are said to be equivalent if there is a vector $\mathbf{a} \in \mathbb{Z}_2^n$ and a permutation of coordinates π such that $C_2 = \{\mathbf{a} + \pi(\mathbf{c}) : \mathbf{c} \in C_1\}$. Two $\mathbb{Z}_{2^s}$-additive codes $\mathcal{C}_1$ and $\mathcal{C}_2$ are said to be permutation equivalent if they differ only by a permutation of coordinates, that is, if there is a permutation of coordinates π such that $\mathcal{C}_2 = \{\pi(\mathbf{c}) : \mathbf{c} \in \mathcal{C}_1\}$.

The Hamming weight of a binary vector $\mathbf{u} \in \mathbb{Z}_2^n$, denoted by $\mathrm{wt}_H(\mathbf{u})$, is the number of nonzero coordinates of $\mathbf{u}$. The Hamming distance of two binary vectors $\mathbf{u}, \mathbf{v} \in \mathbb{Z}_2^n$, denoted by $d_H(\mathbf{u}, \mathbf{v})$, is the number of coordinates in which they differ. Note that $d_H(\mathbf{u}, \mathbf{v}) = \mathrm{wt}_H(\mathbf{v} - \mathbf{u})$. The Lee weight of an element $i \in \mathbb{Z}_{2^s}$ is $\mathrm{wt}_L(i) = \min\{i, 2^s - i\}$ and the Lee weight of a vector $\mathbf{u} = (u_1, u_2, \dots, u_n) \in \mathbb{Z}_{2^s}^n$

This work has been partially supported by the Spanish MINECO under Grants TIN2016-77918-P (AEI/FEDER, UE) and MTM2015-69138-REDT, and by the Catalan AGAUR under Grant 2014SGR-691.

© Springer International Publishing AG 2017
A.I. Barbero et al. (Eds.): ICMCTA 2017, LNCS 10495, pp. 107–117, 2017.
DOI: 10.1007/978-3-319-66278-7_10

is $\mathrm{wt}_L(\mathbf{u}) = \sum_{j=1}^{n} \mathrm{wt}_L(u_j) \in \mathbb{Z}_{2^s}$. The Lee distance of two vectors $\mathbf{u}, \mathbf{v} \in \mathbb{Z}_{2^s}^n$ is $d_L(\mathbf{u}, \mathbf{v}) = \mathrm{wt}_L(\mathbf{v} - \mathbf{u})$. The minimum distance of a $\mathbb{Z}_{2^s}$-additive code $\mathcal{C}$ is $d(\mathcal{C}) = \min\{d_L(\mathbf{u}, \mathbf{v}) : \mathbf{u}, \mathbf{v} \in \mathcal{C}, \mathbf{u} \neq \mathbf{v}\}$ and the minimum distance of a binary code C is $d(C) = \min\{d_H(\mathbf{u}, \mathbf{v}) : \mathbf{u}, \mathbf{v} \in C, \mathbf{u} \neq \mathbf{v}\}$.

In [7], a Gray map from $\mathbb{Z}_4$ to $\mathbb{Z}_2^2$ is defined as $\phi(0) = (0,0)$, $\phi(1) = (0,1)$, $\phi(2) = (1,1)$ and $\phi(3) = (1,0)$. There exist different generalizations of this Gray map, which go from $\mathbb{Z}_{2^s}$ to $\mathbb{Z}_2^{2^{s-1}}$ [4,5]. The one given in [4] is the map $\phi : \mathbb{Z}_{2^s} \to \mathbb{Z}_2^{2^{s-1}}$ defined as follows:

$$\phi(u) = (u_{s-1}, \ldots, u_{s-1}) + (u_0, \ldots, u_{s-2})Y, \tag{1}$$

where $u \in \mathbb{Z}_{2^s}$, $[u_0, u_1, \ldots, u_{s-1}]_2$ is the binary expansion of u, that is $u = \sum_{i=0}^{s-1} 2^i u_i$ ($u_i \in \{0,1\}$), and Y is a matrix of size $(s-1) \times 2^{s-1}$ which columns are the elements of $\mathbb{Z}_2^{s-1}$. Note that $(u_{s-1}, \ldots, u_{s-1})$ and $(u_0, \ldots, u_{s-2})Y$ are binary vectors of length 2^{s-1}. This Gray map can also be defined in terms of a first order Reed-Muller code [8]. Then, we define $\Phi : \mathbb{Z}_{2^s}^n \to \mathbb{Z}_2^{n2^{s-1}}$ as the component-wise Gray map ϕ.

Let $\mathcal{C}$ be a $\mathbb{Z}_{2^s}$-additive code of length n. We say that its binary image $C = \Phi(\mathcal{C})$ is a $\mathbb{Z}_{2^s}$-linear code of length $2^{s-1}n$. Since $\mathcal{C}$ is a subgroup of $\mathbb{Z}_{2^s}^n$, it is isomorphic to an abelian structure $\mathbb{Z}_{2^s}^{t_1} \times \mathbb{Z}_{2^{s-1}}^{t_2} \times \cdots \times \mathbb{Z}_4^{t_{s-1}} \times \mathbb{Z}_2^{t_s}$, and we say that $\mathcal{C}$, or equivalently $C = \Phi(\mathcal{C})$, is of type $(n; t_1, \ldots, t_s)$. Note that $|\mathcal{C}| = 2^{st_1} 2^{(s-1)t_2} \cdots 2^{t_s}$. Unlike linear codes over finite fields, linear codes over a ring do not have a basis, but there exists a generator matrix with minimum number of rows. If $\mathcal{C}$ is a $\mathbb{Z}_{2^s}$-additive code of type $(n; t_1, \ldots, t_s)$, then a generator matrix of $\mathcal{C}$ with minimum number of rows has exactly $t_1 + \cdots + t_s$ rows.

Two structural properties of binary codes are the rank and the dimension of the kernel. The rank of a binary code C is simply the dimension of the linear span, $\langle C \rangle$, of C. The kernel of a binary code C is defined as $\mathrm{K}(C) = \{\mathbf{x} \in \mathbb{Z}_2^n : \mathbf{x} + C = C\}$ [2]. If the all-zero vector belongs to C, then $\mathrm{K}(C)$ is a linear subcode of C. Note also that if C is linear, then $K(C) = C = \langle C \rangle$. We denote the rank of a binary code C as $\mathrm{rank}(C)$ and the dimension of the kernel as $\ker(C)$. These parameters can be used to distinguish between nonequivalent binary codes, since equivalent ones have the same rank and dimension of the kernel.

A binary code of length n, $2n$ codewords and minimum distance $n/2$ is called a Hadamard code. Hadamard codes can be constructed from normalized Hadamard matrices [1,10]. Note that linear Hadamard codes are in fact first order Reed-Muller codes, or equivalently, the dual of extended Hamming codes [10, Chap. 13 Sect. 3]. The $\mathbb{Z}_{2^s}$-additive codes that, under the Gray map Φ, give a Hadamard code are called $\mathbb{Z}_{2^s}$-additive Hadamard codes and the corresponding binary images are called $\mathbb{Z}_{2^s}$-linear Hadamard codes.

For $s = 2$, the $\mathbb{Z}_4$-linear Hadamard codes can be classified by using either the rank or the dimension of the kernel. It is known that for a $\mathbb{Z}_4$-linear Hadamard code C of type $(2^{t-1}; t_1, t_2)$, $\ker(C) = t_1 + t_2 + 1$ if $t_1 > 2$, and $\ker(C) = 2t_1 + t_2$ if $t_1 = 1$ or 2, where $t_2 = t + 1 - 2t_1$ [9,11]. Therefore, for any integer $t \geq 3$ and each $t_1 \in \{1, \ldots, \lfloor (t+1)/2 \rfloor\}$, there is a unique (up to equivalence) $\mathbb{Z}_4$-linear Hadamard code of type $(2^{t-1}; t_1, t+1-2t_1)$, and all these codes are pairwise nonequivalent, except for $t_1 = 1$ and $t_1 = 2$, where the codes are equivalent to the linear Hadamard code [9]. The number of nonequivalent $\mathbb{Z}_4$-linear Hadamard codes of length 2^t is $\lfloor \frac{t-1}{2} \rfloor$ for all $t \geq 3$, and it is 1 for $t = 1$ and for $t = 2$.

In this paper, in order to try to classify $\mathbb{Z}_{2^s}$-linear Hadamard codes for $s > 2$, we establish the kernel and its dimension for these codes, and we point out that this invariant does not provide a complete classification. This correspondence is organized as follows. In Sect. 2, we recall and prove some results related to the generalized Gray map. In Sect. 3, we describe the construction of $\mathbb{Z}_{2^s}$-linear Hadamard codes of type $(n; t_1, \ldots, t_s)$. In Sect. 4, we establish for which types these codes are linear, and we give the kernel and its dimensions whenever they are nonlinear. Through several examples, we show that, unlike for $s = 2$, the dimension of the kernel is not enough to classify completely $\mathbb{Z}_{2^s}$-linear Hadamard codes with $s = 3$ or $s = 4$. Finally, in Sect. 5, we give some conclusions and further research on this topic.

2 Generalized Gray Map

In this section, we present some results about the generalized Gray map in order to show the main results related to $\mathbb{Z}_{2^s}$-linear Hadamard codes.

Let e_i be the vector that has 1 in the ith position and 0 otherwise. Let $u, v \in \mathbb{Z}_{2^s}$ and $[u_0, \ldots, u_{s-1}]_2, [v_0, \ldots, v_{s-1}]_2$ be the binary expansions of u and v, respectively. The operation "$\odot$" on $\mathbb{Z}_{2^s}$ is defined as $u \odot v = \sum_{i=0}^{s-1} 2^i u_i v_i$. Note that the binary expansion of $u \odot v$ is $[u_0 v_0, \ldots, u_{s-1} v_{s-1}]_2$.

Proposition 1. *[12] Let $u, v \in \mathbb{Z}_{2^s}$. Then, $\phi(u) + \phi(v) = \phi(u + v - 2(u \odot v))$.*

Corollary 1. *Let $u \in \mathbb{Z}_{2^s}$ and $0 \leq p \leq s - 1$. Then, $\phi(u) + \phi(2^p) = \phi(u + 2^p - u_p 2^{p+1})$, where $[u_0, u_1 \ldots, u_{s-1}]_2$ is the binary expansion of u.*

Corollary 2. *Let $u \in \mathbb{Z}_{2^s}$. Then, $\phi(u) + \phi(2^{s-1}) = \phi(u + 2^{s+1})$.*

Lemma 1. *Let $u \in \{2^{s-2}, \ldots, 2^{s-1} - 1\} \subset \mathbb{Z}_{2^s}$. Then, $\phi(u + 2^{s-2} + 2^{s-1}) = \phi(u) + \phi(2^{s-2})$.*

Proof. By Proposition 1, we have that $\phi(u) + \phi(2^{s-2}) = \phi(u + 2^{s-2} - 2(u \odot 2^{s-2}))$. The binary expansion of 2^{s-2} is $[0, \ldots, 0, 1, 0]_2$ and, if $u \in \{2^{s-2}, \ldots, 2^{s-1} - 1\}$, the binary expansion of u is $[u_0, \ldots, u_{s-3}, 1, 0]_2$. Then, $-2(u \odot 2^{s-2}) = 2^{s-1}$ and the statement follows. □

Corollary 3. *Let $v \in \{2^{s-2}, 3 \cdot 2^{s-2}\}$ and $U = \{2^{s-2}, \ldots, 2^{s-1} - 1\} \cup \{3 \cdot 2^{s-2}, \ldots, 2^s - 1\} \subset \mathbb{Z}_{2^s}$. Then,*

$$\phi(u) + \phi(v) = \begin{cases} \phi(u + v + 2^{s-1}) & \text{if } u \in U \\ \phi(u + v) & \text{if } u \in \mathbb{Z}_{2^s} \backslash U. \end{cases}$$

Proof. Straightforward from Corollary 1 and Lemma 1. □

Lemma 2. *Let $q_i \in \mathbb{Z}_2$, $i \in \{0, \ldots, s-2\}$. Then, $\sum_{i=0}^{s-2} q_i \phi(2^i) = \phi(\sum_{i=0}^{s-2} q_i 2^i)$, where $2^i \in \mathbb{Z}_{2^s}$.*

Proof. Let y_i be the ith row of Y. By the definition of ϕ given by (1), we know that $\sum_{i=0}^{s-2} q_i \phi(2^i) = \sum_{i=0}^{s-2} q_i e_{i+1} Y = \sum_{i=0}^{s-2} q_i y_{i+1} = \mathbf{q} Y$, where $\mathbf{q} = (q_0, \ldots, q_{s-2})$. Since $[q_0, \ldots, q_{s-2}, 0]_2$ is the binary expansion of $\sum_{i=0}^{s-2} q_i 2^i$, then we have that $\mathbf{q} Y = \phi(\sum_{i=0}^{s-2} q_i 2^i)$.   □

Proposition 2. *[4] Let $u, v \in \mathbb{Z}_{2^s}$. Then, $d_H(\phi(u), \phi(v)) = \mathrm{wt}_H(\phi(u-v))$.*

Lemma 3. *Let $u \in \mathbb{Z}_{2^s}$. Then, $d_H(\phi(u), \phi(2^{s-1})) + d_H(\phi(u), \phi(0)) = 2^{s-1}$.*

Proof. By the properties of the distance, we have that $d_H(\phi(u), \phi(2^{s-1})) + d_H(\phi(u), \phi(0)) = \mathrm{wt}_H(\phi(2^{s-1}) - \phi(u)) + \mathrm{wt}_H(\phi(u))$. Then, since $\phi(2^{s-1}) = \mathbf{1}$, $\mathrm{wt}_H(\phi(2^{s-1}) - \phi(u)) = 2^{s-1} - \mathrm{wt}_H(\phi(u))$, and the result follows.   □

Corollary 4. *Let $u, v \in \mathbb{Z}_{2^s}$. Then, $d_H(\phi(u), \phi(v + 2^{s-1})) + d_H(\phi(u), \phi(v)) = 2^{s-1}$.*

Proof. Straightforward from Corollary 1 and Lemma 3.   □

3 Construction of $\mathbb{Z}_{2^s}$-Linear Hadamard Codes

The description of a generator matrix having minimum number of rows for a $\mathbb{Z}_4$-additive Hadamard code, as long as recursive constructions of these matrices, are given in [9]. In this section, we generalize these results for any $s > 2$ and give another proof of the main theorem, already proved in [8], that establishes that the constructed matrices generate $\mathbb{Z}_{2^s}$-linear Hadamard codes.

Let $T_i = \{j \cdot 2^{i-1} : j \in \{0, 1, \ldots, 2^{s-i+1} - 1\}\}$ for all $i \in \{1, \ldots, s\}$. Note that $T_1 = \{0, \ldots, 2^s - 1\}$. Let $t_1, t_2, \ldots, t_s$ be nonnegative integers with $t_1 \geq 1$. Consider the matrix $A^{t_1, \ldots, t_s}$ whose columns are of the form $\mathbf{z}^T$, $\mathbf{z} \in \{1\} \times T_1^{t_1 - 1} \times T_2^{t_2} \times \cdots \times T_s^{t_s}$. Let $\mathbf{0}, \mathbf{1}, \mathbf{2}, \ldots, \mathbf{2^s} - \mathbf{1}$ be the vectors having the elements $0, 1, 2, \ldots, 2^s - 1$ from $\mathbb{Z}_{2^s}$ repeated in each coordinate, respectively. The order of a vector $\mathbf{u}$ over $\mathbb{Z}_{2^s}$, denoted by $\mathrm{ord}(\mathbf{u})$, is the smallest positive integer m such that $m\mathbf{u} = \mathbf{0}$.

Example 1. For $s = 3$, for example, we have the following matrices:

$$A^{1,0,1} = \begin{pmatrix} 1 & 1 \\ 0 & 4 \end{pmatrix}, \quad A^{1,1,0} = \begin{pmatrix} 1 & 1 & 1 & 1 \\ 0 & 2 & 4 & 6 \end{pmatrix}, \quad A^{2,0,0} = \begin{pmatrix} 1 & 1 & 1 & 1 & 1 & 1 & 1 & 1 \\ 0 & 1 & 2 & 3 & 4 & 5 & 6 & 7 \end{pmatrix},$$

$$A^{1,1,1} = \begin{pmatrix} 1 & 1 & 1 & 1 & 1 & 1 & 1 & 1 \\ 0 & 2 & 4 & 6 & 0 & 2 & 4 & 6 \\ 0 & 0 & 0 & 0 & 4 & 4 & 4 & 4 \end{pmatrix}, \quad A^{2,0,1} = \begin{pmatrix} 1 & 1 & 1 & 1 & 1 & 1 & 1 & 1 & 1 & 1 & 1 & 1 & 1 & 1 & 1 & 1 \\ 0 & 1 & 2 & 3 & 4 & 5 & 6 & 7 & 0 & 1 & 2 & 3 & 4 & 5 & 6 & 7 \\ 0 & 0 & 0 & 0 & 0 & 0 & 0 & 0 & 4 & 4 & 4 & 4 & 4 & 4 & 4 & 4 \end{pmatrix},$$

$$A^{2,1,0} = \begin{pmatrix} 1 & 1 & 1 & 1 & 1 & 1 & 1 & 1 & 1 & 1 & 1 & 1 & 1 & 1 & 1 & 1 & 1 & 1 & 1 & 1 & 1 & 1 & 1 & 1 & 1 & 1 & 1 & 1 & 1 & 1 & 1 & 1 \\ 0 & 1 & 2 & 3 & 4 & 5 & 6 & 7 & 0 & 1 & 2 & 3 & 4 & 5 & 6 & 7 & 0 & 1 & 2 & 3 & 4 & 5 & 6 & 7 & 0 & 1 & 2 & 3 & 4 & 5 & 6 & 7 \\ 0 & 0 & 0 & 0 & 0 & 0 & 0 & 0 & 2 & 2 & 2 & 2 & 2 & 2 & 2 & 2 & 4 & 4 & 4 & 4 & 4 & 4 & 4 & 4 & 6 & 6 & 6 & 6 & 6 & 6 & 6 & 6 \end{pmatrix}.$$

Note that any matrix $A^{t_1,\ldots,t_s}$ can be obtained by applying the following iterative construction. We start with $A^{1,0,\ldots,0} = (1)$. Then, if we have a matrix $A^{t_1,\ldots,t_s}$, we may construct

$$A_i = A^{t'_1,\ldots,t'_s} = \begin{pmatrix} A^{t_1,\ldots,t_s} & A^{t_1,\ldots,t_s} & \cdots & A^{t_1,\ldots,t_s} \\ 0 \cdot 2^{i-1} & 1 \cdot 2^{i-1} & \cdots & (2^{s-i+1} - 1) \cdot 2^{i-1} \end{pmatrix}, \qquad (2)$$

where $i \in \{1,\ldots,s\}$, $t'_j = t_j$ for $j \neq i$ and $t'_i = t_i + 1$.

Example 2. From the matrix $A^{1,0,0} = (1)$, we obtain the matrix $A^{2,0,0}$; and from $A^{2,0,0}$ we can construct $A^{2,0,1}$, where $A^{2,0,0}$ and $A^{2,0,1}$ are the matrices given in Example 1. Note that we can also generate another matrix $A^{2,0,1}$ as follows: from $A^{1,0,0} = (1)$ we construct the matrix $A^{1,0,1}$ given in Example 1, and from $A^{1,0,1}$ we obtain the matrix

$$A^{2,0,1} = \begin{pmatrix} 1 & 1 & 1 & 1 & 1 & 1 & 1 & 1 & 1 & 1 & 1 & 1 & 1 & 1 & 1 & 1 \\ 0 & 4 & 0 & 4 & 0 & 4 & 0 & 4 & 0 & 4 & 0 & 4 & 0 & 4 & 0 & 4 \\ 0 & 0 & 1 & 1 & 2 & 2 & 3 & 3 & 4 & 4 & 5 & 5 & 6 & 6 & 7 & 7 \end{pmatrix},$$

which is different to the previous one. These two matrices $A^{2,0,1}$ generate permutation equivalent codes.

Along this paper, we consider that the matrices $A^{t_1,t_2,\ldots,t_s}$ are constructed recursively starting from $A^{1,0,\ldots,0}$ in the following way. First, we add $t_1 - 1$ rows of order 2^s, up to obtain $A^{t_1,0,\ldots,0}$; then t_2 rows of order 2^{s-1} up to generate $A^{t_1,t_2,0,\ldots,0}$; and so on, until we add t_s rows of order 2 to achieve $A^{t_1,t_2,\ldots,t_s}$.

Let $\mathcal{H}^{t_1,\ldots,t_s}$ be the $\mathbb{Z}_{2^s}$-additive code generated by the matrix $A^{t_1,\ldots,t_s}$, where $t_1,\ldots,t_s \geq 0$ with $t_1 \geq 1$. Let $n = 2^{t-s+1}$, where $t = \left(\sum_{i=1}^{s} (s-i+1) \cdot t_i \right) - 1$. It is easy to see that $\mathcal{H}^{t_1,\ldots,t_s}$ is of length n and has $|\mathcal{H}^{t_1,\ldots,t_s}| = 2^s n = 2^{t+1}$ codewords. Note that this code is of type $(n; t_1, t_2, \ldots, t_s)$. Let $H^{t_1,\ldots,t_s} = \Phi(\mathcal{H}^{t_1,\ldots,t_s})$ be the corresponding $\mathbb{Z}_{2^s}$-linear code.

Remark 1. The code $\mathcal{H}^{1,0,\ldots,0}$ is generated by $A^{1,0,\ldots,0} = (1)$, so $\mathcal{H}^{1,0,\ldots,0} = \mathbb{Z}_{2^s}$. This code has length $n = 1$, cardinality 2^s and minimum distance 1. Thus, $H^{1,0,\ldots,0} = \Phi(\mathcal{H}^{1,0,\ldots,0})$ has length $N = 2^{s-1}$, cardinality $2N = 2^s$ and minimum distance $2^{s-2} = N/2$, so it is a binary linear Hadamard code of length 2^{s-1} [4], or equivalently, the first order Reed-Muller code of length 2^{s-1}, denoted by $RM(1, s-1)$ [10, Chap. 13 Sect. 3].

The result given by Theorem 1 is already proved in [8], where it is also shown that each $\mathbb{Z}_{2^s}$-linear Hadamard code is equivalent to $H^{t_1,\ldots,t_s}$ for some $t_1,\ldots,t_s \geq 0$ with $t_1 \geq 1$. We present a new proof of this theorem, which does not use neither the dual of the $\mathbb{Z}_{2^s}$-additive codes nor another generalization of the Gray map for these dual codes, unlike the proof given in [8].

Let $\mathcal{G}$ be a generator matrix of a $\mathbb{Z}_{2^s}$-additive code $\mathcal{C}$ of length n. Then, $(\mathcal{G} \cdots \mathcal{G})$ is a generator matrix of the r-fold replication code of $\mathcal{C}$, $(\mathcal{C},\ldots,\mathcal{C}) = \{(\mathbf{c},\ldots,\mathbf{c}) : \mathbf{c} \in \mathcal{C}\}$ of length $r \cdot n$.

Theorem 1. *[8] Let $t_1, \ldots, t_s$ be nonnegative integers with $t_1 \geq 1$. The $\mathbb{Z}_{2^s}$-linear code $H^{t_1,\ldots,t_s}$ of type $(n; t_1, t_2, \ldots, t_s)$ is a binary Hadamard code of length 2^t, with $t = (\sum_{i=1}^{s}(s - i + 1) \cdot t_i) - 1$ and $n = 2^{t-s+1}$.*

Proof. We prove this theorem by induction on the integers t_i, $i \in \{1, \ldots, s\}$. First, by Remark 1, the code $H^{1,0,\ldots,0}$ is a Hadamard code.

Let $\mathcal{H} = \mathcal{H}^{t_1,\ldots,t_s}$ be the $\mathbb{Z}_{2^s}$-additive code of length n generated by the matrix $A = A^{t_1,\ldots,t_s}$. We assume that $H = \Phi(\mathcal{H})$ is a Hadamard code of length $N = 2^{s-1}n$. Let $i \in \{1, \ldots, s\}$. Define $A_i = A^{t'_1,\ldots,t'_s}$ as in (2), where $t'_j = t_j$ for $j \neq i$ and $t'_i = t_i + 1$. Let $\mathcal{H}_i = \mathcal{H}^{t'_1,\ldots,t'_s}$ be the $\mathbb{Z}_{2^s}$-additive code generated by the matrix A_i. Now, we shall prove that $H_i = \Phi(\mathcal{H}_i)$ is a Hadamard code.

Note that $\mathcal{H}_i$ can be seen as the union of 2^{s-i+1} cosets of the 2^{s-i+1}-fold replication code of $\mathcal{H}$, $(\mathcal{H}, \ldots, \mathcal{H})$, which are

$$(\mathcal{H}, \ldots, \mathcal{H}) + r \cdot \mathbf{w}_i, \tag{3}$$

for $r \in \{0, \ldots 2^{s-i+1} - 1\}$, where $\mathbf{w}_i = (0, \mathbf{2^{i-1}}, 2 \cdot \mathbf{2^{i-1}}, \ldots, (2^{s-i+1} - 1) \cdot \mathbf{2^{i-1}})$.

The code $\mathcal{H}$ of length n has cardinality $2^s n$. It is easy to see that $\mathcal{H}_i$ has length $n_i = 2^{s-i+1}n$ and cardinality $2^{2s-i+1}n$. Therefore, the length of $H_i = \Phi(\mathcal{H}_i)$ is $N_i = 2^{s-1}n_i$ and the cardinality $2N_i$. Now, we just have to prove that the minimum distance of H_i is $N_i/2$.

By Proposition 2, the minimum distance of H_i is equal to the minimum weight of H_i. Thus, we just have to check that the minimum weight of any coset (3) is $N_i/2$. When $r = 0$, we have that $\mathrm{wt}_H(\Phi((\mathbf{u}, \ldots, \mathbf{u}))) = 2^{s-i+1}\mathrm{wt}_H(\Phi(\mathbf{u})) = 2^{s-i+1}N/2 = N_i/2$. Otherwise, when $r \neq 0$, we consider

$$\mathrm{wt}_H(\Phi((\mathbf{u}, \ldots, \mathbf{u}) + r \cdot \mathbf{w}_i)) = d_H(\Phi((\mathbf{u}, \ldots, \mathbf{u})), \Phi(r \cdot \mathbf{w}_i)). \tag{4}$$

Note that, by construction, the coordinates of any nonnegative multiple of $\mathbf{w}_i$ can be partitioned into two multisets V and V' such that $|V| = |V'| = 2^{s-i}$ and there is a bijection from V to V' mapping any element $\mathbf{v} \in V$ into an element $\mathbf{v}' \in V'$ such that $\mathbf{v}' - \mathbf{v} = \mathbf{2^{s-1}}$. Therefore, (4) can be written as

$$\sum_{\mathbf{v} \in V} d_H(\Phi(\mathbf{u}), \Phi(\mathbf{v})) + \sum_{\mathbf{v}' \in V'} d_H(\Phi(\mathbf{u}), \Phi(\mathbf{v}')) =$$

$$\sum_{\mathbf{v} \in V} d_H(\Phi(\mathbf{u}), \Phi(\mathbf{v})) + d_H(\Phi(\mathbf{u}), \Phi(\mathbf{v} + \mathbf{2^{s-1}})) =$$

$$|V| \cdot 2^{s-1}n = 2^{s-i}2^{s-1}n = N_i/2, \tag{5}$$

where (5) holds by Corollary 4. $\square$

4   Partial Classification of $\mathbb{Z}_{2^s}$-Linear Hadamard Codes

The computation of the kernel and its dimension for $\mathbb{Z}_4$-linear codes are given in [9,11]. In this section, we determine them for $\mathbb{Z}_{2^s}$-linear Hadamard codes with $s > 2$. First, we determine when these codes are linear, and, in the case that they are nonlinear, we construct the kernel and its dimension, which allow us to establish a partial classification of these codes.

Proposition 3. *The $\mathbb{Z}_{2^s}$-linear Hadamard codes $H^{1,0,\ldots,0}$ and $H^{1,0,\ldots,0,1,0}$ are linear.*

Proof. By Remark 1, we know that $H^{1,0\ldots,0}$ is linear.

Recall that the code $\mathcal{H}^{1,0,\ldots,0,1,0}$ is generated by

$$A^{1,0,\ldots,0,1,0} = \begin{pmatrix} 1 & 1 & 1 & 1 \\ 0 & 2^{s-2} & 2^{s-1} & 3 \cdot 2^{s-2} \end{pmatrix}.$$

Let $\boldsymbol{\beta}_i = (2^i, 2^i, 2^i, 2^i)$ for $0 \leq i \leq s-1$, $\boldsymbol{\beta}_s = (0, 2^{s-1}, 0, 2^{s-1})$ and $\boldsymbol{\beta}_{s+1} = (0, 2^{s-2}, 2^{s-1}, 3 \cdot 2^{s-2})$. Let C be the linear code generated by $B = \{\Phi(\boldsymbol{\beta}_i) : 0 \leq i \leq s+1\}$. Now, we prove that $C \subseteq H^{1,0,\ldots,0,1,0}$. Let $c = \sum_{i=0}^{s+1} q_i \Phi(\boldsymbol{\beta}_i) \in C$, where $q_i \in \mathbb{Z}_2$. By Corollary 2, we only have to see that

$$c' = q_{s+1}\Phi(\boldsymbol{\beta}_{s+1}) + \sum_{i=0}^{s-2} q_i \Phi(\boldsymbol{\beta}_i) \in H^{1,0,\ldots,0,1,0}.$$

Since $\sum_{i=0}^{s-2} q_i \Phi(\boldsymbol{\beta}_i) = \Phi(\sum_{i=0}^{s-2} q_i \boldsymbol{\beta}_i)$ by Lemma 2, if $q_{s+1} = 0$, then we have that $c' \in H^{1,0,\ldots,0,1,0}$. Assume $q_{s+1} = 1$. We have that $c' = \Phi((0, 2^{s-2}, 2^{s-1}, 3 \cdot 2^{s-2})) + \Phi((u, u, u, u))$, for $u = \sum_{i=0}^{s-2} q_i 2^i$. Let $U = \{2^{s-2}, \ldots, 2^{s-1}-1\} \cup \{3 \cdot 2^{s-2}, \ldots, 2^s - 1\} \subset \mathbb{Z}_{2^s}$. Then, by Corollary 3, $c' = \Phi((0, 2^{s-2}, 2^{s-1}, 3 \cdot 2^{s-2}) + (u, u, u, u) + (0, 2^{s-1}, 0, 2^{s-1}))$ if $u \in U$, and $c' = \Phi((0, 2^{s-2}, 2^{s-1}, 3 \cdot 2^{s-2}) + (u, u, u, u))$ if $u \in \mathbb{Z}_{2^s} \backslash U$. In both cases, $c' \in H^{1,0,\ldots,0,1,0}$.

Since $|C| = |H^{1,0,\ldots,0,1,0}| = 2^{s+2}$, then $C = H^{1,0,\ldots,0,1,0}$, and thus $H^{1,0,\ldots,0,1,0}$ is linear. $\square$

In [6], it is proved that the codes $H^{1,0,\ldots,0,t_s}$, $t_s \geq 0$, are linear. The next result shows that these codes together with the codes $H^{1,0,\ldots,1,t_s}$ are linear, and they are the only $\mathbb{Z}_{2^s}$-linear Hadamard codes which are linear.

Theorem 2. *The codes $H^{1,0,\ldots,1,t_s}$ and $H^{1,0,\ldots,0,t_s}$, with $t_s \geq 0$, are the unique $\mathbb{Z}_{2^s}$-linear Hadamard codes which are linear.*

Proof. First, we show that these codes are linear by induction on t_s. By Proposition 3, the codes $H^{1,0,\ldots,0}$ and $H^{1,0,\ldots,0,1,0}$ are linear. We assume that $H = \Phi(\mathcal{H})$, where $\mathcal{H} = \mathcal{H}^{1,0,\ldots,0,t_{s-1},t_s}$, $t_{s-1} \in \{0,1\}$ and $t_s \geq 0$, is linear. Now, we prove that the code $H_s = H^{1,0,\ldots,0,t_{s-1},t_s+1}$ is linear. Since H is a linear Hadamard code of length $2^{t_s+2t_{s-1}-1}$, then it is the Reed-Muller code $RM(1, t_s + 2t_{s-1} - 1)$ [10, Chap. 13 Sect. 3]. By the iterative construction (2), we have that $H_s = \{\Phi((\mathbf{h}, \mathbf{h}) + (\mathbf{0}, \mathbf{v})) : \mathbf{h} \in \mathcal{H}, \mathbf{v} \in \{0, 2^{s-1}\}\}$. By Corollary 2, $H_s = \{(\Phi(\mathbf{h}), \Phi(\mathbf{h}) + \Phi(\mathbf{v})) : \mathbf{h} \in \mathcal{H}, \mathbf{v} \in \{0, 2^{s-1}\}\} = \{(\mathbf{h}', \mathbf{h}' + \mathbf{v}') : \mathbf{h}' \in H, \mathbf{v}' \in \{0, 1\}\}$ which corresponds to the Reed-Muller code $RM(1, t_s + 2t_{s-1})$. Therefore, H_s is linear.

Now, we prove the nonlinearity of $H = \Phi(\mathcal{H})$, where $\mathcal{H} = \mathcal{H}^{1,0,\ldots,0,2,0}$. Let $\mathbf{r} = (0, 2^{s-2}, 2^{s-1}, 3 \cdot 2^{s-2})$. Recall that $\mathcal{H}$ has length 16 and is generated by

$$A^{1,0,\ldots,0,2,0} = \begin{pmatrix} 1 & 1 & 1 & 1 \\ \mathbf{r} & \mathbf{r} & \mathbf{r} & \mathbf{r} \\ 0 & 2^{s-2} & 2^{s-1} & 3 \cdot 2^{s-2} \end{pmatrix}.$$

By Corollaries 2 and 3, we have that $\Phi((\mathbf{r}, \mathbf{r}, \mathbf{r}, \mathbf{r})) + \Phi((\mathbf{0}, \mathbf{2^{s-2}}, \mathbf{2^{s-1}}, \mathbf{3 \cdot 2^{s-2}})) = \Phi(z)$, where $z = (0, 2^{s-2}, 2^{s-1}, 3 \cdot 2^{s-2}, 2^{s-2}, 0, 3 \cdot 2^{s-2}, 2^{s-1}, 2^{s-1}, 3 \cdot 2^{s-2}, 0, 2^{s-2}, 3 \cdot 2^{s-2}, 2^{s-1}, 2^{s-2}, 0)$. Since $\mathcal{H}$ is linear over $\mathbb{Z}_{2^s}$, $z \in \mathcal{H}$ if and only if $z + (\mathbf{r}, \mathbf{r}, \mathbf{r}, \mathbf{r}) + (\mathbf{0}, \mathbf{2^{s-2}}, \mathbf{2^{s-1}}, \mathbf{3 \cdot 2^{s-2}}) = z' \in \mathcal{H}$, where

$$z' = (0, 0, 0, 0, 0, 2^{s-1}, 0, 2^{s-1}, 0, 0, 0, 0, 0, 2^{s-1}, 0, 2^{s-1}).$$

Since $\mathrm{wt}_H(\Phi(z')) = 2^{s-1} \cdot 4 = N/4$, where N is the length of H, $\Phi(z') \notin H$, so $\Phi(z) \notin H$. Therefore, $H = H^{1,0,\dots,0,2,0}$ is nonlinear.

We consider $(t_1, \dots, t_s) = (1, 0, \dots, 0, 1, 0)$ and $\mathcal{H}_i = \mathcal{H}^{t'_1, \dots, t'_s}$, $t'_i = t_i + 1$ and $t'_j = t_j$ for $j \neq i$. Next, we prove that the $\mathbb{Z}_{2^s}$-linear Hadamard code $H_i = \Phi(\mathcal{H}_i)$ is nonlinear for any $i \in \{1, \dots, s-2\}$. Note that $\mathcal{H}_i$ is permutation equivalent to the code $\overline{\mathcal{H}_i}$, generated by the matrix A_i constructed as in (2) with $A = A^{1,0,\dots,0,1,0}$. Moreover it is easy to see that the $2^{s-i-1} = (2^{s-i+1}/4)$-fold replication code of $\mathcal{H}^{1,0,\dots,0,2,0}$ is contained in $\overline{\mathcal{H}_i}$. Since $\mathcal{H}^{1,0,\dots,0,2,0}$ is nonlinear, by using the same argument as before, there exist $u, v \in \mathcal{H}^{1,0,\dots,0,2,0}$, such that $\Phi(u) + \Phi(v) = \Phi(z') \notin H^{1,0,\dots,2,0}$, so $\Phi((u, \dots, u)) + \Phi((v, \dots, v)) = \Phi((z', \dots, z')) \notin \Phi(\overline{\mathcal{H}_i})$ and H_i is nonlinear.

Finally, we prove that if $H^{t_1,\dots,t_s}$ is nonlinear, then $H^{t'_1,\dots,t'_s}$ is nonlinear for any $t'_i \geq t_i$, $i \in \{1, \dots, s\}$. Assume that $H^{t'_1,\dots,t'_s}$ is linear. Then, by the iterative construction (2), for any $\mathbf{u}, \mathbf{v} \in \mathcal{H}^{t_1,\dots,t_s}$, we have that $(\mathbf{u}, \dots, \mathbf{u}), (\mathbf{v}, \dots, \mathbf{v}) \in \mathcal{H}^{t'_1,\dots,t'_s}$. Moreover, since $H^{t'_1,\dots,t'_s}$ is linear, $\Phi((\mathbf{u}, \dots, \mathbf{u})) + \Phi((\mathbf{v}, \dots, \mathbf{v})) = \Phi((\mathbf{a}, \dots, \mathbf{a}) + \lambda(0 \cdot 2^{i-1}, 1 \cdot 2^{i-1}, \dots, (2^{s-i+1} - 1) \cdot 2^{i-1})) \in H^{t'_1,\dots,t'_s}$, where $\mathbf{a} \in \mathcal{H}^{t_1,\dots,t_s}$ and $\lambda \in \mathbb{Z}_{2^s}$. Therefore, $\Phi(\mathbf{u}) + \Phi(\mathbf{v}) = \Phi(\mathbf{a}) \in H^{t_1,\dots,t_s}$, and we have that $H^{t_1,\dots,t_s}$ is linear and the result follows. □

Let $A^{t_1,\dots,t_s}$ be the generator matrix of $\mathcal{H}^{t_1,\dots,t_s}$, considered along this paper, and let $\mathbf{w}_i$ be the ith row vector of $A^{t_1,\dots,t_s}$. By construction, $\mathbf{w}_1 = \mathbf{1}$ and $\mathrm{ord}(\mathbf{w}_i) \leq \mathrm{ord}(\mathbf{w}_j)$ if $i > j$. We define $\sigma \in \{1, \dots, s\}$ as the integer such that $\mathrm{ord}(\mathbf{w}_2) = 2^{s+1-\sigma}$. Note that $\sigma = 1$ if $t_1 > 1$, and $\sigma = \min\{i : t_i > 0, i \in \{2, \dots, s\}\}$ if $t_1 = 1$.

Theorem 3. *Let $\mathcal{H} = \mathcal{H}^{t_1,\dots,t_s}$ be a $\mathbb{Z}_{2^s}$-additive Hadamard code of type $(n; t_1, \dots, t_s)$ such that $\Phi(\mathcal{H})$ is nonlinear. Let $\mathcal{H}_b$ be the subcode of $\mathcal{H}$ which contains all codewords of order two. Let $P = \{2^p\}_{p=0}^{\sigma-2}$ if $\sigma \geq 2$, and $P = \emptyset$ if $\sigma = 1$. Then,*

$$\left\langle \Phi(\mathcal{H}_b), \Phi(P), \Phi\left(\sum_{i=0}^{s-2} 2^i\right) \right\rangle = K(\Phi(\mathcal{H})),$$

and $\ker(\Phi(\mathcal{H})) = \sigma + \sum_{i=1}^{s} t_i$.

The following example shows that the dimension of the kernel can not be used to classify completely $\mathbb{Z}_8$-linear Hadamard codes of length 256.

Example 3. The $\mathbb{Z}_8$-linear Hadamard codes of length $2^t = 256$ are the following: $H^{1,0,6}, H^{1,1,4}, H^{1,2,2}, H^{1,3,0}, H^{2,0,3}, H^{2,1,1}$ and $H^{3,0,0}$. The first two are equivalent as they are linear by Theorem 2. The remaining ones have kernels of dimension $7, 6, 6, 5$ and 4, respectively, by Theorem 3. Therefore, by using this invariant, we can say that all of them are nonequivalent, with the exception of $H^{1,3,0}$

and $H^{2,0,3}$ which have the same dimension of the kernel. For these two codes, by using the computer algebra system Magma [3], we have that rank($H^{1,3,0}$) = 12 and rank($H^{2,0,3}$) = 11, so they are also nonequivalent. Actually, all these nonlinear codes have ranks 10, 12, 11, 13 and 17, respectively, so we can use the rank instead of the dimension of the kernel to have a classification of these codes.

Example 4. Table 1 shows the type and the pair (r, k), where r is the rank and k the dimension of the kernel, of all $\mathbb{Z}_{2^s}$-linear Hadamard codes of length 2^t for $s \in \{2, 3, 4, 5\}$ and $t \in \{8, 9\}$. As we can see, for a given ring $\mathbb{Z}_{2^s}$ and length 2^t, all nonlinear codes have a different value of the rank, but not all of them have different values of the dimension of the kernel, when $s = 3$ or 4. Therefore, for these cases, as in Example 3, we have a complete classification by using the rank, and just a partial classification by using the kernel. From this table, we

Table 1. Type, rank and kernel of all $\mathbb{Z}_{2^s}$-linear Hadamard codes of length 2^t.

	$t = 8$		$t = 9$	
	type	(r, k)	type	(r, k)
$\mathbb{Z}_4$	$(2^7; 1, 7)$	(9,9)	$(2^8; 1, 8)$	(10,10)
	$(2^7; 2, 5)$	(9,9)	$(2^8; 2, 6)$	(10,10)
	$(2^7; 3, 3)$	(10,7)	$(2^8; 3, 4)$	(11,8)
	$(2^7; 4, 1)$	(12,6)	$(2^8; 4, 2)$	(13,7)
			$(2^9; 5, 0)$	(16,6)
$\mathbb{Z}_8$	$(2^6; 1, 0, 6)$	(9,9)	$(2^7; 1, 0, 7)$	(10,10)
	$(2^6; 1, 1, 4)$	(9,9)	$(2^7; 1, 1, 5)$	(10,10)
	$(2^6; 1, 2, 2)$	(10,7)	$(2^7; 1, 2, 3)$	(11,8)
	$(2^6; 1, 3, 0)$	(12,6)	$(2^7; 1, 3, 1)$	(13,7)
	$(2^6; 2, 0, 3)$	(11,6)	$(2^7; 2, 0, 4)$	(12,7)
	$(2^6; 2, 1, 1)$	(13,5)	$(2^7; 2, 1, 2)$	(14,6)
	$(2^6; 3, 0, 0)$	(17,4)	$(2^7; 2, 2, 0)$	(17,5)
			$(2^7; 3, 0, 1)$	(18,5)
$\mathbb{Z}_{16}$	$(2^5; 1, 0, 0, 5)$	(9,9)	$(2^6; 1, 0, 0, 6)$	(10,10)
	$(2^5; 1, 0, 1, 3)$	(9,9)	$(2^6; 1, 0, 1, 4)$	(10,10)
	$(2^5; 1, 0, 2, 1)$	(10,7)	$(2^6; 1, 0, 2, 2)$	(11,8)
	$(2^5; 1, 1, 0, 2)$	(11,6)	$(2^6; 1, 0, 3, 0)$	(13,7)
	$(2^5; 1, 1, 1, 0)$	(13,5)	$(2^6; 1, 2, 0, 0)$	(18,5)
	$(2^5; 2, 0, 0, 1)$	(15,4)	$(2^6; 1, 1, 0, 3)$	(12,7)
			$(2^6; 1, 1, 1, 1)$	(14,6)
			$(2^6; 2, 0, 0, 2)$	(16,5)
			$(2^6; 2, 0, 1, 0)$	(20,4)
$\mathbb{Z}_{32}$	$(2^4; 1, 0, 0, 0, 4)$	(9,9)	$(2^5; 1, 0, 0, 0, 5)$	(10,10)
	$(2^4; 1, 0, 0, 1, 2)$	(9,9)	$(2^5; 1, 0, 0, 1, 3)$	(10,10)
	$(2^4; 1, 0, 0, 2, 0)$	(10,7)	$(2^5; 1, 0, 0, 2, 1)$	(11,8)
	$(2^4; 1, 0, 1, 0, 1)$	(11,6)	$(2^5; 1, 0, 1, 0, 2)$	(12,7)
	$(2^4; 1, 1, 0, 0, 0)$	(15,4)	$(2^5; 1, 0, 1, 1, 0)$	(14,6)
			$(2^5; 1, 1, 0, 0, 1)$	(16,5)
			$(2^5; 2, 0, 0, 0, 0)$	(26,3)

Table 2. Number of nonequivalent $\mathbb{Z}_{2^s}$-linear Hadamard codes of length 2^t.

t	3	4	5	6	7	8	9	10	11
$\mathbb{Z}_4$	1	1	2	2	3	3	4	4	5
$\mathbb{Z}_8$	1	1	2	3	4	6	7	9	11
$\mathbb{Z}_{16}$	1	1	1	2	4	5	8	10	14

also have that there are at least 7 nonequivalent $\mathbb{Z}_{2^s}$-linear Hadamard codes of length 2^8, and at least 11 of length 2^9.

We can extend these results for other values of t. Table 2 summarizes these computations by showing how many nonequivalent $\mathbb{Z}_{2^s}$-linear Hadamard codes of length 2^t exist for each $s \in \{2, 3, 4\}$ and $t \in \{3, \ldots, 11\}$.

5   Conclusions

The kernel of $\mathbb{Z}_{2^s}$-linear Hadamard codes has been study. We compute the kernel of these codes and its dimension in order to classify them. Examples 3 and 4 show that we can not use the dimension of the kernel to distinguish between nonequivalent $\mathbb{Z}_{2^s}$-linear Hadamard codes for a given $s > 2$ and length 2^t. However, they also show that the rank seems to be enough to classify them. A further research on this topic would be to determine the rank of $\mathbb{Z}_{2^s}$-linear Hadamard codes, and establish their complete classification by using this invariant.

By using Magma [3], we also noticed that, for lengths 2^8 and 2^9, $\mathbb{Z}_4$-linear and $\mathbb{Z}_8$-linear Hadamard codes with the same rank (and dimension of the kernel) are equivalent. This means that, up to equivalence, the same code can be $\mathbb{Z}_4$-linear and $\mathbb{Z}_8$-linear. Another further research in this sense would be to establish whether all $\mathbb{Z}_{2^s}$-linear Hadamard codes of length 2^t having the same rank (and dimension of the kernel) are equivalent.

References

1. Assmus, E.F., Key, J.D.: Designs and Their Codes. Cambridge University Press, Great Britain (1992)
2. Bauer, H., Ganter, B., Hergert, F.: Algebraic techniques for nonlinear codes. Combinatorica **3**(1), 21–33 (1983)
3. Bosma, W., Cannon, J.J., Fieker, C., Steel, A.: Handbook of Magma Functions, 2.22 edn., p. 5669 (2016). http://magma.maths.usyd.edu.au/magma/
4. Carlet, C.: $\mathbb{Z}_{2^k}$-linear codes. IEEE Trans. Inform. Theory **44**(4), 1543–1547 (1998)
5. Dougherty, S.T., Fernández-Córdoba, C.: Codes over $\mathbb{Z}_{2^k}$, gray map and self-dual codes. Adv. Math. Commun. **5**(4), 571–588 (2011)
6. Gupta, M.K.: On some linear codes over $\mathbb{Z}_{2^s}$. Indian Institute of Technology, Kanpur (1999)
7. Hammons, A.R., Kumar, P.V., Calderbank, A.R., Sloane, N.J.A., Solé, P.: The $\mathbb{Z}_4$-linearity of Kerdock, Preparata, Goethals and related codes. IEEE Trans. Inform. Theory **40**(2), 301–319 (1994)

8. Krotov, D.S.: On $\mathbb{Z}_{2^k}$-dual binary codes. IEEE Trans. Inf. Theory **53**(4), 1532–1537 (2007)
9. Krotov, D.S.: $\mathbb{Z}_4$-linear Hadamard and extended perfect codes. In: International Workshop on Coding and Cryptography. Electronic Notes Discrete Mathematics, vol. 6, pp. 107–112 (2001)
10. MacWilliams, F.J., Sloane, N.J.A.: The Theory of Error-Correcting Codes, vol. 16. Elsevier, Amsterdam (1977)
11. Phelps, K.T., Rifà, J., Villanueva, M.: On the additive ($\mathbb{Z}_4$-linear and non-$\mathbb{Z}_4$-linear) Hadamard codes: rank and kernel. IEEE Trans. Inf. Theory **52**(1), 316–319 (2006)
12. Tapia-Recillas, H., Vega, G.: On $\mathbb{Z}_{2^k}$-linear and quaternary codes. SIAM J. Discrete Math. **17**(1), 103–113 (2003)

Random Network Coding over Composite Fields

Olav Geil[1]([⊠])(iD) and Daniel E. Lucani[2](iD)

[1] Department of Mathematical Sciences, Aalborg University, Aalborg, Denmark
olav@math.aau.dk
[2] Department of Engineering, Aarhus University, Aarhus, Denmark
daniel.lucani@eng.au.dk

Abstract. Random network coding is a method that achieves multicast capacity asymptotically for general networks [1,7]. In this approach, vertices in the network randomly and linearly combine incoming information in a distributed manner before forwarding it through their outgoing edges. To ensure success, the involved finite field needs to be large enough [2,7], which can be an obstacle if some inner (intermediate) nodes have less computational power than others. In this work, we analyze what can be achieved if different nodes are allowed to use different finite fields from a selection of fields all contained in some composite extension finite field [3,5].

Keywords: Composite fields · Random network coding · Success probability

1 Introduction

In the seminal paper [1], Ahlswede, Cai, Li and Yeung showed that if in a network with one sender and several receivers the unicast problem is solvable for each of the receivers, then the multicast problem of sending simultaneously the information to all receivers is also solvable. That is, the multicast problem is solvable if and only if the information rate does not exceed the minimum of the min-cuts of the network, where cuts are with respect to the sender on the one side and to each of the receivers on the other side. The insight is that information should not be treated as a fluid, but as mathematical objects that can be manipulated by intermediate nodes in the network.

Surprisingly, when solvable, the multicast problem can always be dealt with using simple linear algebra, i.e. the vertices of the network linearly combine incoming information before passing it on to other vertices. The linear algebra approach is always possible if the finite field $\mathbb{F}_q$ under consideration is of size at least that of the number of receivers [1,8,11]. In [7], it was further shown that allowing vertices to choose coding coefficients at random from elements of the finite field in a distributed manner will result in successful communication with high probability, provided that the field is large enough. This method now is known as random linear network coding (RLNC). More concretely, lower bounds on the success probability were given in terms of the field's size and certain

© Springer International Publishing AG 2017
A.I. Barbero et al. (Eds.): ICMCTA 2017, LNCS 10495, pp. 118–127, 2017.
DOI: 10.1007/978-3-319-66278-7_11

characteristics of the network, such as the number of edges in a flow system (we formally define flow systems in Sect. 2). The analysis in [7] is algebraic using the insight from [11].

In another direction, a simple polynomial time algorithm was given in [8], which finds a solution whenever it exists. As it is noted by the authors, this method gives as a corollary also a lower bound on the success probability of RLNC. Whereas the method in [8] is purely combinatorial in contrast to [7], it focuses on edges rather than vertices in much the same way as [7] does. In [2], it was recognized that using the combinatorial approach from [8], but focusing instead on vertices, often results in tighter bounds on the success probability.

In the present paper, we follow a trend in more practical network coding [6,9,13,17] to use small fields or even composite fields [3,5,12], i.e. fields of varying sizes, to address the problem of restricted computational power in some network devices. The key intuition is that manipulating objects from a small field requires fewer operations than in a large field and can be computed faster in commercial devices [4,16]. The purpose of our paper is to build a bridge between this more practical point of view and the purely mathematical point of view taken in [2,7,8,11] by revisiting the purely mathematical findings in a setting of varying field sizes (i.e., of composite extension fields). As each vertex in this case is assumed to conduct calculations in a fixed field corresponding to its computational power, we shall apply the method from [2], where the focus is on vertices rather than edges.

To build the bridge, Sect. 2 explains the mathematical problem of multicast in a network and Sect. 3 provides brief summary on how to implement the use of varying field sizes in connection with communication in networks presented in [5]. To derive our main result, Sect. 4 first presents a strongly modified version of the polynomial time algorithm from [8], which in its original form finds a solution if one exists. Instead, our modified algorithm checks if a given random encoding is successful. From that, we obtain estimates on the success probability given information on which field sizes are used by different vertices, following the method in [2].

2   The Multicast Problem

Let $G = (V, E)$ be a cycle free directed graph, where $E = \{1, \ldots, |E|\}$. is the set of edges and V is the set of vertices. Fix a vertex $s \in V$ called a sender and R other vertices $r_1, \ldots, r_R \in V$ called receivers. A message vector $X = (X_1, \ldots, X_h)$ is a variable that takes on values in $\mathbb{F}_q^h$, where $\mathbb{F}_q$ is the finite field with q elements. The edges are considered as channels of capacity 1, and the multicast problem is to get the h messages through to each of the receivers in one time-step, meaning that each edge is allowed to be used at most once during the transmission. Vertices are allowed to linearly combine incoming information before forwarding it to outgoing edges, and receivers will do decoding on the information traveling on their incoming edges to hopefully reconstruct X. To describe the communication problem, we introduce variables $a_{1,j}, \ldots, a_{h,j}$ and

$f_{i,j}$ where $1 \leq i,j \leq |E|$. For $1 \leq m \leq h$ and $1 \leq j \leq |E|$, $a_{m,j} \equiv 0$ if j is not an outgoing edge of s and for $1 \leq i,j \leq |E|$, $f_{i,j} \equiv 0$ if there is no vertex having both i as incoming edge and j as outgoing edge. All variables take on values in $\mathbb{F}_q$. For a vertex v, let in(v) denote the incoming edges and out(v) the outgoing edges. Similarly for an edge $i = (u,v)$ we have in(i) = in(u) and out(i) = out(v). The directed graph being cycle free the vertices can be ordered by an ancestral ordering $\prec^V$, meaning that if $u \prec^V v$ then there does not exist a path from v to u. Similarly, the edges can be ordered by an ancestral ordering $\prec^E$. For simplicity, we shall assume in the following that the sender s has no incoming edges[1].

For each edge in the network, we have a variable $Y(j)$ which takes on values in $\mathbb{F}_q$. For $j =$ out(s), we have

$$Y(j) = \sum_{i=1}^{h} a_{i,j} X_i \tag{1}$$

and for other edges we define

$$Y(j) = \sum_{i \in \text{in}(j)} f_{i,j} Y(i) \tag{2}$$

if there exists a path from s using j. Otherwise, we have $Y(j) \equiv 0$. Observe, that the above is well-defined due to the existence of an ancestral ordering $\prec^E$, and that for each edge $j \in E$

$$Y(j) = c_1(j)X_1 + \cdots + c_h(j)X_h \tag{3}$$

for some $c_1(j), \ldots, c_h(j) \in \mathbb{F}_q$. The task is to choose, if possible, the coefficients $a_{i,j}$ and $f_{i,j}$ in such a way that a receiver from the $Y(j)$'s, as seen on its incoming edges, can reconstruct $X = (X_1, \ldots, X_h)$. Define a flow $\mathcal{F}_\ell$ of size c from s to receiver r_ℓ, $\ell \in \{1, \ldots, R\}$, to be a set of c edge disjoint paths from s to r_ℓ. The union $\mathcal{F} = \cup_{\ell=1}^{R} \mathcal{F}_\ell$ is then called a flow system of size c. It is well known [1,8,11] that the multicast problem is solvable if and only if a flow system of size h exists, and in this case considering only linear encoding (as described above) is no restriction. It is also known that for a solvable multicast problem one can use any field $\mathbb{F}_q$, where $q \geq R$ (the number of receivers).

The purpose of the present work is to reinvestigate previous work on the topic to see which vertices can be allowed to use smaller fields when linearly combining messages from the large field and still obtain a high probability of success. We shall always assume that receivers have full computational power, as they need to perform operations in the big field. Such an analysis is relevant as

---

[1] Neither this nor the assumption that we have only one sender is really a restriction, as one can always add an artificial vertex s' to any cycle free network, assume the messages are generated in s', and for each message add one edge from s' to the source originally generating it. In this case, the $a_{i,j}$'s will need to be chosen in an obvious particular way to reflect the situation.

some vertices may represent physical devices having less computational power than others. To make (1) and (2) still valid in this situation, we will restrict to subfields of the overall field $\mathbb{F}_q$. Before continuing the discussion of the multicast problem, we briefly recall in the next section material from [5] on how to use varying fields sizes in network communication.

3 Calculations in Composite Fields

In the proposed model, we consider linear combinations of elements from a large field but with coefficients coming from a possible smaller subfield. Such calculations of course can be done in the large field, but what we need is a method to perform the calculations using only operations in the small (sub)field. Fortunately, a method to do exactly this was described in [5]. We illustrate the idea by an example.

Example 1. Consider $\mathbb{F}_2 \subset \mathbb{F}_4 \subset \mathbb{F}_{16}$. We write $\mathbb{F}_{16} = \mathbb{F}_4[X_4]/\langle F_4(X_4)\rangle$, where $F_4 \in \mathbb{F}_4[X_4]$ and $\deg F_4 = 2$ and we let $\alpha_4 = [X_4] + \langle F_4\rangle$. Similarly, we write $\mathbb{F}_4 = \mathbb{F}_2[X_2]/\langle F_2(X_2)\rangle$ where $F_2 \in \mathbb{F}_2[X_2]$, $\deg F_2 = 2$, and we let $\alpha_2 = [X_2] + \langle F_2\rangle$. In this way, we obtain

$$\mathbb{F}_{16} = \{a\alpha_4 + b \mid a, b \in \mathbb{F}_4\} \tag{4}$$

$$\mathbb{F}_4 = \{c\alpha_2 + d \mid c, d \in \mathbb{F}_2\} \tag{5}$$

We now identify an element $\beta \in \mathbb{F}_{16}$ with the following tuples (the latter being the one traveling through the network):

$$\begin{pmatrix} \beta_1^{(4)} \\ \beta_2^{(4)} \end{pmatrix} \in \mathbb{F}_4^2, \qquad \begin{pmatrix} \beta_1^{(2)} \\ \beta_2^{(2)} \\ \beta_3^{(2)} \\ \beta_4^{(2)} \end{pmatrix} \in \mathbb{F}_2^4. \tag{6}$$

Here,

$$\beta = \beta_1^{(4)}\alpha_4 + \beta_2^{(4)}, \quad \beta_1^{(4)} = \beta_1^{(2)}\alpha_2 + \beta_2^{(2)}, \quad \text{and} \quad \beta_2^{(4)} = \beta_3^{(2)}\alpha_2 + \beta_4^{(2)},$$

where we used the description of $\mathbb{F}_{16}$ and $\mathbb{F}_4$ from (4) and (5). To perform multiplication with $\gamma \in \mathbb{F}_4$ one simply multiplies each coordinate in the left vector of (6) by γ. These operations take place in $\mathbb{F}_4$. Then, the resulting vector can be translated into a new vector in $\mathbb{F}_2^4$ using a similar approach as above. Multiplying β with $\lambda \in \mathbb{F}_2$ is much simpler: one simply needs to multiply each entry in the right vector in (6) by λ. Finally, addition of elements in $\mathbb{F}_{16}$ is always done by adding the two corresponding vectors in $\mathbb{F}_2^4$. The translation between the different descriptions can either be done algebraically or by look-up tables (or even by a combination).

In the previous example, we considered nested fields, but the method applies to all subfields of any given field $\mathbb{F}_q$. Hence, we could use it for $\mathbb{F}_{64}$ despite the corresponding subfields not being nested. The important fact is that for fields of characteristic 2, which is the relevant case for current applications, it is the binary presentation that travels through the network, and indeed $\mathbb{F}_2$ is contained in any field of characteristic 2. A similar remark of course holds for any finite characteristic p as long as it is the p-ary vector that travels through the network.

Remark 1. Network coding over varying fields as described in [3,5] and illustrated in this section support the use of outer codes in exactly the same way as classical network coding [1,7] does. This being subspace codes [10], rank-metric codes [14] or hybrid codes [15].

4 Success Probability

In this section, we establish a lower bound on the success probability of RLNC when composite extension fields are used. To better explain the results, we present an algorithm inspired by the one proposed by Jaggi et al. in [8], but using, rather than update on edges, updates on vertices as in [2]. We stress that our algorithm is only intended as a tool to obtain estimates on the success probability of random network coding – not as an algorithm to find actual solutions of the network coding problem. In contrast to the algorithm in [8], which finds a solution to a multicast problem whenever it exists (with fixed field size at least that of the number of receivers), our modification simply checks sufficient conditions for given coding coefficients $a_{i,j}$ and $f_{i,j}$ to work successfully, in which case it returns "success". The estimate on the success probability is found by estimating from below the probability that the algorithm will return "success" under the assumption that the coding coefficients $a_{i,j}$ and $f_{i,j}$ not a priori[2] set to 0, are chosen uniformly and independently at random from the respective fields. By this we mean that $a_{i,j}$ is chosen from the field supported by the sender s and that $f_{i,j}$ is chosen from the field supported by the vertex having i as incoming edge (and j as outgoing edge). The algorithm uses the concept of global coding coefficients.

Definition 1. Let $\boldsymbol{X} = (X_1, \ldots, X_h)$ be the information vector introduced in Sect. 2 (recall that the entries are variables). Given chosen values of the coding coefficients $a_{i,j}$ and $f_{i,j}$ of the network, let for an edge $\ell \in E$, $Y(\ell)$ to be as in (3). The global coding coefficient for the edge ℓ now is $c_g(\ell) = (c_1(\ell), \ldots, c_h(\ell))$.

As previously described, the algorithm takes as input some choice of the coding coefficients $a_{i,j}$ and $f_{i,j}$. It also takes as input some positive integer h less than or equal to the minimum of the min-cuts taken with respect to all the receivers. It starts by finding a flow system of size h. The algorithm works with

---

[2] We recall from Sect. 2 that coefficients are set to 0 when they do not correspond to existing connections.

two sets of lists related to this flow system, namely, C_ℓ and B_ℓ, where the index refers to the receiver r_ℓ, $\ell = 1, \ldots, R$. The list C_ℓ is a cut in the flow system with respect to r_ℓ and B_ℓ is the list of global coding vectors of the edges in the cut. The algorithm returns "failure" if in a step some B_ℓ is linearly dependent. The algorithm starts by considering the cuts $C_1, \ldots, C_R$ corresponding to having s in the one vertex set and all other vertices in the other vertex sets. If the corresponding $B_1, \ldots, B_R$ are all linearly independent sets, then it visits the next vertex according to a given ancestral ordering. The cuts $C_1, \ldots, C_R$ are now the cuts of the flow system corresponding to having s and the newly added vertex in the one vertex set and all other vertices in the other vertex sets. Again a check is done to see if all corresponding $B_1, \ldots, B_R$ are linearly independent. The procedure continues in this way until this is no longer the case or until all vertices in the flow system, except the receivers, have been visited. If a flow to some receiver passes through some other receiver, then after having passed the other receiver, the algorithm will continue by working on a flow system with one less flow than before. If the above criteria for returning "failure" is not satisfied during the run, the algorithm ends by returning "success". Clearly, the coding coefficients allow for decoding at all the receivers in this case. Hence, we can estimate from below the success probability of random network coding by giving a lower bound of the algorithm to return "success".

The analysis of success probability is based on the following lemma.

Lemma 1. *Let k, μ be non-negative integers, and let h be a positive integer such that $k + \mu < h$. Given a field $\mathbb{F}_q$ consider a basis $\{b_1, \ldots, b_h\}$ for $\mathbb{F}_q^h$ as a vector space over $\mathbb{F}_q$ and let $b'_{k+1}, \ldots, b'_{k+\mu}$ be such that*

$$V = Span_{\mathbb{F}_q}\{b_1, \ldots b_k, b'_{k+1}, \ldots, b'_{k+\mu}\}$$

is of dimension $k + \mu$ as a vector space over $\mathbb{F}_q$. Given $c \in \mathbb{F}_q^h$ the number of choices of $(a_{k+1}, \ldots, a_h) \in \mathbb{F}_q^{h-k}$ such that $c + a_{k+1}b_{k+1} + \cdots + a_h b_h \in V$ equals q^μ. If $\mathbb{F}_t$ is a proper subfield of $\mathbb{F}_q$ then number of choices of $(a_{k+1}, \ldots, a_h) \in \mathbb{F}_t^{h-k}$ such that $c + a_{k+1}b_{k+1} + \cdots + a_h b_h \in V$ is at most t^μ.

Proof. We choose additional vectors $b'_{k+\mu+1}, \ldots, b'_h$ such that

$$\mathbb{F}_q^h = Span_{\mathbb{F}_q}\{b_1, \ldots, b_k, b'_{k+1}, \ldots, b'_h\}.$$

Without loss of generality we may assume that $\{b'_{k+1}, \ldots, b'_h\}$ is a basis for $U = Span_{\mathbb{F}_q}\{b_{k+1}, \ldots, b_h\}$ as a vector space over $\mathbb{F}_q$.
Write

$$c = c_1 b_1 + \cdots + c_k b_k + c_{k+1}b'_{k+1} + \cdots + c_h b'_h.$$

We have that $x \in U$ satisfies $c + x \in V$ if and only x is of the form

$$x = x_{k+1}b'_{k+1} + \cdots + x_{k+\mu}b'_{k+\mu} - c_{k+\mu+1}b'_{k+\mu+1} - \cdots - c_h b'_h$$

That is, there are exactly q^μ possible choices of $x \in U$ with $c + x \in V$. But then there are exactly q^μ choices of $(a_{k+1}, \ldots, a_h) \in \mathbb{F}_q^{h-k}$ such that $x = a_{k+1}b_{k+1} + \cdots + a_h b_h$ satisfies $c + x \in V$.

To prove the lemma, we must estimate the number of $(a_{k+1}, \ldots, a_h) \in \mathbb{F}_t^{h-k}$ that satisfy

$$\boldsymbol{c} + a_{k+1}\boldsymbol{b}_{k+1} + \cdots + a_h\boldsymbol{b}_h \in V. \tag{7}$$

If there are no solutions, then we are done. Assume therefore that a solution $(d_{k+1}, \ldots, d_h) \in \mathbb{F}_t^{k-h}$ exists. Then, $(e_{k+1}, \ldots, e_h) \in \mathbb{F}_t^{h-k}$ is a solution to the same problem if and only if

$$(d_{k+1} - e_{k+1})\boldsymbol{b}_{k+1} + \cdots + (d_h - e_h)\boldsymbol{b}_h \in \mathrm{Span}_{\mathbb{F}_q}\{\boldsymbol{b}'_{k+1}, \ldots, \boldsymbol{b}'_{k+\mu}\}. \tag{8}$$

But,

$$\{(y_{k+1}, \ldots, y_h) \in \mathbb{F}_t^{h-k} \mid y_{k+1}\boldsymbol{b}_{k+1} + \cdots + y_h\boldsymbol{b}_h \in \mathrm{Span}_{\mathbb{F}_q}\{\boldsymbol{b}'_{k+1}, \ldots, \boldsymbol{b}'_h\}\}$$

is a vector space over $\mathbb{F}_t$, as it is closed under addition and scalar multiplication. Therefore,

$$\{(d_{k+1} - e_{k+1}, \ldots, d_h - e_h) \mid (e_{k+1}, \ldots, e_h) \in \mathbb{F}_t^{h-k}$$
$$\text{is a solution to (7)}\} \tag{9}$$

is a vector space over $\mathbb{F}_t$. Aiming for a contradiction we assume that there are more than t^μ solutions over $\mathbb{F}_t$. Consequently, a basis for (9) as a vector space over $\mathbb{F}_t$ is of size at least $\mu + 1$. Linear independence of vectors in $\mathbb{F}_t^{h-k}$ over $\mathbb{F}_t$ implies that the same vectors are linear independent over any extension field. The span over $\mathbb{F}_q$ of such linear independent vectors by (8) constitutes a subspace of a space of dimension μ. We have reached a contradiction. □

In the following analysis we shall use q_u to denote the field size used in vertex u and ρ_u to denote the number of flows in the flow system passing through u. Recall, that all the fields must be contained in the field $\mathbb{F}_q$ from which the messages $X_1, \ldots, X_h$ are generated, and also recall that all receivers $r_1, \ldots, r_R$ need to be able to work in the same field $\mathbb{F}_q$.

To analyze the probability that the algorithm returns success when taking as input randomly generated coding coefficients, we may start by defining initial values of $B_1, \ldots, B_R$ to be

$$B_1 = \cdots = B_R = \{(1, 0 \ldots, 0), (0, 1, 0, \ldots, 0), \ldots, (0, \ldots, 0, 1)\}$$

(all vectors are of length h) before running the algorithm. This describes the situation, where we have h different sources $X_1, \ldots, X_h$ from the beginning. The global coding vectors traveling on the outgoing edges of the sender s in an obvious way can be understood as linearly combinations of these vectors using the coefficients $a_{i,j}$. Using Lemma 1, we now estimate from below the probability for not returning failure in one update of the algorithm. Recall, that we reach a given step in the algorithm provided that all $B_1, \ldots, B_R$ are linearly independent before the update. Consider a vertex u in which an update takes place. The existing $B_1, \ldots, B_R$ are to be updated by replacing in B_ℓ global coding vectors on incoming edges in the flow $\mathcal{F}_\ell$ with global coding vector on outgoing edges. In the first update where the sender is visited, there are no incoming edges and all vectors should be updated using the coefficients $a_{i,j}$'s. To analyze the probability of not returning "failure", we study the probability that for each of the receivers r_ℓ the new B_ℓ is linearly independent. For a given fixed receiver r_ℓ, let $\boldsymbol{b}_1, \ldots, \boldsymbol{b}_k$ be the global coding vectors in B_ℓ not corresponding to edges that touch u. Let $\boldsymbol{b}_{k+1}, \ldots, \boldsymbol{b}_h$ be the vectors traveling on the incoming edges to u in $\mathcal{F}_\ell$. We shall estimate the probability that vectors traveling on the $h - k$ outgoing edges from u in $\mathcal{F}_\ell$ together with $\boldsymbol{b}_1, \ldots, \boldsymbol{b}_k$ constitute a basis for $\mathbb{F}_q^h$. Adapting the notation from the lemma, the vectors traveling on the outgoing edges from u in $\mathcal{F}_\ell$ will be denoted $\boldsymbol{b}'_{k+1}, \ldots, \boldsymbol{b}'_h$. Hence, we are interested in the probability that $\boldsymbol{b}_1, \ldots, \boldsymbol{b}_k, \boldsymbol{b}'_{k+1}, \ldots, \boldsymbol{b}'_h$ are linearly independent over $\mathbb{F}_q$. Applying first Lemma 1 to estimate the probability that $\boldsymbol{b}'_{k+1}$ is independent from $\boldsymbol{b}_1, \ldots, \boldsymbol{b}_k$, we must choose $\mu = 0$ in the lemma. Hence, the number of local encoding coefficients related to the incoming edges in u of $\mathcal{F}_\ell$ and the particular outgoing edge under consideration, which will result in a failure, is at most $q_u^0 = 1$. Given that this has been successful, the number of unsuccessful choices of local encoding coefficients when updating for the next outgoing edge becomes q_u^1 as now we must use $\mu = 1$ in the lemma. Continuing in this way and assuming the worst case, the probability of success for a given receiver when the incoming edges of a given node u is replaced in the flow $\mathcal{F}_\ell$ with the outgoing edges becomes

$$\prod_{i=1}^{h-k} \frac{q_u^{h-k} - q_u^i}{q_u^{h-k}} \geq 1 - \frac{1}{q_u - 1}.$$

Thus, we conclude that the probability for success to hold for all receivers is at least

$$1 - \frac{\rho_u}{q_u - 1}$$

when updating from one cut in the flow system $\mathcal{F}$ to the next. Hence, we obtain the following theorem.

Theorem 1. *Consider a cycle free network with one sender s and R receivers. Assume that h messages from $\mathbb{F}_q$ are generated in s and that the network contains a flow system $\mathcal{F}$ of size h. Then, the probability of success when choosing coding coefficients independently and uniformly at random is at least*

$$\prod_{\substack{u \in \mathcal{F}, \ u \text{ has} \\ \text{outgoing edges in } \mathcal{F}}} \left(1 - \frac{\rho_u}{q_u - 1}\right) \geq \prod_{u \in \mathcal{F}} \left(1 - \frac{\rho_u}{q_u - 1}\right).$$

provided that all receivers are able to conduct computations in $\mathbb{F}_q$.

From this theorem, it is clear that if the network is sparse in the sense that inner vertices only are contained in few flows of the flow system then these vertices can use small subfields of $\mathbb{F}_q$ and still the probability of success can be high. A similar insight was obtained in [3, p. 363 and Sec. 9.9] using the approach from [8].

5 Concluding Remarks

From the estimate in Theorem 1, it is clear that a solution to the multicast problem can always be found if $\rho_v < q_v - 1$ holds for all vertices. We note that it is actually possible to prove that the problem is solvable under the milder condition $\rho_v \leq q_v$. The proof of this result requires a more direct use of Jaggi et al.'s algorithm. We also note that following the approach from [2] it is possible to obtain improved information on the success probability when the information rate is strictly smaller than the minimum of the min-cuts with respect to all receivers.

Acknowledgments. The first listed author gratefully acknowledge the support from The Danish Council for Independent Research (Grant No. DFF–4002-00367). The second listed author acknowledges the support of the TuneSCode project (Grant No. DFF - 1335-00125) granted by the Danish Council for Independent Research and by the Cisco University Research Program Fund (Project CG No. 593761), Gift No. 2015-146035 (3696).

References

1. Ahlswede, R., Cai, N., Li, S.-Y.R., Yeung, R.W.: Network information flow. IEEE Trans. Inf. Theor. **46**(4), 1204–1216 (2000)
2. Balli, H., Yan, X., Zhang, Z.: On randomized linear network codes and their error correction capabilities. IEEE Trans. Inf. Theor. **55**(7), 3148–3160 (2009)
3. Barbero, A.I., Ytrehus, Ø.: An introduction to network coding for acyclic and cyclic networks. Sel. Topics Inf. Coding Theor. **7**, 339–421 (2010)
4. Guenther, S.M., Riemensberger, M., Utschick, W.: Efficient GF arithmetic for linear network coding using hardware SIMD extensions. In: International Symposium on Network Coding (NetCod), pp. 1–6, June 2014
5. Heide, J., Lucani, D.E.: Composite extension finite fields for low overhead network coding: telescopic codes. In: 2015 IEEE International Conference on Communications (ICC), pp. 4505–4510. IEEE (2015)
6. Heide, J., Pedersen, M.V., Fitzek, F.H.P., Larsen, T.: Network coding for mobile devices - systematic binary random rateless codes. In: 2009 IEEE International Conference on Communications Workshops, pp. 1–6, June 2009
7. Ho, T., Médard, M., Koetter, R., Karger, D.R., Effros, M., Shi, J., Leong, B.: A random linear network coding approach to multicast. IEEE Trans. Inf. Theor. **52**(10), 4413–4430 (2006)
8. Jaggi, S., Sanders, P., Chou, P.A., Effros, M., Egner, S., Jain, K., Tolhuizen, L.M.G.M.: Polynomial time algorithms for multicast network code construction. IEEE Trans. Inf. Theor. **51**(6), 1973–1982 (2005)
9. Jones, A.L., Chatzigeorgiou, I., Tassi, A.: Binary systematic network coding for progressive packet decoding. In: IEEE International Conference on Communications (ICC), pp. 4499–4504, June 2015
10. Koetter, R., Kschischang, F.R.: Coding for errors and erasures in random network coding. IEEE Trans. Inf. Theor. **54**(8), 3579–3591 (2008)
11. Koetter, R., Médard, M.: An algebraic approach to network coding. IEEE/ACM Trans. Netw. (TON) **11**(5), 782–795 (2003)

12. Lucani, D.E., Pedersen, M.V., Ruano, D., Sørensen, C.W., Fitzek, F.H.P., Heide, J., Geil, O.: Fulcrum network codes: a code for fluid allocation of complexity. arXiv preprint arXiv:1404.6620 (2014)
13. Paramanathan, A., Pedersen, M.V., Lucani, D.E., Fitzek, F.H.P., Katz, M.: Lean and mean: network coding for commercial devices. IEEE Wirel. Commun. **20**(5), 54–61 (2013)
14. Silva, D., Kschischang, F.R., Koetter, R.: A rank-metric approach to error control in random network coding. IEEE Trans. Inf. Theor. **54**(9), 3951–3967 (2008)
15. Skachek, V., Milenkovic, O., Nedić, A.: Hybrid noncoherent network coding. IEEE Trans. Inf. Theor. **59**(6), 3317–3331 (2013)
16. Soerensen, C.W., Paramanathan, A., Cabrera, J.A., Pedersen, M.V., Lucani, D.E., Fitzek, F.H.P.: Leaner and meaner: network coding in SIMD enabled commercial devices. In: IEEE Wireless Communications and Networking Conference (WCNC), pp. 1–6, April 2016
17. Trullols-Cruces, O., Barcelo-Ordinas, J.M., Fiore, M.: Exact decoding probability under random linear network coding. IEEE Commun. Lett. **15**(1), 67–69 (2011)

Bounding the Minimum Distance of Affine Variety Codes Using Symbolic Computations of Footprints

Olav Geil[1]([✉])[iD] and Ferruh Özbudak[1,2][iD]

[1] Department of Mathematical Sciences,
Aalborg University, Aalborg, Denmark
olav@math.aau.dk
[2] Department of Mathematics and Institute of Applied Mathematics,
Middle East Technical University, Ankara, Turkey
ozbudak@metu.edu.tr

Abstract. We study a family of primary affine variety codes defined from the Klein quartic. The duals of these codes have previously been treated in [12, Example 3.2]. Among the codes that we construct almost all have parameters as good as the best known codes according to [9] and in the remaining few cases the parameters are almost as good. To establish the code parameters we apply the footprint bound [7,10] from Gröbner basis theory and for this purpose we develop a new method where we inspired by Buchbergers algorithm perform a series of symbolic computations.

Keywords: Affine variety codes · Buchberger's algorithm · Klein curve · Minimum distance

1 Introduction

Affine variety codes [5] are codes defined by evaluating multivariate polynomials at the points of an affine variety. Despite having a simple description such codes constitute the entire class of linear codes [5, Pro. 1]. Given a description of a code as an affine variety code it is easy to determine the length n and dimension k, but no simple general method is known which easily estimates the minimum distance d. Of course such methods exists for particular classes of affine variety codes. For instance the Goppa bound for one-point algebraic geometric codes extends to an improved bound on the more general class of order domain codes [6,11], and in larger generality the Feng-Rao bounds and their variants can be successfully applied to many different types of codes [2–4,6,8,12,13]. In this paper we consider a particular family of primary affine variety codes for which none of the above mentioned bounds provide accurate information. More precisely we consider primary codes defined from the Klein quartic using the same weighted degree lexicographic ordering as in [12, Example 3.2] where they studied the

© Springer International Publishing AG 2017
Á.I. Barbero et al. (Eds.): ICMCTA 2017, LNCS 10495, pp. 128–138, 2017.
DOI: 10.1007/978-3-319-66278-7_12

corresponding dual codes. A common property of the Feng-Rao bound for primary codes and its variants are that they can be viewed [6,8] as consequences of the footprint bound [7,10] from Gröbner basis theory. To establish more accurate information for the codes under consideration it is therefore natural to try to apply the footprint bound in a more direct way, which is exactly what we do in the present paper using ingredients from Buchberger's algorithm and by considering an exhaustive number of special cases. Our analysis reveals that the codes under consideration are in most cases as good as the best known codes according to [9] and for the remaining few cases the minimum distance is only one less than the best known codes of the same dimension.

The paper is organized as follows. In Sect. 2 we introduce the footprint of an ideal and define affine variety codes. We then describe how the footprint bound can be applied to determine the Hamming weight of a code word. Then in Sect. 3 we apply symbolic computations leading to estimates on the minimum distance on each of the considered codes the information of which we collect in Sect. 4.

2   Affine Variety Codes and the Footprint Bound

The footprint (also called the delta-set) is defined as follows:

Definition 1. *Given a field k, a monomial ordering $\prec$ and an ideal $J \subseteq k[X_1, \ldots, X_m]$ the footprint of J is*

$$\Delta_\prec(J) = \{M \mid M \text{ is a monomial which is not leading monomial}$$
$$\text{of any polynomial in } J\}$$

From [1, Property 7, Sect. 5.3] we have the following well-known result.

Theorem 1. *Let the notation be as in the above definition. The set*

$$\{M + J \mid M \in \Delta_\prec(J)\}$$

is a basis for $k[X_1, \ldots, X_m]/J$ as a vector space over k.

Recall that by definition a Gröbner basis is a finite basis for the ideal J from which one can easily determine the footprint. Concretely a monomial is a leading monomial of some polynomial in the ideal if and only if it is divisible by a leading monomial of some polynomial in the Gröbner basis. The following corollary is an instance of the more general footprint bound [10].

Corollary 1. *Let $I \subseteq \mathbb{F}_q[X_1, \ldots, X_m]$ be an ideal and $I_q = I + \langle X_1^q - X_1, \ldots, X_m^q - X_m \rangle$. The variety of I_q is of size $\#\Delta_\prec(I_q)$ for any monomial ordering $\prec$.*

Proof. Let the variety of I_q be $\{P_1, \ldots, P_n\}$ with $P_i \neq P_j$ for $i \neq j$. The field $\mathbb{F}_q$ being perfect, the ideal I_q is radical because it contains a univariate square-free polynomial in each variable and by the ideal-variety correspondence therefore

I_q is in fact the vanishing ideal of $\{P_1, \ldots, P_n\}$. Therefore the evaluation map $\mathrm{ev} : \mathbb{F}_q[X_1, \ldots, X_m]/I_q \to \mathbb{F}_q^n$ given by $\mathrm{ev}(F + I_q) = (F(P_1), \ldots, F(P_n))$ is injective. On the other hand the evaluation map is also surjective which is seen by applying Lagrange interpolation. We have demonstrated that ev is a bijection and the corollary follows from Theorem 1.   □

We are now ready to define primary affine variety codes formally.

Definition 2. *Let the notation be as in the proof of Corollary 1. Given an ideal $I \subseteq \mathbb{F}_q[X_1, \ldots, X_m]$ and a monomial ordering $\prec$ choose $L \subseteq \Delta_\prec(I_q)$. Then*

$$C(I, L) = \mathrm{Span}_{\mathbb{F}_q}\{\mathrm{ev}(M + I_q) \mid M \in L\}$$

is called a primary affine variety code.

From the above discussion it is clear that $C(I, L)$ is a code of length $n = \#\Delta_\prec(I_q)$ and dimension $k = \#L$. Given a code word $\mathbf{c} = \mathrm{ev}(F + I_q)$ then by Corollary 1 we have

$$w_H(\mathbf{c}) = n - \#\Delta_{\prec_w}(\langle F \rangle + I_q) = \#\Delta_{\prec_w}(I_q) \cap \mathrm{lm}(\langle F \rangle + I_q) = \#\square_{\prec_w}(F),$$

where $\square_{\prec_w}(F) := \Delta_{\prec_w}(I_q) \cap \mathrm{lm}(\langle F \rangle + I_q)$. Reducing a polynomial modulo a Gröbner basis for I_q one obtains a (unique) polynomial which has support in the footprint $\Delta(I_q)$ (this is the result behind Theorem 1). Hence we shall always assume that F is of this form. In the rest of the paper we concentrate on estimating for a concrete class of codes $\#\square_\prec(F)$ using only information on the leading monomial.

3   Code Words from the Klein Curve

In the remaining part of the paper I will always be the ideal

$$I = \langle Y^3 + X^3Y + X \rangle \subseteq \mathbb{F}_8[X, Y]$$

and consequently $I_8 = \langle Y^3 + X^3Y + X, X^8 + X, Y^8 + Y \rangle$. The corresponding variety is of size 22, hence we write it as $\{P_1, \ldots, P_{22}\}$. The evaluation map then becomes $\mathrm{ev}(F + I_8) = (F(P_1), \ldots, F(P_{22}))$.

As monomial ordering we choose the same ordering as in [12, Example 3.2], namely the weighted degree lexicographic ordering $\prec_w$ given by the rule that $X^\alpha Y^\beta \prec_w X^\gamma Y^\delta$ if either (i) or (ii) below holds

(i) $2\alpha + 3\beta < 2\gamma + 3\delta$,   (ii) $2\alpha + 3\beta = 2\gamma + 3\delta$ but $\beta < \delta$.

By inspection $\{Y^3 + X^3Y + X, X^8 - X, X^7Y + Y\}$ is a Gröbner basis for I_8 with respect to $\prec_w$. Hence, the footprint $\Delta_{\prec_w}(I_8)$ and the corresponding weights are as in Fig. 1. We remind the reader that for $L \subseteq \Delta_{\prec_w}(I_8)$ the code $C(I, L)$ equals $\mathrm{ev}(\mathrm{Span}_{\mathbb{F}_8}(L) + I_8)$ which is of length $n = 22$ and dimension $k = \#L$.

$$
\begin{array}{ccccccc}
Y^2 & XY^2 & X^2Y^2 & X^3Y^2 & X^4Y^2 & X^5Y^2 & X^6Y^2 \\
Y & XY & X^2Y & X^3Y & X^4Y & X^5Y & X^6Y \\
1 & X & X^2 & X^3 & X^4 & X^5 & X^6 & X^7
\end{array}
$$

$$
\begin{array}{ccccccc}
6 & 8 & 10 & 12 & 14 & 16 & 18 \\
3 & 5 & 7 & 9 & 11 & 13 & 15 \\
0 & 2 & 4 & 6 & 8 & 10 & 12 & 14
\end{array}
$$

Fig. 1. The footprint $\Delta_{\prec_w}(I_8)$ with corresponding weights.

Our method to estimate $\#\square_{\prec_w}(F)$ (which corresponds to estimating the Hamming weight of the corresponding code word) consists in two parts. First we observe that all monomials in $\Delta_{\prec_w}(I_8)$ divisible by the leading monomial of F are in $\square_{\prec_w}(F)$. In the second part we then for a number of exhaustive special cases find more monomials in $\square_{\prec_w}(F)$ by establishing clever combinations of polynomials that we already know are in $\langle F \rangle + I_q$. To describe how such combinations are derived we will need the following notation. Consider polynomials $S(X,Y)$, $D(X,Y)$ and $R(X,Y)$. By

$$
S(X,Y) \overset{D(X,Y)}{\longrightarrow} R(X,Y) \tag{1}
$$

we shall indicate that $R(X,Y) = S(X,Y) - Q(X,Y)D(X,Y)$ for some polynomial $Q(X,Y)$. The important fact – which we shall use frequently throughout the paper – is that $R(X,Y) \in \langle S(X,Y), D(X,Y) \rangle$.

Remark 1. The Feng-Rao bound can be applied to any affine variety code; but it works most efficiently when the ideal I and the monomial ordering $\prec$ under consideration satisfy the order domain conditions [6, Sect. 7]. That is,

1. The ordering $\prec$ must be a weighted degree lexicographic ordering (or in larger generality a generalized weighted degree ordering [6, Definition 8]).
2. A Gröbner basis for I must exist with the property that any polynomial in it contains in its support (exactly) two monomials of the highest weight.
3. No two different monomials in $\Delta_{\prec}(I)$ are of the same weight.

In such cases the method often establishes many more monomials in $\square_{\prec}(F)$ than those divisible by the leading monomial of F. In [8] an improved Feng-Rao bound was presented which treats in addition efficiently certain families of cases where the conditions 1. and 2. are satisfied, but 3. is not. Even though the ideal and monomial ordering studied in the present section exactly satisfy conditions 1. and 2., but not 3, the improved Feng-Rao bound produces the same information as the Feng-Rao bound in this case. By inspection both methods only "detect" monomials divisible by the leading monomial of F as being members of $\square_{\prec_w}(F)$.

In the following, for simplicity, we shall in our calculations always assume that the leading coefficient of F is 1.

3.1   Leading Monomial Equal to Y

Consider $c = \mathrm{ev}(F + I_8)$ where $F(X,Y) = Y + a_1X + a_2$. Clearly

$$\{Y, Y^2, XY, XY^2, \ldots, X^6Y, X^6Y^2\} \subset \square_{\prec_w}(F).$$

We next establish more monomials in $\square_{\prec_w}(F)$ under different conditions on the coefficients a_1, a_2. Consider

$$Y^2 F(X,Y)$$
$$\overset{Y^3 + X^3Y + X}{\longrightarrow} X^3Y + a_1XY^2 + a_2Y^2 + X$$
$$\overset{F(X,Y)}{\longrightarrow} a_1X^4 + (a_1^3 + a_2)X^3 + a_1^2a_2X^2 + (a_1a_2^2 + 1)X + a_2^3.$$

If $a_1 \neq 0$ then we have

$$\{X^4, X^5, X^6, X^7\} \subset \square_{\prec_w}(F).$$

Next assume $a_1 = 0$. If $a_2 \neq 0$ then we obtain

$$\{X^3, X^4, X^5, X^6, X^7\} \subset \square_{\prec_w}(F).$$

Finally, assume $a_1 = a_2 = 0$ in which case we have

$$\{X, X^2, X^3X^4, X^5, X^6, X^7\} \subset \square_{\prec_w}(F).$$

In conclusion we have shown that $\square_{\prec_w}(F)$ contains at least $14 + 4 = 18$ elements which implies $w_H(c) \geq 18$.

3.2   Leading Monomial Equal to Y^2

Consider a codeword $c = \mathrm{ev}(F + I_8)$ where

$$F(X,Y) = Y^2 + a_1X^3 + a_2XY + a_3X^2 + a_4Y + a_5X + a_6.$$

Independently of the coefficients $a_1, \ldots, a_6$ we see that

$$\{Y^2, XY^2, \ldots, X^6Y^2\} \subset \square_{\prec_w}(F). \tag{2}$$

We next consider an exhaustive series of conditions under which we establish more monomials in $\square_{\prec_w}(F)$. We have

$$YF(X,Y)$$
$$\overset{Y^3 + X^3Y + X}{\longrightarrow} (a_1 + 1)X^3Y + a_2XY^2 + a_3X^2Y$$
$$+ a_4Y^2 + a_5XY + a_6Y + X. \tag{3}$$

If $a_1 \neq 1$ then the leading monomial of the last polynomial becomes X^3Y and consequently

$$\{X^3Y, X^4Y, X^5Y, X^6Y\} \in \square_{\prec_w}(F). \tag{4}$$

Continuing the calculations for this case we obtain:

$$Y((a_1 + 1)X^3Y + a_2XY^2 + a_3X^2Y + a_4Y^2 + a_5XY + a_6Y + X)$$

$$\xrightarrow{F(X,Y)} (a_1 + 1)(a_1X^6 + a_2X^4Y + a_3X^5 + a_4X^3Y + a_5X^4 + a_6X^3)$$

$$+ a_2XY^3 + a_3X^2Y^2 + a_4Y^3 + a_5XY^2 + a_6Y^2 + XY.$$

If $a_1 \neq 0$ then we also have

$$\{X^6, X^7\} \subset \square_{\prec_w}(F).$$

Assuming next that $a_1 = 0$ the above expression becomes

$$a_2X^4Y + a_3X^5 + a_4X^3Y + a_5X^4 + a_6X^3 + a_2XY^3$$

$$+ a_3X^2Y^2 + a_4Y^3 + a_5XY^2 + a_6Y^2 + XY$$

$$\xrightarrow{Y^3 + X^3Y + X} a_3X^5 + a_4X^3Y + a_5X^4 + a_6X^3 + a_3X^2Y^2 + a_4Y^3$$

$$+ a_5XY^2 + a_6Y^2 + XY + a_2X^2$$

$$\xrightarrow{F(X,Y)} a_3X^5 + a_5X^4 + a_6X^3 + a_3a_2X^3Y + a_3^2X^4 + a_3a_4X^2Y$$

$$+ a_3a_5X^3 + a_3a_6X^2 + a_5XY^2 + a_6Y^2 + XY + a_2X^2.$$

If $a_3 \neq 0$ then

$$\{X^5, X^6, X^7\} \subset \square_{\prec_w}(F).$$

Hence, continuing under the assumption $a_3 = 0$ we are left with

$$a_5X^4 + a_6X^3 + a_5XY^2 + a_6Y^2 + XY + a_2X^2$$

$$\xrightarrow{F(X,Y)} a_5X^4 + a_6X^3 + a_5a_2X^2Y + a_5a_4XY + a_5^2X^2 + a_5a_6X + a_6Y^2$$

$$+ XY + a_2X^2.$$

if $a_5 \neq 0$ then

$$\{X^4, X^5, X^6, X^7\} \subset \square_{\prec_w}(F).$$

Hence, assume $a_5 = 0$ and we are left with

$$a_6X^3 + a_6Y^2 + XY + a_2X^2$$

$$\xrightarrow{F(X,Y)} a_6X^3 + a_6a_2XY + a_6a_4Y + a_6^2 + XY + a_2X^2.$$

If $a_6 \neq 0$ then

$$\{X^3, X^4, X^5, X^6, X^7\} \subset \square_{\prec_w}(F).$$

If on the other hand $a_6 = 0$ then we are left with $XY + a_2X^2$ in which case we obtain

$$\{XY, X^2Y\} \subset \square_{\prec_w}(F).$$

In conclusion, for the case $a_1 \neq 1$ we obtained in addition to the elements in (2) the elements in (4) and at least 2 more. That is, in addition to the elements in (2) at least 6 more.

Assume in the following that $a_1 = 1$ and continue the reduction from (3)

$$\overset{F(X,Y)}{\longrightarrow} a_2X^4 + (a_3 + a_2^2)X^2Y + (a_2a_3 + a_4)X^3 + a_5XY$$
$$+(a_2a_5 + a_3a_4)X^2 + (a_6 + a_4^2)Y + (1 + a_4a_5)X + a_4a_6. \qquad (5)$$

If $a_2 \neq 0$ then

$$\{X^4, X^5, X^7, X^4Y, X^5Y, X^6Y\} \subset \square_{\prec_w}(F).$$

Next assume $a_2 = 0$. If $a_3 \neq 0$ then

$$\{X^2Y, X^3Y, X^4Y, X^5Y, X^6Y\} \subset \square_{\prec_w}(F). \qquad (6)$$

Continuing the reduction under the assumption $a_3 \neq 0$ we multiply (5) by Y and continue the reduction:

$$a_3X^2Y^2 + a_4X^3Y + a_5XY^2 + a_3a_4X^2Y + (a_6 + a_4^2)Y$$
$$+(1 + a_4a_5)X + a_4a_6$$
$$\overset{F(X,Y)}{\longrightarrow} a_3X^5 + a_3^2X^4 + a_3a_4X^2Y + a_3a_5X^3 + a_3a_6X^2 + a_4X^3Y + a_5XY^2$$
$$+a_3a_4X^2Y + (a_6 + a_4^2)Y + (1 + a_4a_5)X + a_4a_6.$$

As $a_3 \neq 0$ we obtain in addition to (2) and (6) that

$$\{X^5, X^6, X^7\} \subset \square_{\prec_w}(F).$$

That is, in addition to (2) we found in total 8 more elements in $\square_{\prec_w}(F)$.

Next assume $a_3 = 0$ and continue from (5). If $a_4 \neq 0$ then

$$\{X^3, X^4, X^5, X^6, X^7, X^3Y, X^4Y, X^5Y, X^6Y\} \subset \square_{\prec_w}(F).$$

Next assume $a_4 = 0$. if $a_5 \neq 0$ then

$$\{XY, X^2Y, X^3Y, X^4Y, X^5Y, X^6Y\} \subset \square_{\prec_w}(F).$$

Hence, assume $a_5 = 0$. If $a_6 \neq 0$ then

$$\{Y, XY, X^2Y, X^3Y, X^4Y, X^5Y, X^6Y\} \subset \square_{\prec_w}(F).$$

Finally, assume $a_6 = 0$. But then

$$\{X, X^2, X^3, X^4, X^5, X^6, X^7, XY, X^2Y, X^3Y, X^4Y, X^5Y, X^6Y\} \subset \square_{\prec_w}(F).$$

In conclusion, we have at least $7 + \min\{6, 6, 8, 9, 6, 7, 13\} = 13$ monomials in $\square_{\prec_w}(F)$ and therefore $w_H(c) \geq 13$.

3.3   Leading Monomial Equal to XY

Consider $c = \text{ev}(F + I_8)$ where

$$F(X,Y) = XY + a_1 X^2 + a_2 Y + a_3 X + a_4.$$

For sure

$$\{XY, X^2Y, X^3Y, X^4Y, X^5Y, X^6Y,$$
$$XY^2, X^2Y^2, X^3Y^2, X^4Y^2, X^5Y^2, X^6Y^2\} \subset \square_{\prec_w}(F). \qquad (7)$$

We next consider an exhaustive series of conditions under which we establish more monomials in $\square_{prec_w}(F)$. We have

$$Y^2 F(X,Y)$$

$$\xrightarrow{Y^3+X^3Y+X} a_1 X^5 + a_3 X^4 + a_4 X^3$$
$$+a_1(a_1^2 X^4 + a_2^2 Y^2 + a_3^2 X^2 + a_4^2)$$
$$+a_3 XY^2 + a_4 Y^2 + X^2 + a_2 X.$$

If $a_1 \neq 0$ then

$$\{X^5, X^6, X^7\} \subset \square_{\prec_w}(F).$$

Hence, assume $a_1 = 0$ and continue the reduction:

$$\xrightarrow{F(X,Y)} a_3 a_2 Y^2 + a_3^2 XY + a_3 a_4 Y + a_3 X^4 + a_4 X^3 + a_4 Y^2 + X^2 + a_2 X.$$

If $a_3 \neq 0$ then

$$\{X^4, X^5, X^6, X^7\} \subset \square_{\prec_w}(F).$$

Hence, assume $a_3 = 0$ in which case the above becomes

$$a_4 Y^2 + a_4 X^3 + X^2 + a_2 X.$$

If $a_4 = 0$ then

$$\{X^2, X^3, X^4, X^5, X^6, X^7\} \subset \square_{\prec_w}(F).$$

Hence, assume $a_4 \neq 0$, in which case we have

$$\{Y^2\} \subset \square_{\prec_w}(F).$$

We continue the calculations to add more elements. We have:

$$X^2(a_4 Y^2 + a_4 X^3 + X^2 + a_2 X) \xrightarrow{F(X,Y)} a_4 X^5 + a_2^2 Y^2 + a_2 X + a_4^2.$$

But then

$$\{X^5, X^6, X^7\} \subset \square_{\prec_w}(F).$$

That is, for the case $a_4 \neq 0$ $\square_{\prec_w}(F)$ contains in addition to (7) at least $1 + 3 = 4$ more monomials.

In conclusion $w_H(c) \geq 12 + \min\{3, 4, 6, 4\} = 15$, and if $a_1 = 0$ then $w_H(c) \geq 16$.

3.4   The Remaining Cases

Using a similar approach as above we can treat the remaining 19 possible cases corresponding to other choices of leading monomial. For the cases that the leading monomial is in $\{XY^2, X^2Y, X^2Y^2, X^3Y, X^3Y^2\}$ our analysis is as involved – or a little more involved – than in the last three subsections. Due to lack of space these calculations are not included in the present paper. For the last 14 out of the 19 cases the analysis is simply done by using the fact that all monomials in $\Delta_{\prec_w}(I_8)$ divisible by the leading monomial of F must be in $\Box_{\prec_w}(F)$. Our findings are collected in Fig. 2.

$$
\begin{array}{ccccccc}
13 & 10 & 7 & 5 & 3 & 2 & 1 \\
18 & 15 & 12 & 9 & 6 & 4 & 2 \\
22 & 19 & 16 & 13 & 10 & 7 & 4 & 1
\end{array}
$$

Fig. 2. Lower bounds on $\#\Box_{\prec_w}(F)$ where $\mathrm{lm}(F)$ are as in Fig. 1

4   Code Parameters

As code construction we use

$$
\mathrm{Span}_{\mathbb{F}_8}\{\mathrm{ev}(M + I_8) \mid M \in \Delta_{\prec_w}(I_8), \delta(M) \geq s\},
$$

where $\delta(M)$ are the estimates of $\#\Box_{\prec_w}(F)$ as depicted in Fig. 2. In this way we obtain the best possible codes, according to our estimates. The resulting parameters are shown in Table 1. In almost all cases, given a dimension in the table, then the corresponding estimate on the minimum distance equals the best value known to exist according to [9]. The only exceptions are the dimensions $4, 14, 15$ and 18 where the best minimum distances known to exist are one more than we obtain.

Table 1. Parameters $[n, k, d]_8$ of codes from the Klein quartic. Here, n and k are sharp values, whereas d represents a lower bound estimate.

$[22, 1, 22]_8$	$[22, 2, 19]_8$	$[22, 3, 18]_8$
$[22, 4, 16]_8$	$[22, 5, 15]_8$	$[22, 7, 13]_8$
$[22, 8, 12]_8$	$[22, 10, 10]_8$	$[22, 11, 9]_8$
$[22, 13, 7]_8$	$[22, 14, 6]_8$	$[22, 15, 5]_8$
$[22, 17, 4]_8$	$[22, 18, 3]_8$	$[22, 20, 2]_8$

5 Concluding Remarks

In [12, Example 3.2] the authors estimated the minimum distances of the duals of the codes studied in the present paper using the Feng-Rao bound for dual codes. We believe that is should be possible to improve (possibly even drastic) upon their estimates of the minimum distance in the same way as we in this paper improved upon the Feng-Rao bound for primary codes. We leave this question for future research. The method of the present paper also applies to estimate higher weights (possible relative). We leave it for future research to establish examples where this gives improved information compared to what can be derived from the Feng-Rao bound. In the light of Remark 1 and the information established in Sect. 3, evidently our new method sometimes significantly improves upon the previous known methods. We stress that our method is very general in that it can be applied to any primary affine variety code. In particular it works for any monomial ordering and consequently also without any of the order domain conditions (Remark 1). Finding more families of good affine variety codes using our method is subject to future work.

Acknowledgments. The authors gratefully acknowledge the support from The Danish Council for Independent Research (Grant No. DFF–4002-00367). They are also grateful to Department of Mathematical Sciences, Aalborg University for supporting a one-month visiting professor position for the second listed author. The research of Ferruh Özbudak has been funded by METU Coordinatorship of Scientific Research Projects via grant for projects BAP-01-01-2016-008 and BAP-07-05-2017-007.

References

1. Cox, D.A., Little, J., O'Shea, D.: Ideals, Varieties, and Algorithms: An Introduction to Computational Algebraic Geometry and Commutative Algebra, vol. 4. Springer, New York (2015)
2. Feng, G.L., Rao, T.R.N.: Decoding algebraic-geometric codes up to the designed minimum distance. IEEE Trans. Inform. Theory **39**(1), 37–45 (1993)
3. Feng, G.L., Rao, T.R.N.: A simple approach for construction of algebraic-geometric codes from affine plane curves. IEEE Trans. Inform. Theory **40**(4), 1003–1012 (1994)
4. Feng, G.L., Rao, T.R.N.: Improved geometric Goppa codes part I: basic theory. IEEE Trans. Inform. Theory **41**(6), 1678–1693 (1995)
5. Fitzgerald, J., Lax, R.F.: Decoding affine variety codes using Gröbner bases. Des. Codes Crypt. **13**(2), 147–158 (1998)
6. Geil, O.: Evaluation codes from an affine variety code perspective. In: Martínez-Moro, E., Munuera, C., Ruano, D. (eds.) Advances in Algebraic Geometry Codes. Coding Theory and Cryptology, vol. 5, pp. 153–180. World Scientific, Singapore (2008)
7. Geil, O., Høholdt, T.: Footprints or generalized Bezout's theorem. IEEE Trans. Inform. Theory **46**(2), 635–641 (2000)
8. Geil, O., Martin, S.: An improvement of the Feng-Rao bound for primary codes. Des. Codes Crypt. **76**(1), 49–79 (2015)

9. Grassl, M.: Bounds on the minimum distance of linear codes and quantum codes (2007). http://www.codetables.de. Accessed 20 Apr 2017
10. Høholdt, T.: On (or in) Dick Blahut's' footprint'. Codes, Curves and Signals, pp. 3–9 (1998)
11. Høholdt, T., van Lint, J.H., Pellikaan, R.: Algebraic geometry codes. In: Pless, V.S., Huffman, W.C. (eds.) Handbook of Coding Theory, vol. 1, pp. 871–961. Elsevier, Amsterdam (1998)
12. Kolluru, M.S., Feng, G.L., Rao, T.R.N.: Construction of improved geometric Goppa codes from Klein curves and Klein-like curves. Appl. Algebra Engrg. Comm. Comput. **10**(6), 433–464 (2000)
13. Salazar, G., Dunn, D., Graham, S.B.: An improvement of the Feng-Rao bound on minimum distance. Finite Fields Appl. **12**, 313–335 (2006)

On Binary Matroid Minors and Applications to Data Storage over Small Fields

Matthias Grezet[✉], Ragnar Freij-Hollanti, Thomas Westerbäck,
and Camilla Hollanti

Department of Mathematics and Systems Analysis,
Aalto University, Espoo, Finland
{matthias.grezet,ragnar.freij-hollanti,
thomas.westerback,camilla.hollanti}@aalto.fi

Abstract. Locally repairable codes for distributed storage systems have gained a lot of interest recently, and various constructions can be found in the literature. However, most of the constructions result in either large field sizes and hence too high computational complexity for practical implementation, or in low rates translating into waste of the available storage space.

In this paper we address this issue by developing theory towards code existence and design over a given field. This is done via exploiting recently established connections between linear locally repairable codes and matroids, and using matroid-theoretic characterisations of linearity over small fields. In particular, nonexistence can be shown by finding certain forbidden uniform minors within the lattice of cyclic flats. It is shown that the lattice of cyclic flats of binary matroids have additional structure that significantly restricts the possible locality properties of $\mathbb{F}_2$-linear storage codes. Moreover, a collection of criteria for detecting uniform minors from the lattice of cyclic flats of a given matroid is given, which is interesting in its own right.

Keywords: Binary matroids · Distributed storage systems · Lattice of cyclic flats · Locally repairable codes · Uniform minors

1 Introduction

The need for large-scale data storage is continuously increasing. Within the past few years, *distributed storage systems* (DSSs) have revolutionised our traditional ways of storing, securing, and accessing data. Storage node failure is a frequent obstacle, making repair efficiency an important objective. Network coding techniques for DSSs were considered in [6], characterising a storage space–repair bandwidth tradeoff.

A bottle-neck for repair efficiency, measured by the notion of *locality* [13], is the number of contacted nodes needed for repair. To this end, our motivation in this paper comes from *locally repairable codes* (LRCs), which are, informally speaking, storage systems where a small number of failing nodes can be recovered

© Springer International Publishing AG 2017
A.I. Barbero et al. (Eds.): ICMCTA 2017, LNCS 10495, pp. 139–153, 2017.
DOI: 10.1007/978-3-319-66278-7_13

by boundedly many other (close-by) nodes. Repair-efficient LRCs are already implemented on HDFS-Xorbas used by Facebook [14] and Windows Azure storage [11]. Here, the field size is not yet a huge concern, as the coding is done over a small number (<20) of nodes. Nevertheless, if we wish for more flexibility in terms of the code parameters, the field size quickly becomes a critical issue. Hence, a question arises as to how and when can we maintain a small field size, regardless of the number of the storage nodes. Some explicit constructions of LRCs over small fields can be found in the literature, e.g., [4,8,12,16,17].

Let us denote by (n, k, d, r, δ), respectively, the code length, dimension, global minimum distance, locality, and local minimum distance. In terms of a storage system employing an (n, k, d, r, δ)-LRC, this means that we encode k information symbols into n code symbols that are then stored on n storage nodes, and can globally tolerate $d - 1$ node failures while still being able to repair by contacting k nodes. Locally, if we lose at most $\delta - 1$ nodes ($\delta \leq d$), we can repair those by contacting at most $r < k$ close-by nodes.

It was shown in [18] that the $(r, \delta = 2)$-locality of a linear LRC is a matroid invariant. The connection between matroid theory and linear LRCs was examined in more detail in [20]. In addition, the parameters (n, k, d, r, δ) for linear LRCs were generalised to matroids, and new results for both matroids and linear LRCs were given therein.

In this paper, we develop theory towards matroids that are representable over the binary field or some other fixed-sized field. We show that well-known matroid-theoretic criteria on binary linear codes give conditions on the lattice of cyclic flats, which govern the locality properties of the associated storage codes. As a consequence, we get stronger structural constraints on binary LRCs compared to those previously known for general LRCs. Moreover, a collection of criteria for detecting uniform minors from the lattice of cyclic flats of a given matroid is given, which is interesting in its own right.

2   Preliminaries on LRCs and Matroids

To study LRCs in more detail, we consider punctured codes $C|Y$, where $Y \subseteq E$ is a set of coordinates of the code C. For a fixed code C, we denote by d_Y the minimum Hamming distance of the punctured code $C|Y$. As is common practice, we say that C is an (n, k, d)-code if it has length n, dimension k and minimum Hamming distance d. A linear (n, k, d)-code C over a field is a *non-degenerate storage code* if $d \geq 2$ and there is no zero column in a generator matrix of C.

Definition 1. *Here, a linear (n, k, d, r, δ)-LRC over a finite field $\mathbb{F}$ is a non-degenerate linear (n, k, d)-code C over $\mathbb{F}^E$ such that any coordinate $x \in E$ of C has locality (r, δ), meaning that there is a subset R of E, called repair set of x, such that $x \in R$, $|R| \leq r + \delta - 1$ and $d_R \geq \delta$.*

The parameters (n, k, d, r, δ) can immediately be defined and studied for matroids in general, as in [18,20].

2.1   Matroid Fundamentals

Matroids were first introduced by Whitney in 1935, to capture and generalise the notion of linear dependence in purely combinatorial terms. Indeed, the combinatorial setting is general enough to also capture many other notions of dependence occurring in mathematics, such as cycles or incidences in a graph, non-transversality of algebraic varieties, or algebraic dependence of field extensions. Of special interest for linear LRCs is the connection between linear algebra and matroids.

Matroids have many equivalent definitions in the literature. Here, we choose to present matroids via their rank functions. Much of the contents in this section can be found in more detail in [9].

Definition 2 (Matroid). *A (finite) matroid* $M = (\rho, E)$ *is a finite set* E *together with a rank function* $\rho : 2^E \to \mathbb{Z}$ *such that for all subsets* $X, Y \subseteq E$

$$(R.1)\ 0 \le \rho(X) \le |X|,$$
$$(R.2)\ X \subseteq Y \quad \Rightarrow \quad \rho(X) \le \rho(Y),$$
$$(R.3)\ \rho(X) + \rho(Y) \ge \rho(X \cup Y) + \rho(X \cap Y).$$

A subset $X \subseteq E$ is called *independent* if $\rho(X) = |X|$. If X is independent and $\rho(X) = \rho(E)$, then X is called a *basis*. Strongly related to the rank function is the *nullity function* $\eta : 2^E \to \mathbb{Z}$, defined by $\eta(X) = |X| - \rho(X)$ for $X \subseteq E$.

Any matrix G over a field $\mathbb{F}$ generates a matroid $M_G = (\rho, E)$, where E is the set of columns of G, and $\rho(X)$ is the rank of $G(X)$ over $\mathbb{F}$, where $G(X)$ denotes the submatrix of G formed by the columns indexed by X. As elementary row operations preserve the row space of $G(X)$ for all $X \subseteq E$, it follows that row-equivalent matrices generate the same matroid.

Thus, there is a straightforward connection between linear codes and matroids. Let C be a linear code over a field $\mathbb{F}$. Then any two different generator matrices of C will have the same row space by definition, so they will generate the same matroid. Therefore, without any inconsistency, we can denote the matroid associated to these generator matrices by $M_C = (\rho_C, E)$. The rank function ρ_C can be defined directly from the code without referring to a generator matrix, via $\rho_C(X) = \dim(C|X)$ for $X \subseteq E$.

Example 1. Let C be the linear code generated by the following matrix G over $\mathbb{F}_2$:

$$
G = \begin{array}{c} \begin{array}{cccccc} 1\ 2\ 3\ 4\ 5\ 6 \end{array} \\ \left[\begin{array}{cccccc} 1 & 0 & 1 & 0 & 1 & 1 \\ 0 & 1 & 1 & 0 & 1 & 1 \\ 0 & 0 & 0 & 1 & 1 & 1 \end{array} \right] \end{array}
$$

Then, for the matroid $M_C = (\rho_C, \{1, 2, 3, 4, 5, 6\})$,

$$\rho_C(\emptyset) = 0,\ \rho_C(\{1, 2, 3\}) = \rho_C(\{3, 4, 5\}) = 2,\ \rho_C(\{1, 2, 3, 4, 5, 6\}) = 3.$$

Two matroids $M_1 = (\rho_1, E_1)$ and $M_2 = (\rho_2, E_2)$ are *isomorphic* if there exists a bijection $\psi : E_1 \to E_2$ such that $\rho_2(\psi(X)) = \rho_1(X)$ for all subsets $X \subseteq E_1$.

Definition 3. *A matroid that is isomorphic to M_G for some matrix G over $\mathbb{F}$ is said to be* representable *over $\mathbb{F}$. We also say that such a matroid is $\mathbb{F}$-representable. A* binary *matroid is a matroid that is $\mathbb{F}_2$-representable.*

Definition 4. *The* uniform matroid *$U_n^k = (\rho, [n])$ is a matroid with a ground set $[n] = \{1, 2, \ldots, n\}$ and a rank function $\rho(X) = \min\{|X|, k\}$ for $X \subseteq [n]$.*

The following straightforward observation gives a characterisation of maximum distance separable (MDS) codes and also shows that uniform matroids constitute a subclass of representable matroids.

Proposition 1. *A linear code C is an $(n, k, n - k + 1)$-MDS code if and only if M_C is the uniform matroid U_n^k.*

There are several elementary operations that are useful for explicit constructions of matroids, as well as for analysing their structure. The operations that we will need for this paper are dualisation, contraction, and deletion.

Definition 5. *Let $M = (\rho, E)$ be a matroid and $X, Y \subseteq E$, and denote by $\bar{X} = E - X$ for any $X \subseteq E$. Then*

(i) *The* restriction *of M to Y is the matroid $M|Y = (\rho_{|Y}, Y)$, where $\rho_{|Y}(A) = \rho(A)$ for $A \subseteq Y$.*
(ii) *The* contraction *of M by X is the matroid $M/X = (\rho_{/X}, \bar{X})$, where $\rho_{/X}(A) = \rho(A \cup X) - \rho(X)$ for $A \subseteq \bar{X}$.*
(iii) *A* minor *of M is the matroid $M|Y/X = (\rho_{|Y/X}, Y - X)$ obtained from M by restriction to Y and contraction by X. Observe that this does not depend on the order in which the restriction and contraction are performed.*
(iv) *The* dual *of M is the matroid $M^* = (\rho^*, E)$, where $\rho^*(A) = |A| + \rho(\bar{A}) - \rho(E)$ for $A \subseteq E$.*

The restriction operation to Y is also referred to as *deletion* of the set $E - Y$.

Let $M = M_C$ be a representable matroid. Then, restriction, contraction, and dualisation of M correspond to puncturing, shortening, and orthogonal complement of the code C, respectively.

Example 2. Let C be the code generated by the matrix G in Example 1. Then for the matroid $M_C = (\rho = \rho_C, [6])$, $X = \{1\}$, and $Y = \{2, 3, 4, 5\}$,

$$\rho_{/X}(\{2, 3, 4\}) = 2, \quad \rho_{|Y}(\{2, 3, 4\}) = 3, \quad \text{and} \quad \rho^*(\{2, 3, 4\}) = 2.$$

The minors of uniform matroids are very easily described:

Lemma 1. *Let $U_n^k = (\rho, [n])$ be a uniform matroid, and let $X \subseteq Y \subseteq E$. Then the minor $U_n^k|Y/X$ is isomorphic to $U_{n'}^{k'}$, where $k' = \max\{0, k - |X|\}$ and $n' = |Y| - |X|$. In particular, M is a minor of U_n^k if and only if $M \cong U_{n'}^{k'}$, for some $0 \le k' \le k$ and $0 \le n' - k' \le n - k$.*

In general there is no simple criterion to determine if a matroid is representable. However, there is a simple criterion for when a matroid is binary.

Theorem 1 ([19]). *Let $M = (\rho, E)$ be a matroid. The following two conditions are equivalent.*

1. *M is linearly representable over $\mathbb{F}_2$.*
2. *There are no sets $X \subseteq Y \subseteq E$ such that $M|Y/X$ is isomorphic to the uniform matroid U_4^2.*

In essence, this means that the only obstruction that needs to be overcome in order to be representable over the binary alphabet, is that no more than three nonzero points can fit in the same plane. Clearly, if M is representable over $\mathbb{F}$, then so are all its minors. The following result, Theorem 2, is a far-going extension of Theorem 1, that was first conjectured by Gian-Carlo Rota in 1970. A proof of this conjecture was announced by Geelen, Gerards, and Whittle in 2014, but the details of the proof still remain to written up [10].

Theorem 2 ([10]). *For any finite field $\mathbb{F}$, there is a finite set $L(\mathbb{F})$ of matroids such that any matroid M is representable if and only if it contains no element from $L(\mathbb{F})$ as a minor.*

Since the 1970's, it has been known that a matroid is representable over $\mathbb{F}_3$ if and only if it avoids the uniform matroids U_5^2, U_5^3, the Fano plane $\mathbb{P}^2(\mathbb{F}_2)$, and its dual $\mathbb{P}^2(\mathbb{F}_2)^*$ as minors. The list $L(\mathbb{F}_4)$ has seven elements, and was given explicitly in 2000. For larger fields, the explicit list is not known, and there is little hope to even find useful bounds on its size.

By the Critical Theorem [5], the matroid M_C determines the supports of a linear code C. Consequently, since binary codes are determined uniquely by the support of the codewords, binary matroids are in one-to-one correspondence with binary codes. This is in sharp contrast to linear codes over larger fields, where many interesting properties are not determined by the associated matroid. An important example of such a property is the covering radius [3].

2.2   Fundamentals on Cyclic Flats

The main tool from matroid theory in this paper are the cyclic flats. We will define them using the closure and cyclic operator.

Let $M = (\rho, E)$ be a matroid. The *closure* operator $\mathrm{cl} : 2^E \to 2^E$ and *cyclic* operator $\mathrm{cyc} : 2^E \to 2^E$ are defined by

$$(i)\ \ \mathrm{cl}(X) = X \cup \{e \in E - X : \rho(X \cup e) = \rho(X)\},$$
$$(ii)\ \ \mathrm{cyc}(X) = \{e \in X : \rho(X - e) = \rho(X)\}.$$

A subset $X \subseteq E$ is a *flat* if $\mathrm{cl}(X) = X$ and a *cyclic set* if $\mathrm{cyc}(X) = X$. Therefore, X is a *cyclic flat* if

$$\rho(X \cup y) > \rho(X) \quad \text{and} \quad \rho(X - x) = \rho(X)$$

for all $y \in E - X$ and $x \in X$. The collection of flats, cyclic sets, and cyclic flats of M are denoted by $\mathcal{F}(M)$, $\mathcal{U}(M)$, and $\mathcal{Z}(M)$, respectively.

It is easy to verify, as in [2], that the closure operator induces flatness and preserves cyclicity, and that the cyclic operator induces cyclicity and preserves flatness. Thus we can write

$$\mathrm{cl} : \begin{cases} 2^E \to \mathcal{F}(M) \\ \mathcal{U}(M) \to \mathcal{Z}(M), \end{cases} \quad \text{and} \quad \mathrm{cyc} : \begin{cases} 2^E \to \mathcal{U}(M) \\ \mathcal{F}(M) \to \mathcal{Z}(M). \end{cases}$$

In particular, for any set $X \subseteq E$, we have $\mathrm{cyc}(\mathrm{cl}(X)) \in \mathcal{Z}(M)$ and $\mathrm{cl}(\mathrm{cyc}(X)) \in \mathcal{Z}(M)$. Some more fundamental properties of flats, cyclic sets, and cyclic flats are given in [2].

The cyclic flats of a linear code C of $\mathbb{F}^E$ can be described as sets $X \subseteq E$ such that

$$C|(X \cup y) \supsetneq C|X \quad \text{and} \quad C|(X - x) = C|X$$

for all $y \in E - X$ and $x \in X$. Thus, we have the following immediate proposition.

Proposition 2. *Let $M = (\rho, E)$ be a matroid. The following are equivalent:*

(i) M is the uniform matroid U_n^k
(ii) $\mathcal{Z} = \mathcal{Z}(M)$ is the two element lattice with bottom element $0_{\mathcal{Z}} = \emptyset$, top element $1_{\mathcal{Z}} = E$ and $\rho(1_{\mathcal{Z}}) = k$

Non-degeneracy of a truncated code can be observed immediately from the lattice of cyclic flats of the associated matroid, as follows.

Proposition 3. *Let C be a linear code over $\mathbb{F}^E$ and $X \subseteq E$. Then $C|X$ is non-degenerate if and only if $1_{\mathcal{Z}(M_C|X)} = X$ and $0_{\mathcal{Z}(M_C|X)} = \emptyset$.*

The minimum distances and the ranks of punctured codes can be computed from the lattice of cyclic flats via the following theorem. In [20], this was used to construct matroids and linear LRCs with prescribed parameters (n, k, d, r, δ).

Theorem 3 ([20]). *Let C be a linear code over $\mathbb{F}^E$ and $X \subseteq E$. Then, if $C|X$ is non-degenerate, it has dimension $k_X = \rho(1_{\mathcal{Z}(M_C|X)})$ and minimum distance $d_X = \eta(X) + 1 - \max\{\eta(Y) : Y \in \mathcal{Z}(M_C|X) \text{ and } Y \neq X\}$.*

Example 3. Let $M_C = (\rho_C, E = [6])$ be the matroid associated to the linear code C generated by the matrix G given in Example 1. The lattice of cyclic flats $(\mathcal{Z}, \subseteq)$ of M_C is given in Fig. 1, where the cyclic flat is given inside the node and its rank is labelled outside the node on the right.

From the lattice of cyclic flats given above, we can conclude that C is a $(6, 3, 2)$-code.

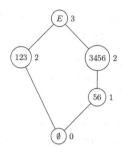

Fig. 1. Lattice of cyclic flats of M_C.

3 Sufficient Conditions for Uniformity

Our main goal is to study criteria for when $M|Y/X$ is uniform, for $X \subseteq Y \subseteq E(M)$. As a preparation, observe that for a uniform matroid $M \cong U_n^k$, the only cyclic flats are $\emptyset$ and $E(M)$, with $|E(M)| = n$ and $\rho(E) = k$.

Our first interest is in the case when X and Y are themselves cyclic flats. Then we have a straightforward characterisation of cyclic flats in $M|Y/X$, via the following two lemmas:

Lemma 2. *Let M be a matroid, and let $X \subseteq Y \subseteq E(M)$ be two sets with $Y \in \mathcal{F}(M)$. Then $\mathcal{F}(M|Y/X) = \{F \subseteq Y - X, F \cup X \in \mathcal{F}(M)\}$.*

Proof. A set S is flat in $M|Y/X$ precisely if $\rho(S \cup X \cup i) > \rho(S \cup X)$ for all $i \in (Y - X) - S$. Since Y is flat, the inequality $\rho(S \cup X \cup i) > \rho(S \cup X)$ will hold for all $i \in \bar{Y}$ regardless of S. Thus, S is flat in $M|Y/X$ if and only if $S \cup X$ is flat in M.

Lemma 3. *Let M be a matroid, and let $X \subseteq Y \subseteq E(M)$ be two sets with $X \in \mathcal{U}(M)$. Then $\mathcal{U}(M|Y/X) = \{U \subseteq Y - X, U \cup X \in \mathcal{U}(M)\}$.*

This is the dual statement, and thus an immediate consequence, of Lemma 2. We write out the proof explicitly only for illustration.

Proof. A set S is cyclic in $M|Y/X$ precisely if $\rho((S \cup X) - i) = \rho(S \cup X)$ for all $i \in S$. For $i \in X$, this will hold regardless of S, since X is cyclic. Thus, S is cyclic in $M|Y/X$ if and only if $S \cup X$ is cyclic in M.

The previous lemmas give the following immediate corollary:

Corollary 1. *Let $M = (E, \rho)$ be a matroid, and let $X \subseteq Y \subseteq E(M)$ be two sets with $X \in \mathcal{U}(M)$ and $Y \in \mathcal{F}(M)$. Then $\mathcal{Z}(M|Y/X) = \{Z \subseteq Y - X, Z \cup X \in \mathcal{Z}(M)\}$, with the rank function $\rho_{|Y/X}(Z) = \rho(Z \cup X) - \rho(X)$.*

As a consequence, we get a sufficient condition for uniformity of minors, that only depends on the Hasse diagram of $\mathcal{Z}(M)$. Recall that v is said to *cover* u in a poset P if $u < v$ and there is no w with $u < w < v$. If this is the case, then we write $u \lessdot_P v$. This is equivalent to (u, v) being an upwards directed edge in the Hasse diagram of P.

Theorem 4. *Let X and Y be two cyclic flats in M with $X \lessdot_{\mathcal{Z}(M)} Y$. Let $n = |Y| - |X|$ and $k = \rho(Y) - \rho(X)$. Then $M|Y/X \cong U_n^k$.*

Proof. By Corrolary 1, we have $\mathcal{Z}(M|Y/X) = \{\emptyset, Y - X\}$ as there are no cyclic flats with $X \subset Z \subset Y$. Again by Corrolary 1, we have $\rho(Y - X) = \rho(Y) - \rho(X) = k$. Thus, by Proposition 2, $M|Y/X$ is the uniform matroid U_n^k. $\square$

Corollary 2. *Let M be a matroid that contains no U_n^k minors. Then, for every edge $X \lessdot_{\mathcal{Z}(M)} Y$ in the Hasse diagram of $\mathcal{Z}(M)$, we have $\rho(Y) - \rho(X) < k$ or $\eta(Y) - \eta(X) < n - k$.*

Proof. Assume for a contradiction that $X \lessdot_{\mathcal{Z}(M)} Y$ has $\rho(Y) - \rho(X) = k' \geq k$ or $\eta(Y) - \eta(X) = n' - k' \geq n - k$. Then by Theorem 4, $M|Y/X \cong U_{n'}^{k'}$, and so contains U_n^k as a minor by Lemma 1. $\square$

Now, we are going to need formulas for how to compute the lattice operators in $\mathcal{Z}(M|Y/X)$ in terms of the corresponding operators in $\mathcal{Z}(M)$. These can be derived from corresponding formulas for the closure and cyclic operator. To derive these, we will need to generalise Corollary 1 to the setting where the restriction and contraction are not necessarily performed at cyclic flats.

Theorem 5. *For $X \subseteq Y \subseteq E$, we have*

1. $\mathcal{Z}(M|Y) = \{\operatorname{cyc}(Z \cap Y) : Z \in \mathcal{Z}(M)\}$
2. $\mathcal{Z}(M/X) = \{\operatorname{cl}(X \cup Z) - X : Z \in \mathcal{Z}(M)\}$
3. $\mathcal{Z}(M|Y/X) = \left\{\operatorname{cl}\left(X \cup \operatorname{cyc}(Z \cap Y)\right) \cap (Y - X) : Z \in \mathcal{Z}(M)\right\} = \left\{\operatorname{cyc}\left(\operatorname{cl}(X \cup Z) \cap Y\right) - X : Z \in \mathcal{Z}(M)\right\}$

Proof 1. First, observe that the cyclic operator in $M|Y$ is the same as that in M, and that the flats in $M|Y$ are $\{F \cap Y : F \in \mathcal{F}(M)\}$. Thus we have

$$\mathcal{Z}(M|Y) = \{\operatorname{cyc}(F \cap Y) : F \in \mathcal{F}(M)\} \supseteq \{\operatorname{cyc}(Z \cap Y) : Z \in \mathcal{Z}(M)\}.$$

On the other hand, let $A \in \mathcal{Z}(M|Y)$, so $\operatorname{cyc}(A) = A$ and $\operatorname{cl}(A) \cap Y = A$. But the closure operator preserves cyclicity, so $\operatorname{cl}(A) \in \mathcal{Z}(M)$. We then observe that

$$A = \operatorname{cyc}(\operatorname{cl}(A) \cap Y) \in \{\operatorname{cyc}(Z \cap Y) : Z \in \mathcal{Z}(M)\}.$$

This proves the reverse inclusion

$$\mathcal{Z}(M|Y) = \{\operatorname{cyc}(F \cap Y) : F \in \mathcal{F}(M)\} \subseteq \{\operatorname{cyc}(Z \cap Y) : Z \in \mathcal{Z}(M)\}.$$

2. This is the dual statement of 1., and so follows immediately by applying 1. to the matroid $M^*|\bar{X}/\bar{Y}$.

3. We first apply 2. and then 1. to the restricted matroid $M|Y$, and get

$$\mathcal{Z}(M|Y/X) = \{\operatorname{cl}_{|Y}(X \cup \operatorname{cyc}(Z \cap Y)) - X : Z \in \mathcal{Z}(M)\}.$$

But if $T \subseteq Y$, then $\mathrm{cl}_{|Y}(T) = \mathrm{cl}(T) \cap Y$. Then,

$$\mathcal{Z}(M|Y/X) = \{\mathrm{cl}(X \cup \mathrm{cyc}(Z \cap Y)) \cap (Y - X) : Z \in \mathcal{Z}(M)\}$$

For the second equality of 3. we need to study the operator $\mathrm{cyc}_{/X}$. Suppose $T \subseteq E - X$. Using duality and the formula for $cl_{|\bar{X}}$, we find that

$$\mathrm{cyc}_{/X}(T) = \mathrm{cyc}(X \cup T) - X.$$

Now we are ready to prove the last equality. Applying first 1. and then 2. to the contracted matroid M/X, we get

$$\mathcal{Z}(M|Y/X) = \{\mathrm{cyc}_{/X}((\mathrm{cl}(X \cup Z) - X) \cap Y) : Z \in \mathcal{Z}(M)\}$$

Applying the formula for $\mathrm{cyc}_{/X}$, we obtain

$$\begin{aligned}
\mathcal{Z}(M|Y/X) &= \{\mathrm{cyc}((\mathrm{cl}(X \cup Z) \cap Y - X) \cup X) - X : Z \in \mathcal{Z}(M)\} \\
&= \{\mathrm{cyc}(\mathrm{cl}(X \cup Z) \cap Y) - X : Z \in \mathcal{Z}(M)\},
\end{aligned}$$

where the last equality follows as $X \subseteq cl(X \cup Z)$. This concludes the proof.

4 Criteria for Uniformity via Cyclic Flats

We can use Theorem 5 to find conditions for a minor to be isomorphic to a uniform matroid. The idea is to detect when $\mathcal{Z}(M|Y/X)$ as calculated in Theorem 5 is precisely $\{\emptyset, Y - X\}$. Using this, we will be able to find some conditions on the sets X and Y, as well as on the matroid itself.

4.1 Minors Given by Restriction or Contraction only

We will begin by considering a simpler case when the minor is the result of a restriction only, $i.e.$, when the minor is given by $M|Y$. So, let $M = (E, \rho)$ be a matroid and Y an arbitrary subset of E. We can use Corollary 1 to restrict the amount of information we need to consider. Indeed, by properties of minors, we have

$$M|Y = M|\mathrm{cl}(Y) \setminus (\mathrm{cl}(Y) - Y). \tag{1}$$

Then, Theorem 5 states that the cyclic flats of $M|Y$ depend only on the cyclic flats of $M|\mathrm{cl}(Y)$. Furthermore, according to Corollary 1 the cyclic flats of $M|\mathrm{cl}(Y)$ are exactly the cyclic flats of M contained in $\mathrm{cl}(Y)$. Hence, we can restrict the study to the case when $M = (E, \rho)$ is a matroid and Y is a subset of full rank. Define $k := \rho(Y)$ and $n := |Y|$. With this setup, we obtain the following theorem.

Theorem 6. Let $M = (E, \rho)$ be a matroid and Y a subset of full rank. $M|Y$ is isomorphic to the uniform matroid U_n^k if and only if either Y is a basis of M or the following two conditions are satisfied:

1. Y *is a cyclic set of* M.
2. *For all* $Z \in \mathcal{Z}(M)$ *with* $\rho(Z) < k$, *we have that* $Z \cap Y$ *is independent in* M.

Before stating the proof, we will need one useful lemma about the properties of the closure and cyclic operator.

Lemma 4. *Let* $M = (E, \rho)$ *be a matroid and* $Y \subseteq E$. *Then*

1. $\mathrm{cl}(\mathrm{cyc}(Y)) \cap Y = \mathrm{cyc}(Y)$.
2. $\mathrm{cyc}(\mathrm{cl}(Y)) \cup Y = \mathrm{cl}(Y)$.

The proof of Lemma 4 is straightfrorward from the definition of the operators together with the submodularity of the rank function. Details of the proof can be found in [2]. We now present the proof of Theorem 6.

Proof. We know that $M|Y \cong U_n^k$ if and only if $\mathcal{Z}(M|Y) = \{\emptyset, Y\}$. On the other hand, we know by Theorem 5, that $\mathcal{Z}(M|Y) = \{\mathrm{cyc}(Z \cap Y) : Z \in \mathcal{Z}(M)\}$. Then we have

$$M|Y \cong U_n^k \text{ if and only if } \mathrm{cyc}(Z \cap Y) \in \{\emptyset, Y\} \text{ for all } Z \in \mathcal{Z}(M).$$

Now consider the cyclic flat $Z_Y := \mathrm{cl}(\mathrm{cyc}(Y))$. Using Lemma 4, we have $\mathrm{cyc}(Z_Y \cap Y) = \mathrm{cyc}(Y)$. Two cases can occur. If $\mathrm{cyc}(Y) = \emptyset$ then Y was a basis of M and we end up with a minor isomorphic to U_k^k. If not, then $\mathrm{cyc}(Y)$ must be equal to Y. So, we have that $Y \in \mathcal{Z}(M|Y)$ if and only if Y is a cyclic set and we obtain the first condition. Since Y already has full rank, there is only one cyclic flat that contains Y, namely $\mathrm{cl}(\mathrm{cyc}(Y)) = E$. Therefore, for every other cyclic flat Z, i.e., for all Z with $\rho(Z) < k$, we have $\mathrm{cyc}(Z \cap Y) \subseteq Z \cap Y \neq Y$. But, by Theorem 5, $\mathrm{cyc}(Z \cap Y)$ is a cyclic flat of $M|Y$. Thus, for all $Z \in \mathcal{Z}(M)$ with $\rho(Z) < k$, we have $\mathrm{cyc}(Z \cap Y) = \emptyset$, or equivalently, $Z \cap Y$ is independent. Notice that, combined with the first condition, this implies immediately that $\mathcal{Z}(M|Y) = \{\emptyset, Y\}$. This concludes the proof.

Corollary 3. *Under the above assumptions, if* $M|Y$ *is isomorphic to* U_n^k *then the ground set* E *must be a cyclic flat, i.e.,* $E \in \mathcal{Z}(M)$.

Example 4. We will look at the matroid M_C arising from the binary matrix in Example 1. Since it is a binary matroid, we cannot find a minor isomorphic to U_4^2. Using the lattice of cyclic flats of Fig. 1, we can find a minor isomorphic to U_3^2 if we look at, e.g., $M_C|\{1, 2, 3\}$. We can also find a minor isomorphic to U_4^3. To this end, we look at $M_C|\{1, 2, 4, 5\}$. The set $\{1, 2, 4, 5\}$ is indeed a cyclic set because $\mathrm{cyc}(\{1, 2, 4, 5\}) = \{1, 2, 4, 5\}$. Furthermore, there is no cyclic flat properly contained in $\{1, 2, 4, 5\}$, and every intersection of a cyclic flat different from E with the set $\{1, 2, 4, 5\}$ gives us an independent set. Thus, the conditions from the previous theorem are met and we get a minor isomorphic to U_4^3.

Now we can do the same for M/X and use duality to get back to the restriction case. By minor properties, we have

$$M/X = M/\mathrm{cyc}(X)/(X - \mathrm{cyc}(X)). \tag{2}$$

Then, Corollary 1 states that the cyclic flats of $M/\mathrm{cyc}(X)$ are the cyclic flats of M that contain $\mathrm{cyc}(X)$. Thus, we will consider a matroid $M = (E, \rho)$ and X an independent subset of E. Define $k := \rho(E) - \rho(X)$ and $n := |E - X|$. Then, we have the following dual statement of Theorem 6.

Theorem 7. *Let $M = (E, \rho)$ be a matroid and X an independent subset of E. M/X is isomorphic to the uniform matroid U_n^k if and only if either X is a basis of M or the following two conditions are satisfied:*

1. *X is a flat of M.*
2. *For all $Z \in \mathcal{Z}(M)$ with $\rho(Z) > 0$, we have $\mathrm{cl}(X \cup Z) = E$.*

Corollary 4. *Under the above assumptions, if M/X is isomorphic to U_n^k then the empty set must be a cyclic flat, i.e., $\emptyset \in \mathcal{Z}(M)$.*

4.2 Minors Given by both Restriction and Contraction

This part combines the two previous situations into a more general statement. We will see that, when we allow both a restriction and a contraction to occur, we lose some conditions on the matroid that are then replaced by conditions on the sets used in the minor.

Let $M = (E, \rho)$ be a matroid and $X \subset Y \subseteq E$ two sets. Combining minor properties (1) and (2) and Corollary 1, it is sufficient to only consider the cyclic flats between $\mathrm{cyc}(X)$ and $\mathrm{cl}(Y)$. In addition, we want to avoid some known cases, namely when Y is a basis (we will obtain U_k^k) and when X has full rank (we will obtain U_0^0). Define $k := \rho(E) - \rho(X)$ and $n := |Y - X|$. We get the following theorem.

Theorem 8. *Let $M = (E, \rho)$ be a matroid and $X \subset Y \subseteq E$ two sets such that Y is a dependent full-rank set and X is an independent set with $\rho(X) < \rho(E)$. The minor $M|Y/X$ is isomorphic to a uniform matroid U_n^k if and only if*

1. *$\mathrm{cl}(X) \cap Y = X$,*
2. *$Y - X \subseteq \mathrm{cyc}(Y)$,*
3. *for all $Z \in \mathcal{Z}(M)$ either $Z \cap Y$ is independent or $\mathrm{cl}(X \cup \mathrm{cyc}(Z \cap Y)) = E$.*

Proof. Using Theorem 5, we have that $M|Y/X \cong U_n^k$ if and only if

$$\mathrm{cl}(X \cup \mathrm{cyc}(Z \cap Y)) \cap (Y - X) \in \{\emptyset, Y - X\} \text{ for all } Z \in \mathcal{Z}(M).$$

In particular, it holds for $Z = 0_{\mathcal{Z}}$. Let $Z'_0 := \mathrm{cl}(X \cup \mathrm{cyc}(0_{\mathcal{Z}} \cap Y))$. Using the properties of the closure, we have

$$\mathrm{cl}(X) \subseteq Z'_0 \subseteq \mathrm{cl}(X \cup 0_{\mathcal{Z}}) = \mathrm{cl}(X).$$

Thus, we have a chain of equalities and, in particular, $\rho(Z'_0) < \rho(E)$. This means that $Z'_0 = \emptyset$ in order to have $\emptyset \in \mathcal{Z}(M|Y/X)$. But, since $Z'_0 = \mathrm{cl}(X)$ then $Z'_0 = \emptyset$ is equivalent to $\mathrm{cl}(X) \cap Y = X$ and Condition 1 is proved.

Now, consider the cyclic flat $Z_Y := \text{cl}(\text{cyc}(Y))$. First, using again Lemma 4, we have $\text{cyc}(Z_Y \cap Y) = \text{cyc}(Y)$. Since Y is a dependent subset, $\text{cyc}(Y) \neq \emptyset$ and thus $X \subsetneq X \cup \text{cyc}(Y)$. Then, the closure cannot be contained in X and we must have

$$\text{cl}(X \cup \text{cyc}(Y)) \cap (Y - X) = Y - X.$$

Define $Z'_Y := \text{cl}(X \cup \text{cyc}(Y))$. The above equality means that $Y - X \subseteq Z'_Y$. On the other hand, we have that $X \subseteq Z'_Y$. Then, $Y \subseteq Z'_Y$ and $Z'_Y = E$. In particular, we must have $Y - X \subseteq \text{cyc}(Y)$. Indeed, assume by contradiction that there exists $a \in Y - X$ and $a \notin \text{cyc}(Y)$. Then, by definition of the cyclic operator, $\rho(Y - a) < \rho(Y)$. But since $a \notin \text{cyc}(Y)$ and $a \notin X$ then $X \cup \text{cyc}(Y) \subseteq Y - a$. This implies that $\rho(X \cup \text{cyc}(Y)) \leq \rho(Y - a) < \rho(Y)$ which is a contradiction. The condition $Y - X \subseteq \text{cyc}(Y)$ is also sufficient to guarantee that $Y - X \in \mathcal{Z}(M|Y/X)$. This proves Condition 2.

Finally, for every other cyclic flat of M, $\text{cl}(X \cup \text{cyc}(Z \cap Y)) \cap (Y - X) = \emptyset$ if and only if $\text{cyc}(Z \cap Y) \subseteq X$. But since X is independent, this is equivalent to $Z \cap Y$ being independent. On the other hand, $\text{cl}(X \cup \text{cyc}(Z \cap Y)) \cap (Y - X) = Y - X$ if and only if $\text{cl}(X \cup \text{cyc}(Z \cap Y)) = E$. This concludes the proof.

5   Stuctural Properties of Binary LRCs

From a practical viewpoint, storage systems over alphabets of bounded size are of special interest. The field size is important both because it governs the complexity of the computations involved in repair and retrieval, and because it restricts the size of the data items stored. We are therefore interested in understanding the matroidal structure of LRCs that are linearly representable over the finite field $\mathbb{F}_q$, where q is small.

Assuming the MDS conjecture [15], a matroid M that is linearly representable over $\mathbb{F}_q$ must avoid U_{q+2}^k as a minor, for $k = 2$, $4 \leq k \leq q - 2$, and $k = q$. If q is odd, M must also avoid U_{q+2}^3 and U_{q+2}^{q-1} minors. The MDS conjecture is widely believed to be true, and is proven when q is prime [1].

By Corollary 2, matroids avoiding U_n^k minors have a rather special structure in their lattice of cyclic flats. In particular, it tells us that whenever $X <_{\mathcal{Z}} Y$, we cannot simultaneously have $\rho(Y) - \rho(X) > k$ and $\eta(Y) - \eta(X) > n - k$. This observation can be exploited to bound locality parameters of an LRC in terms of the field size.

Of special interest are binary storage codes that are linear over $\mathbb{F}_2$. It is known that a matroid is representable over $\mathbb{F}_2$ if and only if it avoids U_4^2 as a minor. In this case, Corollary 2 tells us that we cannot simultaneously have $\rho(Y) - \rho(X) > 1$ and $\eta(Y) - \eta(X) > 1$. On the other hand, we know by Theorem 3.2 in [2] or by direct calculation, that we always have $\rho(Y) - \rho(X) \geq 1$ and $\eta(Y) - \eta(X) \geq 1$. Thus, if M is representable over $\mathbb{F}_2$, then every edge $X <_{\mathcal{Z}} Y$ in the Hasse diagram of $\mathcal{Z}(M)$ satisfies exactly one of the following:

(i) $\rho(Y) - \rho(X) = l > 1$. We call such an edge a *rank edge*, and label it $\rho = l$. Such an edge corresponds to a U_{l+1}^l minor in M.

(ii) $\eta(Y) - \eta(X) = l > 1$. We call such an edge a *nullity edge*, and label it $\eta = l$. Such an edge corresponds to a U^1_{l+1} minor in M.

(iii) $\rho(Y) - \rho(X) = 1$ and $\eta(Y) - \eta(X) = 1$. We call such an edge an *elementary edge*. Such an edge corresponds to a U^1_2 minor in M.

As an example, the matroid from Examples 1 and 2 gets an edge labelling as illustrated in Fig. 2:

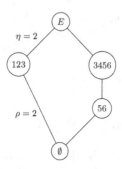

Fig. 2. Lattice of cyclic flats of M_C.

It is clear that this representation is enough to reconstruct the so-called *configuration* of the matroid, *i.e.*, the isomorphism type of the lattice of cyclic flats, together with the cardinality and rank of the cyclic flats. However, this data does not uniquely determine the matroid, as is shown in [7].

As is proven in [20], the configuration of a representable matroid determines the minimum distance of the corresponding code, via the formula

$$d_C = \eta(E) + 1 - \max_{Z \in \mathcal{Z}(M) - \{E\}} \eta(Z).$$

In particular, for a binary code C with $d_C > 2$, all edges $Z \lessdot E$ on the "top level" of $\mathcal{Z}(M_C)$ must be nullity edges. Moreover, the minimum distance is then one higher than the smallest label of a top level edge in $\mathcal{Z}(M_C)$.

Recall that an (n, k, d)-storage code is said to have locality (r, δ) if every storage node $i \in [n]$ is contained in a set X with $|X| \leq r + \delta - 1$ and $d_X \geq \delta$. This is equivalent to that every node $i \in [n]$ is contained in a set X with $s_X := |X| - d_X + 1 \leq r$ and $d_X \geq \delta$. Now, notice that $s_X = \rho(X) + \max_{Z < _Z X} \eta(Z)$, and so s_X does not increase when replacing X by its closure, which is a cyclic flat as X is cyclic. This gives a lattice-theoretic description of $\mathcal{Z}(M_C)$, when C is a binary code with (r, δ)-locality.

Surprisingly, the descriptions are qualitatively different depending on whether $\delta = 2$ or $\delta > 2$, *i.e.*, whether one or more erasures can be corrected locally. This is in sharp contrast to the case when the field size is ignored, as in [20]. We conclude this paper by formulating the locality criteria for binary storage codes in terms of lattices of cyclic flats. Exploiting this description to obtain quantitative bounds on the parameters (n, k, d, r, δ) is left for future research. Such bounds are also likely to suggest explicit constructions of extremal LCFs satisfying the conditions of Theorem 9.

Theorem 9. *Let $d > 2$ and let C be a linear (n, k, d, r, δ)-LRC over $\mathbb{F}_2$. Then $\mathcal{Z} = \mathcal{Z}(M_C)$ satisfies the following:*

1. $\emptyset$ *and $[n]$ are cyclic flats.*
2. *Every covering relation $Z \lessdot_{\mathcal{Z}} [n]$ is a nullity edge labeled with a number $\geq d-1$.*
3. *If $\delta = 2$, then for every $i \in [n]$, there is $X \in \mathcal{Z}$ with $i \in X$ such that $\rho(X) \leq r$.*
4. *If $\delta > 2$, then for every $i \in [n]$, there is $X \in \mathcal{Z}$ with $i \in X$ such that*
   (i) Every covering relation $Y \lessdot_{\mathcal{Z}} X$ is a nullity edge labeled with a number $\geq \delta - 1$.
   (ii) Every cyclic flat Y with $Y \lessdot_{\mathcal{Z}} X$ has size $\leq r - 1$.

6   Conclusions and Future Work

We have studied the lattice of cyclic flats of matroids, with a special emphasis on identifying uniform minors, and with applications to locally repairable codes over small fields. Necessary and sufficient criteria for a specified minor $M|Y/X$ to be uniform are derived in general, and in the special case of U_4^2-minors, necessary global criteria for U_4^2-avoidance are given. Finally, it is shown how these criteria dictate the structure of binary storage codes with prescribed locality parameters. Future work include translating these structural results to quantitative parameter bounds. Similar arguments are likely to be applicable when studying storage codes over other small fields, although new methods would then be needed to identify other minors than uniform ones.

Acknowledgment. The authors gratefully acknowledge the financial support from the Academy of Finland (grants #276031 and #303819).

References

1. Ball, S.: On sets of vectors of a finite vector space in which every subset of basis size is a basis. J. Eur. Math. Soc. **14**, 733–748 (2012)
2. Bonin, J.E., de Mier, A.: The lattice of cyclic flats of a matroid. Ann. Comb. **12**, 155–170 (2008)
3. Britz, T., Rutherford, C.G.: Covering radii are not matroid invariants. Discrete Math. **296**, 117–120 (2005)
4. Cadambe, V., Mazumdar, A.: An upper bound on the size of locally recoverable codes. In: International Symposium on Network Coding, pp. 1–5 (2013)
5. Crapo, H., Rota, G.C.: On the Foundations of Combinatorial Theory: Combinatorial Geometries, Preliminary edition edn. MIT Press, Cambridge (1970)
6. Dimakis, A., Godfrey, P.B., Wu, Y., Wainwright, M.J., Ramchandran, K.: Network coding for distributed storage systems. IEEE Trans. Inf. Theory **56**(9), 4539–4551 (2010)

7. Eberhardt, J.: Computing the tutte polynomial of a matroid from its lattice of cyclic flats. Electron. J. Comb. **21**, 12 (2014)
8. Ernvall, T., Westerbäck, T., Freij-Hollanti, R., Hollanti, C.: Constructions and properties of linear locally repairable codes. IEEE Trans. Inf. Theory **62**, 5296–5315 (2016)
9. Freij-Hollanti, R., Hollanti, C., Westerbäck, T.: Matroid theory and storage codes: bounds and constructions (2017), arXiv: 1704.0400
10. Geelen, J., Gerards, B., Whittle, G.: Solving Rota's conjecture. Not. Am. Math. Soc. **61**, 736–743 (2014)
11. Huang, C., Simitci, H., Xu, Y., Ogus, A., Calder, B., Gopalan, P., Li, J., Yekhanin, S.: Erasure coding in Windows Azure storage. In: Proceedings of the USENIX Annual Technical Conference, pp. 15–26 (2012)
12. Huang, P., Yaakobi, E., Uchikawa, H., Siegel, P.H.: Binary linear locally repairable codes. IEEE Trans. Inf. Theory **62**, 5296–5315 (2016)
13. Papailiopoulos, D., Dimakis, A.: Locally repairable codes. In: International Symposium on Information Theory, pp. 2771–2775. IEEE (2012)
14. Sathiamoorthy, M., Asteris, M., Papailiopoulos, D., Dimakis, A.G., Vadali, R., Chen, S., Borthakur, D.: Xoring elephants: novel erasure codes for Big Data. Proc. VLDB. **6**, 325–336 (2013)
15. Segre, B.: Curve razionali normali e k-archi negli spazi finiti. Annali di Matematica **39**, 357–359 (1955)
16. Silberstein, N., Zeh, A.: Optimal binary locally repairable codes via anticodes (2015), arXiv: 1501.07114v1
17. Tamo, I., Barg, A., Frolov, A.: Bounds on the parameters of locally recoverable codes. IEEE Trans. Inf. Theory **62**(6), 3070–3083 (2016)
18. Tamo, I., Papailiopoulos, D., Dimakis, A.: Optimal locally repairable codes and connections to matroid theory. IEEE Trans. Inf. Theory **62**, 6661–6671 (2016)
19. Tutte, W.: A homotopy theorem for matroids, I, II. Trans. Am. Math. Soc. **88**, 148–178 (1958)
20. Westerbäck, T., Freij-Hollanti, R., Ernvall, T., Hollanti, C.: On the combinatorics of locally repairable codes via matroid theory. IEEE Trans. Inf. Theory **62**, 5296–5315 (2016)

Absorbing Set Analysis of Codes
from Affine Planes

Kathryn Haymaker[(✉)]

Villanova University, 800 E. Lancaster Ave, Villanova, PA 19085, USA
`kathryn.haymaker@villanova.edu`

Abstract. We examine the presence of absorbing sets, fully absorbing sets, and elementary absorbing sets in low-density parity-check (LDPC) codes arising from certain classes of finite geometries. In particular, we analyze the absorbing set spectra of LDPC codes from finite Euclidean planes. For some parameters, we classify the absorbing sets present and give exact counts on their multiplicities.

Keywords: Absorbing sets · Fully absorbing sets · Finite geometry LDPC codes · Euclidean geometries · Finite planes

1 Introduction

Codes based on low-density parity-check (LDPC) matrices have been in the forefront of research in coding theory due to their low-complexity efficient decoders and near capacity performance at long block lengths. A geometric approach to constructing these codes was given in [13,14]. The resulting finite geometry LDPC (FG-LDPC) codes are based on the points and lines of finite Euclidean and projective geometries, the structure of which can be used to prove parameters of the code. Stopping sets and pseudocodewords of FG-LDPC codes were studied in [11,22]. Trapping sets of FG-LDPC codes were studied in [5,16], while absorbing sets of LDPC codes from finite planes were analyzed in [17]. The research in this paper is inspired by the results in [17], particularly the authors' discussion describing the difficulty of dealing with a subgeometry of EG(2, q). In this paper we analyze the absorbing set structures of the EG(2, q) classes of FG-LDPC codes, where we consider an incidence matrix of the entire Euclidean geometry instead of a subgeometry.

Research into the error floor phenomenon that occurs in the bit error rate (BER) curves for structured families of LDPC codes under message-passing iterative decoders has shown that certain graphical substructures contribute to the persistence of the error floor. Specifically, structural properties of the Tanner graph of the code are tied to pseudocodewords, absorbing sets, trapping sets, and stopping sets [4,7,12,18]. While trapping sets depend on the decoder, absorbing sets are a combinatorial substructure of the Tanner graph that are independent of the channel and in certain cases are also stable under bit-flipping decoding. For EG(2, q) classes of FG-LDPC codes, the error performance of the codes is

© Springer International Publishing AG 2017
A.I. Barbero et al. (Eds.): ICMCTA 2017, LNCS 10495, pp. 154–162, 2017.
DOI: 10.1007/978-3-319-66278-7_14

good under the sum-product and other iterative decoding algorithms [14]. This paper proves theoretical results on the non-existence of small absorbing sets in these codes, which is consistent with the decoding simulations of EG codes [13,14].

This paper is organized as follows. In Sect. 2, we give the necessary background and notation for absorbing sets, and define the classes of finite geometry codes that we consider in this paper. In Sect. 3 we present results for codes based on finite Euclidean geometries. Section 4 concludes the paper.

2 Preliminaries

In [14], Kou, Lin, and Fossorier describe families of cyclic or quasi-cyclic LDPC codes with parity-check matrices determined by the incidence structure of finite Euclidean and projective geometries. The Euclidean geometry constructions involve defining a subgeometry without the origin point, and creating incidence matrices of points and lines for these families of subgeometries. These matrices alone can be used as parity-check matrices of LDPC codes; they can also be extended by a column splitting process that results in a code of longer length. The cyclic or quasi-cyclic structure of these codes is an advantage, however the parity-check matrices can have extra redundancy in the number of rows. Higher redundancy in the parity-check matrices results in increased decoding complexity, but it also has a positive effect on the decoding performance of the codes [8,11,19].

We recall the basic properties of finite projective and Euclidean geometries, starting with the definitions of affine and projective space [3].

Definition 1. *A **linear space** is a collection of points and lines such that any line has at least two points, and two points are on precisely one line. A hyperplane of a linear space is a maximal proper subspace. A **projective plane** is a linear space in which any two lines meet, and there exists a set of four points, no three of which are collinear. (A projective plane has dimension 2.) A **projective space** is a linear space in which any two-dimensional subspace is a projective plane. An **affine space** is a projective space with one hyperplane removed.*

Like the Euclidean space $\mathbb{R}^n$, the set of points formed by m-tuples with entries from the finite field $\mathbb{F}_q$ forms an affine space, called a finite Euclidean geometry. A finite Euclidean geometry satisfies the axioms listed in Definition 1, and comprises the family of finite geometries that we will consider in this paper. In the case of $m = 2$, lines are sets of points (x, y) satisfying an equation $y = mx + b$ or $x = a$, where $m, b, a \in \mathbb{F}_q$. The m-dimensional finite Euclidean geometry $\mathrm{EG}(m, p^s)$ has the following parameters. There are p^{ms} points and the number of lines is

$$\frac{p^{s(m-1)}(p^{ms} - 1)}{p^s - 1}.$$

Each line contains p^s points and each point is on $\dfrac{p^{ms} - 1}{p^s - 1}$ lines. Any two points have exactly one line in common and any two lines either have one point in common or are parallel (i.e., have no points in common).

A μ-dimensional subspace of a finite geometry is called a μ-flat.

An LDPC code can be formed from an m-dimensional finite geometry by taking the incidence matrix of μ_1-flats and μ_2-flats, where $0 \le \mu_1 < \mu_2 \le m$. Taking $\mu_1 = 0$ and $\mu_2 = 1$ gives the incidence matrix of points and lines in a finite geometry, which is one of the constructions presented in [14]. However, in [14], the authors eliminate the origin point and all lines incident to it in the Euclidean geometry before creating the incidence matrix, because it results in a cyclic or quasi-cyclic code. In this paper, we include the origin point, in order to use the full geometric structure for absorbing set analysis (see [17] for the subgeometry viewpoint). Keeping the origin point also allows us to view the finite plane codes as a special case of the Gallager-like codes constructed in [20].

We use the notation $\mathcal{H}_{\mathrm{EG}}(m, p^s)$ to denote a parity-check matrix formed as the incidence matrix of points and lines in $\mathrm{EG}(m, p^s)$. Points correspond to columns in $\mathcal{H}_{\mathrm{EG}}(m, p^s)$, and lines in the geometry correspond to rows in the parity-check matrix. Notice that $\mathcal{H}_{\mathrm{EG}}(m, p^s)$ has entries from $\mathbb{F}_2$. The code defined by $\mathcal{H}_{\mathrm{EG}}(m, p^s)$ is denoted $\mathcal{C}_{\mathrm{EG}}(m, p^s)$. When we refer to the EG code formed by puncturing the geometry on the origin point, we denote that subgeometry code by $\mathcal{C}^*_{\mathrm{EG}}(m, p^s)$. In this paper we will focus on codes of form $\mathcal{C}_{\mathrm{EG}}(2, q)$. The minimum distance of such codes is given by: $d \ge q + 2$. (To see this, consider a point in the geometry and the $q + 1$ lines containing it.)

Let $G = (V, W; E)$ be a bipartite Tanner graph corresponding to an LDPC code, where V and W denote the sets of variable nodes and check nodes, respectively, and E is the set of edges. For a subset S of V, let $G_S = (S, W_S; E_S)$ denote the subgraph induced by S in G. In the induced subgraph, W_S is the set of constraint neighbors of S, and E_S is the set of edges between S and W_S.

Definition 2 (Dolecek et al., [7]). *An (a, b) **absorbing set** is a subset S of V where $|S| = a$, and there are b odd degree vertices in W_S, with the property that every vertex $v \in S$ has more even-degree than odd-degree neighbors in G_S. Let $\mathcal{O}(S)$ (resp., $\mathcal{E}(S)$) denote the vertices in W_S with odd degree (even degree) in G_S. If in addition, all variable nodes in $V \backslash S$ have strictly more neighbors in $W \backslash \mathcal{O}(S)$ than in $\mathcal{O}(S)$, then S is a **fully absorbing set**. An **elementary absorbing set** is an absorbing set in which all vertices in W_S have degree one or two in G_S [1].*

Although an absorbing set is a special subgraph of the Tanner graph of $\mathcal{H}$, for brevity we will refer to an absorbing set of $\mathcal{H}$.

Constraint nodes in the Tanner graph of $\mathcal{H}_{\mathrm{EG}}(2, q)$ correspond to lines in the geometry, so we say a line is *odd* if its corresponding constraint node is in $\mathcal{O}(S)$ with respect to some fixed set S. An *even line* has a constraint node in $\mathcal{E}(S)$.

The *smallest* (a, b) absorbing set of the Tanner graph of a code is the size of the smallest a, and the corresponding smallest b for that given a, for which an absorbing set exists. Small absorbing sets may provide information that can be used to design low error-floor decoders [15], and the structure inherent in FG-LDPC codes allows for enumeration and explicit description of these structures.

There are a wide variety of finite geometry codes, including codes for which a parity-check matrix is the incidence matrix of lines and points (the transpose of $\mathcal{H}$), along with many other variations [14]. Creative constructions of codes using other finite incidence structures such as generalized quadrangles and latin squares have also been studied extensively [10,11,21].

We now define an important geometry substructure that has been used to find absorbing sets in other work [17].

Definition 3. *A (k,d)-arc in a finite affine or projective plane is a set of k points, such that d of them are collinear, and any collection of $d+1$ are not collinear. Often $(k,2)$-arcs are simply referred to as k-**arcs**.*

Arcs can be defined for higher-dimensional finite geometries as well; these geometric structures are the subject of ongoing research in geometry [2,9].

Remark 1. The following rephrases Lemma 1 of [6] in terms of codes from finite Euclidean planes, $\mathcal{C}_{\mathrm{EG}}(2,q)$.

Lemma 1 (Dolecek, [6]). *The parameters (a^*, b^*) of the smallest absorbing sets for $\mathcal{H}_{\mathrm{EG}}(2,q)$ satisfy $a^* \geq 2 + \left\lfloor \frac{q}{2} \right\rfloor$, and $b^* \geq a^* \cdot \left\lfloor \frac{q}{2} \right\rfloor$.*

3 Finite Euclidean LDPC Codes

Our first result concerns small absorbing sets in finite Euclidean geometries.

Proposition 1. *The only nontrivial EG-LDPC code that has a $(3,3)$ absorbing set is the code with parity-check matrix $\mathcal{H}_{\mathrm{EG}}(2,2)$. There are four such sets in $\mathcal{H}_{\mathrm{EG}}(2,2)$.*

Proof. First we show that if a set of three points of $EG(m,q)$ forms an absorbing set, then $m = q = 2$. Consider a subset $\mathcal{A}$ of three points in a nontrivial finite geometry $(m,q \geq 2)$. In order for $\mathcal{A}$ to be an absorbing set, each point in $\mathcal{A}$ must be on more even lines than odd lines. The number of even lines containing $p \in \mathcal{A}$ can be at most two, since there are two other points in $\mathcal{A}$, and at best they each contribute a distinct even line. Thus, the total number of lines that p can be on is three. There is one finite Euclidean geometry in which $m, q \geq 2$ and each point is on three lines: $EG(2,2)$.

Next we will demonstrate a $(3,3)$ absorbing set in the code constructed from $EG(2,2)$ (see Fig. 1). In the Tanner graph for $\mathcal{H}_{\mathrm{EG}}(2,2)$ there are four variable nodes and six check nodes. Since any subset of three variable nodes will form a $(3,3)$ absorbing set, there are $\binom{4}{3} = 4$ such sets. □

Next we show that the smallest absorbing sets of $\mathcal{H}_{\mathrm{EG}}(2,4)$ are $(4,8)$ absorbing sets, which are the minimal parameters according to the bound in Lemma 1.

The four points in the absorbing set $\mathcal{A}$ must be chosen from two parallel lines, two on each line, as shown by the X's in Fig. 2. Then each pair of points

Fig. 1. The Euclidean geometry EG(2, 2). Any subset of three points forms a (3, 3) absorbing set of $\mathcal{H}_{EG}(2, 2)$.

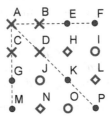

Fig. 2. The set $\{A, B, C, D\}$ forms a (4, 8) absorbing set in EG(2, 4). The only 'lines' represented are those containing the point A—three are shown by dotted lines. The fourth line containing A is denoted by open circles and the fifth by diamonds.

has a unique line containing them, which is even with respect to $\mathcal{A}$. Since each point is on five lines, with three even and two odd, the set is an absorbing set.

The general approach in [17] for finding the smallest absorbing sets in $\mathcal{C}^*_{EG}(2, q)$ is to first find a k-arc, where k is the minimal possible value for a, $k = 2 + \lfloor \frac{q}{2} \rfloor$. However, we immediately run into problems with this approach, as the full geometry EG(2, q) has no such k-arcs for sufficiently large q.

Lemma 2. *When $q \geq 6$, no k-arc exists in* EG(2, q) *for $k \geq \lfloor \frac{q}{2} \rfloor + 2$.*

Proof. Consider a k-arc in EG(2, q), denoted $\mathcal{A}$. Since no three points in $\mathcal{A}$ are collinear, we have that any line through two points in $\mathcal{A}$ has all other points outside of $\mathcal{A}$. Since every pair of points in EG(2, q) has a unique line that contains those points, there are $\binom{|\mathcal{A}|}{2}$ lines through pairs of points in $\mathcal{A}$, each of which contains $q - 2$ points not in $\mathcal{A}$. Therefore there are at least

$$\binom{|\mathcal{A}|}{2} \cdot (q - 2)$$

points outside of $\mathcal{A}$. When $q \geq 6$, this number is less than $q^m - |\mathcal{A}|$ only if $k < \lfloor \frac{q}{2} \rfloor + 2$. □

Since the approach of finding small absorbing sets by starting with k-arcs will not work for $q \geq 6$, we will use other geometric substructures, such as parallel bundles of lines. In the following theorem, we see that the parameters of smallest absorbing sets of $\mathcal{H}_{EG}(2, 8)$ are strictly larger than the lower bound given in Lemma 1. For comparison, in [17] the authors show that the smallest absorbing sets for both projective geometry codes and the Euclidean subgeometry codes (in which a point is deleted) all meet the bound in Lemma 1.

Theorem 1. *The smallest absorbing set of $\mathcal{H}_{\mathrm{EG}}(2,8)$ has $a = 8$, $b = 32$. There are $9 \cdot \binom{9}{2} \cdot \binom{8}{4}^{2} = 1,587,600$ distinct smallest absorbing sets in $\mathcal{H}_{\mathrm{EG}}(2,8)$.*

Proof. By the bound in Lemma 1, the smallest (a, b) absorbing set in $\mathcal{C}_{\mathrm{EG}}(2,8)$ has the following: $a \geq 6$ and $b \geq 24$.

Consider the case when $a = 6$. By Lemma 2, we have that there is no 6-arc in $\mathrm{EG}(2,8)$. Thus, any collection of six points must have at least three collinear. Suppose that $\mathcal{A}$ is a subset of the points of $\mathrm{EG}(2,8)$, where $|\mathcal{A}| = 6$. Then there exists a line L in $\mathrm{EG}(2,8)$ such that $3 \leq |L \cap \mathcal{A}| \leq 6$. In the Tanner graph, denote the subgraph induced by the variable nodes corresponding to $\mathcal{A}$ as $(\mathcal{A}, W_{\mathcal{A}}; E_{\mathcal{A}})$.

There are three subcases:

1. Suppose that $|L \cap \mathcal{A}| = 3$. Then L contains three points of $\mathcal{A}$, so $L \in \mathcal{O}(\mathcal{A})$. Consider a point $p \in L \cap \mathcal{A}$. There are nine lines in $\mathrm{EG}(2,8)$ that contain p, and we claim that at least six of them must be odd with respect to $\mathcal{A}$. Indeed, since p is on L, there are three other points of $\mathcal{A}$ that are not on L, each of which lies on a line with p. If each of those lines is even, p is still incident to six odd lines. Thus $\mathcal{A}$ is not an absorbing set.

2. Suppose $|L \cap \mathcal{A}| = 4$. Consider a point $p \in L \cap \mathcal{A}$. The number of even lines in $W_{\mathcal{A}}$ containing p can be at most three—L and the lines containing p and the points in $\mathcal{A} \setminus L$. Therefore the number of odd lines incident to p is at least six.

3. Suppose $|L \cap \mathcal{A}| = 5$ or 6. Then for a point p on L, the number of odd lines containing p in $W_{\mathcal{A}}$ is eight.

In all cases above $\mathcal{A}$ is not an absorbing set. A similar analysis shows that a set $\mathcal{A}$ of size 7 cannot be an absorbing set.

The existence of an $(8, 32)$ absorbing set in $\mathcal{H}_{\mathrm{EG}}(2,8)$ is demonstrated by the following construction. Consider two parallel lines L_1 and L_2 in $\mathrm{EG}(2,8)$. Let the set $\mathcal{A}$ consist of four points on L_1 and four points on L_2. We claim that $\mathcal{A}$ is an absorbing set. Consider a point $p \in L_1 \cap \mathcal{A}$. The number of even lines in $W_{\mathcal{A}}$ that contain p is five—L_1, along with the four unique lines through p and the points in $\mathcal{A} \cap L_2$. Notice that a line containing p and a point q in $\mathcal{A} \cap L_2$ cannot contain any other points in $\mathcal{A}$, since a third point r would have to lie on either L_1 or L_2. Suppose r lies on L_1. There is a unique line through the points p and r, so that line would have to be L_1. But the set $\mathcal{A}$ was chosen so that $q \notin L_1$.

Similar reasoning shows that any point in $\mathcal{A}$ is on five even lines and four odd lines, and so $\mathcal{A}$ is an absorbing set of $\mathcal{H}_{\mathrm{EG}}(2,8)$.

The size of $\mathcal{O}(A)$ is 32, since all eight points of $\mathcal{A}$ are on four distinct odd lines, and all 32 of these lines are distinct.

Every absorbing set with $a = 8$ must be arranged as two sets of four points on parallel lines in order to guarantee that every point in the absorbing set has a majority of even line neighbors. First choose a parallel bundle in $\mathrm{EG}(2,8)$, then two lines in that bundle, then four points on each of those lines. $\qquad \square$

Corollary 1. *The absorbing set construction in Theorem 1 generalizes to show the existence of at least* $(2^s + 1) \cdot \binom{2^s + 1}{2} \cdot \binom{2^s}{2^{s-1}}^2$ *absorbing sets with parameters* $(2^s, 2^{2s+1})$ *in* $\mathcal{H}_{EG}(2, 2^s)$.

Proof. Start with two parallel lines in $EG(2, 2^s)$. Let $\mathcal{A}$ consist of 2^{s-1} points on each line. Then $\mathcal{A}$ is an absorbing set with the parameters given above.   □

In fact, the parallel line construction leads to results on (a, b) absorbing sets in $\mathcal{C}_{EG}(2, q)$ for a range of values of a and q.

Proposition 2. *Let a be an even integer, $a = 2\alpha$.*

If $a \equiv 0 \pmod 4$ *and* $\left\lceil \frac{q}{2} \right\rceil \le \alpha \le q$, *then there are at least* $(q+1) \cdot \binom{q}{2} \cdot \binom{q}{\alpha}^2$ *type (a, b) absorbing sets in* $\mathcal{H}_{EG}(2, q)$.

If a is even but $a \not\equiv 0 \pmod 4$, and $\left\lceil \frac{q}{2} \right\rceil + 1 \le \alpha \le q$, *then there are at least* $(q + 1) \cdot \binom{q}{2} \cdot \binom{q}{\alpha}^2$ *type (a, b) absorbing sets in* $\mathcal{H}_{EG}(2, q)$.

Proof. Parallel bundles of lines are the basis for this construction of (a, b) absorbing sets. Let $a = 2\alpha$ be an integer such that $\lceil \frac{q}{2} \rceil \le a \le 2q$, where $a = \lceil \frac{q}{2} \rceil$ only if $a \equiv 0 \pmod 4$. Choose two lines, L_1 and L_2, from a parallel bundle. Designate α points on each line, and call this set of points $\mathcal{A}$. Then $\mathcal{A}$ forms an absorbing set with a vertices. Indeed, for a given point $p \in \mathcal{A}$ (say $p \in L_1$), p is on a total of $q + 1$ lines. The lines joining points in $\mathcal{A} \cap L_1$ to points in $\mathcal{A} \cap L_2$ are all even, since they each have exactly two points from $\mathcal{A}$ on them. The point p is on α such lines. Moreover, if α is even, the line L_1 is even with respect to $\mathcal{A}$. Therefore any point p is on more even lines than odd, so $\mathcal{A}$ is an absorbing set. By counting the parallel classes and choice of positions for the points in the absorbing set, we obtain the enumeration in the proposition.   □

Conjecture 1. *The smallest absorbing set in* $\mathcal{H}_{EG}(2, 2^s)$ *has* $a > 2^{s-1} + 2$, *for* $s \ge 3$.

Using the computational system GAP, we have verified cases showing the non-existence of an absorbing set with $a = 12$ or $a = 14$ for $EG(2, 16)$, which provides some evidence for the conjecture. Moreover, the construction in Proposition 2 shows that a $(2^s, 2^{2s-1})$ absorbing set always exists in $\mathcal{H}_{EG}(2, 2^s)$.

3.1   Fully Absorbing Sets and Elementary Absorbing Sets

The $(4, 8)$ absorbing set in $\mathcal{H}_{EG}(2, 4)$ is a fully absorbing set. Figure 2 shows the absorbing set points denoted with X's, and the five lines through the point A. Consider a point outside of the absorbing set $\mathcal{A}$, say L. There is a line (not drawn) containing the points B, C, O, L, which is in $\mathcal{E}(\mathcal{A})$. Since L is on five lines and only two of them are in $\mathcal{O}(\mathcal{A})$, it must be that L is on more lines corresponding to check nodes in $W \setminus \mathcal{O}(\mathcal{A})$ than $\mathcal{O}(\mathcal{A})$. A similar argument applies to all other points in $V \setminus \mathcal{A}$. Therefore $\mathcal{A}$ is a fully absorbing set. Any absorbing set with $a = 4$ must have two points on parallel lines, and so the symmetry of the geometry would yield the same result regardless of the choice of parallel bundle.

The $(4, 8)$ absorbing sets in $\mathcal{H}_{\mathrm{EG}}(2, 4)$ are also all elementary absorbing sets, since each line of the geometry contains either $0, 1$, or 2 points in $\mathcal{A}$.

However, we now show that codes from finite Euclidean plane geometries over larger fields do not have elementary absorbing sets.

Proposition 3. *There are no elementary absorbing sets in $\mathcal{H}_{\mathrm{EG}}(2, q)$ for $q \geq 6$.*

Proof. Recall that an elementary absorbing set given by the subgraph $G_S = (S, W_S; E_S)$ is an absorbing set in which all vertices in W_S have degree one or two in G_S. Lemma 2 shows that for $q \geq 6$ and $|S| \geq 5$, there must be at least three collinear points in S, so some line corresponds to a check node in W_S with degree at least three. Combined with Lemma 1, which shows that an absorbing set in $\mathcal{H}_{\mathrm{EG}}(2, q)$ has $|S| \geq 2 + \left\lfloor \frac{q}{2} \right\rfloor$, we conclude that there are no elementary absorbing sets in $\mathcal{H}_{\mathrm{EG}}(2, q)$ for $q \geq 6$. $\square$

Remark 2. The fact that the minimal absorbing sets in codes from the geometries $\mathrm{EG}(2, q)$ have relatively large values of a^* and b^* indicates that these combinatorial structures may not impede iterative decoding processes. While knowledge of the structure of these minimal absorbing sets may not be necessary to optimize BP or SPA decoding of these codes, the bounds are useful in contributing to the theoretical basis for the robustness of the error-performance of the codes.

4   Conclusions

We constructed and enumerated certain absorbing sets for codes from finite affine planes. We showed that k-arcs do not exist in $\mathrm{EG}(2, q)$ for $k \geq \left\lfloor \frac{q}{2} \right\rfloor + 2$ when $q \geq 6$. We enumerated the smallest absorbing sets in $\mathcal{H}_{\mathrm{EG}}(2, 8)$ and described their structure, which demonstrates that a sharper lower bound than Lemma 1 on the size of the smallest absorbing sets in $\mathcal{H}_{\mathrm{EG}}(2, q)$ is possible for this class of codes. We also gave a general construction and enumeration for absorbing sets in $\mathcal{H}_{\mathrm{EG}}(2, q)$ for a range of values of a and q. Ongoing work includes obtaining an improved lower bound for the size of minimal (a, b) absorbing sets in $\mathcal{H}_{\mathrm{EG}}(2, q)$, as well as extending these ideas to codes from geometries with dimension larger than two.

References

1. Amiri, B., Lin, C.W., Dolecek, L.: Asymptotic distribution of absorbing sets and fully absorbing sets for regular sparse code ensembles. IEEE Trans. Commun. **61**(2), 455–464 (2013)
2. Ball, S., Weiner, Z.: An introduction to finite geometry. Preprint (2011)

3. Batten, L.M.: Combinatorics of Finite Geometries, 2nd edn. Cambridge University Press, Cambridge (1997)
4. Di, C., Proietti, D., Telatar, I.E., Richardson, T.J., Urbanke, R.L.: Finite-length analysis of low-density parity-check codes on the binary erasure channel. IEEE Trans. Inform. Theory 48(6), 1570–1579 (2002)
5. Diao, Q., Tai, Y.Y., Lin, S., Abdel-Ghaffar, K.: Trapping set structure of LDPC codes on finite geometries. In: Information Theory and Applications Workshop (ITA), pp. 1–8. IEEE (2013)
6. Dolecek, L.: On absorbing sets of structured sparse graph codes. In: Information Theory and Applications Workshop (ITA), pp. 1–5. IEEE (2010)
7. Dolecek, L., Zhang, Z., Anantharam, V., Wainwright, M., Nikolic, B.: Analysis of absorbing sets for array-based LDPC codes. In: IEEE International Conference on Communications, pp. 6261–6268 (2007)
8. Feldman, J., Wainright, M.J., Karger, D.R.: Using linear programming to decode binary linear codes. IEEE Trans. Inform. Theory 51(3), 954–972 (2005)
9. Hirschfeld, J.W., Storme, L.: The packing problem in statistics, coding theory and finite projective spaces: update 2001. In: Finite Geometries, pp. 201–246. Springer, US (2001)
10. Johnson, S.J., Weller, S.R.: Codes for iterative decoding from partial geometries. IEEE Trans. Commun. 52(2), 236–243 (2004)
11. Kelley, C.A., Sridhara, D., Rosenthal, J.: Tree-based construction of LDPC codes having good pseudocodeword weights. IEEE Trans. Inform. Theory 53(4), 1460–1478 (2007)
12. Koetter, R., Vontobel, P.O.: Graph covers and iterative decoding of finite-length codes. In: Proceedings of the 3rd International Symposium on Turbo Codes and Related Topics, Brest, France, pp. 75–82 (2003)
13. Kou, Y., Lin, S., Fossorier, M.: Construction of low density parity check codes: a geometric approach. In: 2nd IEEE International Symposium on Turbo Codes and Related Topics, Brest, France, pp. 137–140 (2000)
14. Kou, Y., Lin, S., Fossorier, M.: Low density parity-check codes based on finite geometries: a rediscovery and new results. IEEE Trans. Inform. Theory 47(7), 2711–2736 (2001)
15. Kyung, G.B., Wang, C.C.: Finding the exhaustive list of small fully absorbing sets and designing the corresponding low error-floor decoder. IEEE Trans. Commun. 60(6), 1487–1498 (2012)
16. Landner, S., Milenkovic, O.: Algorithmic and combinatorial analysis of trapping sets in structured ldpc codes. In: Wireless Networks, Communications, and Mobile Computing, vol. 1, pp. 630–635. IEEE (2005)
17. Liu, H., Li, Y., Ma, L., Chen, J.: On the smallest absorbing sets of LDPC codes from finite planes. IEEE Trans. Inform. Theory 58(6) (2012)
18. Richardson, T.: Error floors of LDPC codes. In: Proceedings of the Annual Allerton Conference on Communication, Control, and Computing, vol. 41 (2003)
19. Schwartz, M., Vardy, A.: On the stopping distance and the stopping redundancy of codes. IEEE Trans. Inform. Theory 52, 922–932 (2006)
20. Tang, H., Xu, J., Lin, S., Abdel-Ghaffar, K.A.S.: Codes on finite geometries. IEEE Trans. Inform. Theory 51(2), 572–596 (2005)
21. Tanner, R.M.: Explicit concentrators from generalized n-gons. SIAM J. Algebraic Discrete Methods 5(3), 287–293 (1984)
22. Xia, S.T., Fu, F.W.: On the stopping distance of finite geometry LDPC codes. IEEE Commun. Lett. 10(5), 381–383 (2006)

Asymptotic Bounds for the Sizes of Constant Dimension Codes and an Improved Lower Bound

Daniel Heinlein$^{(\boxtimes)}$ and Sascha Kurz

University of Bayreuth, Bayreuth, Germany
{Daniel.Heinlein,Sascha.Kurz}@uni-bayreuth.de

Abstract. We study asymptotic lower and upper bounds for the sizes of constant dimension codes with respect to the subspace or injection distance, which is used in random linear network coding. In this context we review known upper bounds and show relations between them. A slightly improved version of the so-called linkage construction is presented which is e.g. used to construct constant dimension codes with subspace distance $d = 4$, dimension $k = 3$ of the codewords for all field sizes q, and sufficiently large dimensions v of the ambient space. It exceeds the MRD bound, for codes containing a lifted MRD code, by Etzion and Silberstein.

Keywords: Constant dimension codes · Subspace distance · Injection distance · Random network coding

1 Introduction

Let $V \cong \mathbb{F}_q^v$ be a v-dimensional vector space over the finite field $\mathbb{F}_q$ with q elements. By $\begin{bmatrix} V \\ k \end{bmatrix}$ we denote the set of all k-dimensional subspaces in V, where $0 \leq k \leq v$, which has size $\begin{bmatrix} v \\ k \end{bmatrix}_q := \prod_{i=1}^{k} \frac{q^{v-k+i}-1}{q^i-1}$. More general, the set $P(V)$ of all subspaces of V forms a metric space with respect to the subspace distance defined by $d_s(U, W) = \dim(U + W) - \dim(U \cap W) = \dim(U) + \dim(W) - 2\dim(U \cap W)$, see [32], and the injection distance defined by $d_i(U, W) = \max\{\dim(U), \dim(W)\} - \dim(U \cap W)$, see [40]. Coding Theory on $P(V)$ is motivated by Kötter and Kschischang [32] via error correcting random network coding, see [4]. In this context it is natural to consider codes $\mathcal{C} \subseteq P(V)$ where each codeword, i.e., each element of $\mathcal{C}$, has the same dimension k, called *constant dimension code*, since this knowledge can be exploited by decoders. For constant dimension codes we have $d_s(U, W) = 2d_i(U, W)$, so that we will only consider the subspace distance in this paper. By $(v, N, d; k)_q$ we denote a constant dimension code in V with minimum (subspace) distance d and size N, where the dimensions of each codeword is $k \in \{0, 1, \ldots, v\}$. As usual, a constant dimension

The work was supported by the ICT COST Action IC1104 and grants KU 2430/3-1, WA 1666/9-1 – "Integer Linear Programming Models for Subspace Codes and Finite Geometry" – from the German Research Foundation.

© Springer International Publishing AG 2017
Á.I. Barbero et al. (Eds.): ICMCTA 2017, LNCS 10495, pp. 163–191, 2017.
DOI: 10.1007/978-3-319-66278-7_15

code $\mathcal{C}$ has the *minimum distance* d, if $d \le \mathrm{d_s}(U, W)$ for all $U \ne W \in \mathcal{C}$ and equality is attained at least once. If $\#\mathcal{C} = 1$, we set the minimum distance to ∞. The corresponding maximum size is denoted by $A_q(v, d; k)$, where we allow the minimum distance to be larger than d. The authors of [32] provided lower and upper bounds for $A_q(v, d; k)$ which are less than a factor of 4 apart. For increasing field size q this factor tends to 1. Here, we tighten the corresponding analysis and end up in a factor of less than 2 for the binary field $q = 2$ and a strictly better factor for larger values of q. With respect to lower bounds, we slightly generalize the so-called linkage construction by Gluesing-Luerssen, Troha, Morrison [21, 22] and Silberstein, Trautmann [38]. This improvement then gives the best known lower bounds for $A_q(v, d; k)$ for many parameters, cf. the online tables http://subspacecodes.uni-bayreuth.de associated with [23]. For codes containing a lifted maximum rank distance (lifted MRD) code as a subcode an upper bound on the size has been presented in [16] for some infinite series of parameters. Codes larger than this MRD bound are very rare. Based on the improved linkage construction we give an infinite series of such examples.

In this context we mention the following asymptotic result based on the non-constructive probabilistic method. If the subspace distance d and the dimension k of the codewords is fixed, then the ratio of the lower and upper bound tends to 1 as the dimension v of the ambient space approaches infinity, see [18, Theorem 4.1], which is implied by a more general result of Frankl and Rödl on hypergraphs. The same result, with an explicit error term, was also obtained in [8, Theorem 1]. If d and $v - k$ is fixed we have the same result due to the orthogonal code. If the parameter k can vary with the dimension v, then our asymptotic analysis implies there is still a gap of almost 2 between the lower and the upper bound of the code sizes for $d = 4$ and $k = \lfloor v/2 \rfloor$, which is the worst case.

The remaining part of the paper is organized as follows. In Sect. 2 we collect the basic facts and definitions for constant dimension codes. Upper bounds on the achievable code sizes are reviewed in Sect. 3. Here, we extend the current knowledge on the relation between these bounds. While most of them were known around 2008, see [2, 17, 29, 32, 43], there are some recent improvements for the subclass of partial spreads, where $d = 2k$, which we summarize in Subsect. 3.1. In Sect. 4 we present the mentioned improvement of the linkage construction. Asymptotic bounds for the ratio between lower and upper bounds for code sizes are studied in Sect. 5. We continue with the upper bound for constant dimension codes containing a lifted MRD code in Sect. 6, including some numerical results, before we draw a short conclusion in Sect. 7.

2 Preliminaries

For the remainder of the paper we set $V \cong \mathbb{F}_q^v$, where q is a prime power. By v we denote the dimension of V. Using the language of projective geometry, we will call the 1-dimensional subspaces of $\mathbb{F}_q^v$ points and the 2-dimensional subspaces lines. First, we observe that the q-binomial coefficient $\begin{bmatrix} v \\ k \end{bmatrix}_q$ indeed gives the cardinality of $\begin{bmatrix} V \\ k \end{bmatrix}$. To this end, we associate with a subspace $U \in \begin{bmatrix} V \\ k \end{bmatrix}$ a unique $k \times v$ matrix X_U in row reduced echelon form (rref) having the property that $\langle X_U \rangle = U$ and denote the corresponding bijection

$$\tau : \begin{bmatrix} \mathbb{F}_q^v \\ k \end{bmatrix} \to \{X_U \in \mathbb{F}_q^{k \times v} \mid \mathrm{rk}(X_U) = k, X_U \text{ is in rref}\}.$$

An example is given by $X_U = \left(\begin{smallmatrix} 1 & 0 & 0 \\ 0 & 1 & 1 \end{smallmatrix} \right) \in \mathbb{F}_2^{2 \times 3}$, where $U = \tau^{-1}(X_U) \in \begin{bmatrix} \mathbb{F}_2^3 \\ 2 \end{bmatrix}$ is a line that contains the three points $(1,0,0)$, $(1,1,1)$, and $(0,1,1)$. Counting those matrices gives

$$\#\begin{bmatrix} V \\ k \end{bmatrix} = \prod_{i=0}^{k-1} \frac{q^v - q^i}{q^k - q^i} = \prod_{i=1}^{k} \frac{q^{v-k+i} - 1}{q^i - 1} = \begin{bmatrix} v \\ k \end{bmatrix}_q$$

for all integers $0 \le k \le v$. Especially, we have $\begin{bmatrix} v \\ v \end{bmatrix}_q = \begin{bmatrix} v \\ 0 \end{bmatrix}_q = 1$. Given a non-degenerate bilinear form, we denote by $U^\perp$ the orthogonal subspace of a subspace U, which then has dimension $v - \dim(U)$. Then, we have $d_s(U, W) = d_s(U^\perp, W^\perp)$, so that $\begin{bmatrix} v \\ k \end{bmatrix}_q = \begin{bmatrix} v \\ v-k \end{bmatrix}_q$. The recurrence relation for the usual binomial coefficients generalize to $\begin{bmatrix} v \\ k \end{bmatrix}_q = q^k \begin{bmatrix} v-1 \\ k \end{bmatrix}_q + \begin{bmatrix} v-1 \\ k-1 \end{bmatrix}_q$. In order to remove the restriction $0 \le k \le v$, we set $\begin{bmatrix} a \\ b \end{bmatrix}_q = 0$ for $a \in \mathbb{N}_{\ge 0}$ and $b \in \mathbb{Z}$, whenever $b < 0$ or $a < b$. This extension goes in line with the interpretation of the number of b-dimensional subspaces of $\mathbb{F}_q^a$ and respects the orthogonality relation. In order to write $\sum_{j=0}^{v-1} q^j = \begin{bmatrix} v \\ 1 \end{bmatrix}_q$ for positive integers q in later formulas, we apply the definition of $\begin{bmatrix} v \\ k \end{bmatrix}_q$ also in cases where q is not a prime power and set $\begin{bmatrix} v \\ k \end{bmatrix}_1 = \binom{v}{k}$ for $q = 1$.

Using the bijection τ we can express the subspace distance between two k-dimensional subspaces $U, W \in \begin{bmatrix} V \\ k \end{bmatrix}$ via the rank of a matrix:

$$d_s(U, W) = 2 \dim(U + W) - \dim(U) - \dim(W) = 2 \left(\mathrm{rk} \begin{pmatrix} \tau(U) \\ \tau(W) \end{pmatrix} - k \right). \quad (1)$$

Using $\begin{bmatrix} V \\ k \end{bmatrix}$ as vertex set, we obtain the so-called Grassmann graph, where two vertices are adjacent iff the corresponding subspaces intersect in a space of dimension $k - 1$. It is well-known that the Grassmann graph is distance regular. The injection distance $d_i(U, W)$ corresponds to the graph distance in the Grassmann graph. Considered as an association scheme one speaks of the q-Johnson scheme.

If $\mathcal{C} \subseteq \begin{bmatrix} V \\ k \end{bmatrix}$ is a constant dimension code with minimum subspace distance d, we speak of a $(v, \#\mathcal{C}, d; k)$ constant dimension code. In the special case of $d = 2k$ one speaks of so-called partial spreads, i.e., collections of k-dimensional subspaces with pairwise trivial intersection.

Besides the injection and the subspace distance we will also consider the Hamming distance $d_h(u, w) = \#\{i \mid u_i \ne w_i\}$, for two vectors $u, w \in \mathbb{F}_2^v$, and the rank distance $d_r(U, W) = \mathrm{rk}(U - W)$, for two matrices $U, W \in \mathbb{F}_q^{m \times n}$. The latter is indeed a metric, as observed in [20]. A subset $\mathcal{C} \subseteq \mathbb{F}_q^{m \times n}$ is called a rank metric code. If the minimum rank-distance of $\mathcal{C}$ is given by d_r, we will also speak of an $(m \times n, \#\mathcal{C}, d_r)_q$ rank metric code in order to specify its parameters. A rank metric code $\mathcal{C} \subseteq \mathbb{F}_q^{m \times n}$ is called linear if $\mathcal{C}$ forms a subspace of $\mathbb{F}_q^{m \times n}$, which implies that $\#\mathcal{C}$ has to be a power of the field size q.

Theorem 1. *(see [20]) Let $m, n \geq d$ be positive integers, q a prime power, and $\mathcal{C} \subseteq \mathbb{F}_q^{m \times n}$ be a rank metric code with minimum rank distance d. Then, $\#\mathcal{C} \leq q^{\max\{n,m\}\cdot(\min\{n,m\}-d+1)}$.*

Codes attaining this upper bound are called maximum rank distance (MRD) codes. They exist for all (suitable) choices of parameters, which remains true if we restrict to linear rank metric codes, see [20]. If $m < d$ or $n < d$, then only $\#\mathcal{C} = 1$ is possible, which can be achieved by a zero matrix and may be summarized to the single upper bound $\#\mathcal{C} \leq \lceil q^{\max\{n,m\}\cdot(\min\{n,m\}-d+1)} \rceil$. Using an $m \times m$ identity matrix as a prefix one obtains the so-called lifted MRD codes.

Theorem 2. *[39, Proposition 4] For positive integers k, d, v with $k \leq v$, $d \leq 2\min\{k, v-k\}$, and d even, the size of a lifted MRD code in $\begin{bmatrix} V \\ k \end{bmatrix}$ with subspace distance d is given by*

$$M(q, k, v, d) := q^{\max\{k,v-k\}\cdot(\min\{k,v-k\}-d/2+1)}.$$

If $d > 2\min\{k, v-k\}$, then we have $M(q, k, v, d) := 1$.

The Hamming distance can be used to lower bound the subspace distance between two codewords (of the same dimension). To this end let $p : \{M \in \mathbb{F}_q^{k \times v} \mid \mathrm{rk}(M) = k, M \text{ is in rref}\} \to \{x \in \mathbb{F}_2^v \mid \sum_{i=1}^{v} x_i = k\}$ denote the pivot positions of the matrix in rref. For our example X_U we we have $p(X_U) = (1, 1, 0)$. Slightly abusing notation we also write $p(U)$ for subspaces $U \in \begin{bmatrix} V \\ k \end{bmatrix}$ instead of $p(\tau(U))$.

Lemma 1. *[15, Lemma 2] For two subspaces $U, W \leq \mathbb{F}_q^v$, we have $d_s(U, W) \geq d_h(p(U), p(W))$.*

3   Upper Bounds

In this section we review and compare known upper bounds for the sizes of constant dimension codes. Here we assume that v, d, and k are integers with $2 \leq k \leq v - 2$, $4 \leq d \leq 2\min\{k, v-k\}$, and d even in all subsequent results. The bound $0 \leq k \leq v$ just ensures that $\begin{bmatrix} V \\ k \end{bmatrix}$ is non-empty. Note that $d_s(U, W) \leq 2\min\{k, v-k\}$ and $d_s(U, W)$ is even for all $U, W \in \begin{bmatrix} V \\ k \end{bmatrix}$. Restricting to the set case, we trivially have $A_q(v, d; k) = \#\begin{bmatrix} V \\ k \end{bmatrix} = \begin{bmatrix} v \\ k \end{bmatrix}_q$ for $d \leq 2$ or $k \leq 1$, so that we assume $k \geq 2$ and $d \geq 4$, which then implies $k \leq v - 2$ and $v \geq 4$. We remark that some of the latter bounds are also valid for parameters outside the ranges of non-trivial parameters considered by us. Since the maximum size of a code with certain parameters is always an integer and some of the latter upper bounds can produce non-integer values, we may always round them down. To ease the notation we will commonly omit the final rounding step.

The list of known bounds has not changed much since [29], see also [17]. Comparisons of those bounds are scattered among the literature and partially hidden in comments, see e.g. [6]. Additionally some results turn out to be wrong or need a reinterpretation at the very least.

Counting k-dimensional subspaces having a *large* intersection with a fixed m-dimensional subspace gives:

Lemma 2. *For integers* $0 \le t \le k \le v$ *and* $k - t \le m \le v$ *we have*

$$\# \left\{ U \in \begin{bmatrix} V \\ k \end{bmatrix} \mid \dim(U \cap W) \ge k - t \right\} = \sum_{i=0}^{t} q^{(m+i-k)i} \begin{bmatrix} m \\ k-i \end{bmatrix}_q \begin{bmatrix} v-m \\ i \end{bmatrix}_q,$$

where $W \le V$ *and* $\dim(W) = m$.

Proof. Let us denote $\dim(U \cap W)$ by $k - i$, where $\max\{0, k - m\} \le i \le \min\{t, v - m\}$. With this, the number of choices for U is given by

$$\frac{(q^m - q^0) \cdot (q^m - q^1) \cdots (q^m - q^{k-i-1}) \cdot (q^v - q^{m+1}) \cdots (q^v - q^{m+i-1})}{(q^k - q^0) \cdot (q^k - q^1) \cdots (q^k - q^{k-1})}$$

$$= \begin{bmatrix} m \\ k-i \end{bmatrix}_q \cdot \frac{(q^m)^i}{(q^{k-i})^i} \cdot \begin{bmatrix} v-m \\ i \end{bmatrix}_q = q^{(m+i-k)i} \begin{bmatrix} m \\ k-i \end{bmatrix}_q \begin{bmatrix} v-m \\ i \end{bmatrix}_q.$$

Finally apply the convention $\begin{bmatrix} a \\ b \end{bmatrix}_q = 0$ for integers with $b < 0$ or $b > a$. $\square$

Note that $\dim(U \cap W) \ge k - t$ is equivalent to $d_s(U, W) \le m - k + 2t$. The fact that the Grassmann graph is distance-regular implies:

Theorem 3. *(Sphere-packing bound) [32, Theorem 6]*

$$A_q(v, d; k) \le \frac{\begin{bmatrix} v \\ k \end{bmatrix}_q}{\sum_{i=0}^{\lfloor (d/2-1)/2 \rfloor} q^{i^2} \begin{bmatrix} k \\ i \end{bmatrix}_q \begin{bmatrix} v-k \\ i \end{bmatrix}_q}$$

We remark, that we can obtain the denominator of the formula of Theorem 3 by setting $m = k$, $2t = d/2 - 1$ in Lemma 2 and applying $\begin{bmatrix} k \\ k-i \end{bmatrix}_q = \begin{bmatrix} k \\ i \end{bmatrix}_q$. The right hand side is symmetric with respect to orthogonal subspaces, i.e., the mapping $k \mapsto v - k$ leaves it invariant.

By defining a puncturing operation one can decrease the dimension of the ambient space and the codewords. Since the minimum distance decreases by at most two, we can iteratively puncture $d/2 - 1$ times, so that $A_q(v, d; k) \le \begin{bmatrix} v-d/2+1 \\ k-d/2+1 \end{bmatrix}_q = \begin{bmatrix} v-d/2+1 \\ v-k \end{bmatrix}_q$ since $A_q(v', 2; k') = \begin{bmatrix} v' \\ k' \end{bmatrix}_q$. Considering either the code or its orthogonal code gives:

Theorem 4. *(Singleton bound) [32, Theorem 9]*

$$A_q(v, d; k) \le \begin{bmatrix} v-d/2+1 \\ \max\{k, v-k\} \end{bmatrix}_q$$

Referring to [32] the authors of [29] state that even a relaxation of the Singleton bound is always stronger than the sphere packing bound for non-trivial codes. However, for $q = 2$, $v = 8$, $d = 6$, and $k = 4$, the sphere-packing bound gives an upper bound of $200787/451 \approx 445.20399$ while the Singleton bound gives an upper bound of $\begin{bmatrix} 6 \\ 4 \end{bmatrix}_2 = 651$. For $q = 2$, $v = 8$, $d = 4$, and $k = 4$ it is just the other way round, i.e., the Singleton bound gives $\begin{bmatrix} 7 \\ 3 \end{bmatrix}_2 = 11811$ and the sphere-packing bound gives $\begin{bmatrix} 8 \\ 4 \end{bmatrix}_2 = 200787$. Examples for the latter case are easy to find. For $d = 2$ both bounds coincide and for $d = 4$ the Singleton bound is always stronger than the sphere-packing bound since $\begin{bmatrix} v-1 \\ k \end{bmatrix}_q < \begin{bmatrix} v \\ k \end{bmatrix}_q$. The asymptotic bounds [32, Corollaries 7 and 10], using normalized parameters, and [32, Fig. 1] suggest that there is only a small range of parameters where the sphere-packing bound can be superior to the Singleton bound.[1]

Given an arbitrary metric space X, an anticode of diameter e is a subset whose elements have pairwise distance at most e. Since the q-Johnson scheme is an association scheme the Anticode bound of Delsarte [12] can be applied. As a standalone argument we go along the lines of [2] and consider bounds for codes on transitive graphs. By double-counting the number of pairs $(a, g) \in A \times \mathrm{Aut}(\Gamma)$, where $g(a) \in B$, we obtain:

Lemma 3. *[2, Lemma 1], cf. [3, Theorem 1'] Let $\Gamma = (V, E)$ be a graph that admits a transitive group of automorphisms $\mathrm{Aut}(\Gamma)$ and let A, B be arbitrary subsets of the vertex set V. Then, there exists a group element $g \in \mathrm{Aut}(\Gamma)$ such that*

$$\frac{|g(A) \cap B|}{|B|} \geq \frac{|A|}{|V|}.$$

Corollary 1. *[2, Corollary 1], cf. [3, Theorem 1] Let $\mathcal{C}_D \subseteq \begin{bmatrix} V \\ k \end{bmatrix}$ be a code with (injection or graph) distances from $D = \{d_1, \ldots, d_s\} \subseteq \{1, \ldots, v\}$. Then, for an arbitrary subset $\mathcal{B} \subseteq \begin{bmatrix} V \\ k \end{bmatrix}$ there exists a code $\mathcal{C}_D^*(\mathcal{B}) \subseteq \mathcal{B}$ with distances from D such that*

$$\frac{|\mathcal{C}_D^*(\mathcal{B})|}{|\mathcal{B}|} \geq \frac{|\mathcal{C}_D|}{\begin{bmatrix} v \\ k \end{bmatrix}_q}.$$

If $\mathcal{C}_D \subseteq \begin{bmatrix} V \\ k \end{bmatrix}$ is a constant dimension code with minimum injection distance d, i.e., $D = \{d, \ldots, v\}$, and $\mathcal{B}$ is an anticode with diameter $d - 1$, we have $\#\mathcal{C}_D^*(\mathcal{B}) = 1$, so that we obtain Delsarte's Anticode bound

$$\#\mathcal{C}_D \leq \frac{\begin{bmatrix} v \\ k \end{bmatrix}_q}{\#\mathcal{B}}. \tag{2}$$

The set of all elements of $\begin{bmatrix} V \\ k \end{bmatrix}$ which contain a fixed $(k - d/2 + 1)$-dimensional subspace is an anticode of diameter $d - 2$ with $\begin{bmatrix} v-k+d/2-1 \\ d/2-1 \end{bmatrix}_q$ elements. By orthogonality, the set of all elements of $\begin{bmatrix} V \\ k \end{bmatrix}$ which are contained in a fixed $(k + d/2 - 1)$-dimensional subspace is also an anticode of diameter $d - 2$ with

---

[1] By a tedious computation one can check that the sphere-packing bound is strictly tighter than the Singleton bound iff $q = 2$, $v = 2k$ and $d = 6$.

$\begin{bmatrix} k+d/2-1 \\ k \end{bmatrix}_q = \begin{bmatrix} k+d/2-1 \\ d/2-1 \end{bmatrix}_q$ elements. Frankl and Wilson proved in [19, Theorem 1] that these anticodes have the largest possible size, which implies:

Theorem 5. *(Anticode bound)*

$$A_q(v, d; k) \leq \frac{\begin{bmatrix} v \\ k \end{bmatrix}_q}{\begin{bmatrix} \max\{k, v-k\}+d/2-1 \\ d/2-1 \end{bmatrix}_q}$$

Using different arguments, Theorem 5 was proved in [42, Theorem 5.2] by Wang, Xing, Safavi-Naini in 2003. Codes that can achieve the (unrounded) value $\begin{bmatrix} v \\ k \end{bmatrix}_q / \begin{bmatrix} \max\{k,v-k\}+d/2-1 \\ d/2-1 \end{bmatrix}_q$ are called Steiner structures. It is a well-known and seemingly very hard problem to decide whether a Steiner structure for $v = 7$, $d = 4$, and $k = 3$ exists. For $q = 2$ the best known bounds are $333 \leq A_2(7, 4; 3) \leq 381$. Additionally it is known that a code attaining the upper bound can have automorphisms of at most order 2, see [30]. So far, the only known Steiner structure corresponds to $A_2(13, 4; 3) = 1597245$ [9]. The reduction to Delsarte's Anticode bound can be found e.g. in [17, Theorem 1].

Since the sphere underlying the proof of Theorem 3 is also an anticode, Theorem 3 is implied by Theorem 5. For $d = 2$ both bounds coincide. In [43, Sect. 4] Xia and Fu verified that the Anticode bound is always stronger than the Singleton bound for the ranges of parameters considered by us.

Mimicking a classical bound of Johnson on binary error-correcting codes with respect to the Hamming distance, see [28, Theorem 3] and also [41], Xia and Fu proved:

Theorem 6. *(Johnson type bound I)* *[43, Theorem 2]*
If $\left(q^k - 1\right)^2 > \left(q^v - 1\right)\left(q^{k-d/2} - 1\right)$, then

$$A_q(v, d; k) \leq \frac{\left(q^k - q^{k-d/2}\right)\left(q^v - 1\right)}{\left(q^k - 1\right)^2 - \left(q^v - 1\right)\left(q^{k-d/2} - 1\right)}.$$

However, the required condition of Theorem 6 is rather restrictive and can be simplified considerably.

Proposition 1. *For $0 \leq k < v$, the bound in Theorem 6 is applicable iff $d = 2\min\{k, v - k\}$ and $k \geq 1$. Then, it is equivalent to*

$$A_q(v, d; k) \leq \frac{q^v - 1}{q^{\min\{k, v-k\}} - 1}.$$

Proof. If $k = 0$ we have $\left(q^k - 1\right)^2 = 0$, so that we assume $k \geq 1$ in the following. If $k \leq v - k$ and $d \leq 2k - 2$, then

$$(q^v - 1)\left(q^{k-d/2} - 1\right) \geq \left(q^{2k} - 1\right)(q - 1) \geq q^{2k} - 1 \overset{q \geq 2, k \geq 1}{>} q^{2k} - 2q^k + 1 = \left(q^k - 1\right)^2.$$

If $k \geq v - k + 1$ and $d \leq 2v - 2k - 2$, then

$$(q^v - 1)\left(q^{k-d/2} - 1\right) \geq (q^v - 1)\left(q^2 - 1\right) \overset{q \geq 2, v \geq 1}{>} \left(q^{(v+1)/2} - 1\right)^2 \geq \left(q^k - 1\right)^2.$$

If $d = 2\min\{k, v - k\}$, $q \geq 2$, and $k \geq 1$, then it can be easily checked that the condition of Theorem 6 is satisfied and we obtain the proposed formula after simplification. □

For $k = v$ Theorem 6 gives $A_q(v, d; v) \leq 1$ which is trivially satisfied with equality. In Subsect. 3.1 we will provide tighter upper bounds for the special case where $d = 2k$, i.e., partial spreads. Indeed, the bound stated in Proposition 1 corresponds to the most trivial upper bounds for partial spreads that is tight iff k divides v, as we will see later on. So, due to orthogonality, Theorem 6 is dominated by the partial spread bounds discussed later on.

While the previously mentioned generalization of a classical bound of Johnson on binary error-correcting codes yields the rather weak Theorem 6, generalizing [28, Inequality (5)] yields a very strong upper bound:

Theorem 7. (Johnson type bound II) *[43, Theorem 3], [17, Theorem 4,5]*

$$A_q(v, d; k) \leq \frac{q^v - 1}{q^k - 1} A_q(v - 1, d; k - 1) \tag{3}$$

$$A_q(v, d; k) \leq \frac{q^v - 1}{q^{v-k} - 1} A_q(v - 1, d; k) \tag{4}$$

Note that for $d = 2k$ Inequality (3) gives $A_q(v, 2k; k) \leq \left\lfloor \frac{q^v - 1}{q^k - 1} \right\rfloor$ since we have $A_q(v - 1, 2k; k - 1) = 1$ by definition. Similarly, for $d = 2(v - k)$, Inequality (4) gives $A_q(v, 2v - 2k; k) \leq \left\lfloor \frac{q^v - 1}{q^{v-k} - 1} \right\rfloor$.

Some sources like [43, Theorem 3] list just Inequality 3 and omit Inequality 4. This goes in line with the treatment of the classical Johnson type bound II for binary error-correcting codes, see e.g. [35, Theorem 4 on p. 527], where the other bound is formulated as Problem (2) on page 528 with the hint that ones should be replaced by zeros. Analogously, we can consider orthogonal codes:

Proposition 2. *Inequality (3) and Inequality (4) are equivalent using orthogonality, cf. [17, Sect. 3, esp. Lemma 13].*

Proof. We have

$$A_q(v, d; k) = A_q(v, d; v - k) \overset{(3)}{\leq} \frac{q^v - 1}{q^{v-k} - 1} A_q(v - 1, d; v - k - 1)$$

$$= \frac{q^v - 1}{q^{v-k} - 1} A_q(v - 1, d; k),$$

which is Inequality (4), and

$$A_q(v, d; k) = A_q(v, d; v - k) \overset{(4)}{\leq} \frac{q^v - 1}{q^k - 1} A_q(v - 1, d; v - k)$$

$$= \frac{q^v - 1}{q^k - 1} A_q(v - 1, d; k - 1),$$

which is Inequality (3). □

Of course, the bounds in Theorem 7 can be applied iteratively. In the classical Johnson space the optimal choice of the corresponding inequalities is unclear, see e.g. [35, Research Problem 17.1]. Denoting the maximum size of a binary constant-weight block code of length n, Hamming distance d and weight k by $A(n, d, w)$, the two corresponding variants of the inequalities in Theorem 7 are $A(n, d, w) \leq \lfloor n/w \cdot A(n-1, d, w-1) \rfloor$ and $A(n, d, w) \leq \lfloor n/(n-w) \cdot A(n-1, d, w) \rfloor$. Applying the first bound yields

$$A(28, 8, 13) \leq \lfloor 28/13 \cdot A(27, 8, 12) \rfloor \leq \lfloor 28/13 \cdot 10547 \rfloor = 22716$$

while applying the second bound yields

$$A(28, 8, 13) \leq \lfloor 28/15 \cdot A(27, 8, 13) \rfloor \leq \lfloor 28/15 \cdot 11981 \rfloor = 22364$$

using the numerical bounds from

http://webfiles.portal.chalmers.se/s2/research/kit/bounds/cw.html, cf. [1].

The authors of [17,29] state that the optimal choice of Inequality (3) or Inequality (4) is unclear, too. However, this question is much easier to answer for constant dimension codes.

Proposition 3. *For $k \leq v/2$ we have*

$$\left\lfloor \frac{q^v - 1}{q^k - 1} A_q(v - 1, d; k - 1) \right\rfloor \leq \left\lfloor \frac{q^v - 1}{q^{v-k} - 1} A_q(v - 1, d; k) \right\rfloor,$$

where equality holds iff $v = 2k$.

Proof. By considering orthogonal codes we obtain equality for $v = 2k$. Now we assume $k < v/2$ and show

$$\frac{q^v - 1}{q^k - 1} A_q(v - 1, d; k - 1) + 1 \leq \frac{q^v - 1}{q^{v-k} - 1} A_q(v - 1, d; k), \tag{5}$$

which implies the proposed statement. Considering the size of the lifted MRD code we can lower bound the right hand side of Inequality (5) to

$$\frac{q^v - 1}{q^{v-k} - 1} A_q(v - 1, d; k) \geq \frac{q^v - 1}{q^{v-k}} \cdot q^{(v-k-1)(k-d/2+1)}.$$

Since

$$\frac{\begin{bmatrix} v-1 \\ k-1 \end{bmatrix}_q}{\begin{bmatrix} v-k+d/2-1 \\ d/2-1 \end{bmatrix}_q} = \frac{\prod_{i=1}^{k-1} \frac{q^{v-k+i}-1}{q^i-1}}{\prod_{i=1}^{d/2-1} \frac{q^{v-k+i}-1}{q^i-1}} \leq \prod_{i=d/2}^{k-1} \frac{q^{v-k+i}}{q^i - 1} = q^{(v-k)(k-d/2)} \prod_{i=d/2}^{k-1} \frac{1}{1 - q^{-i}}$$

we can use the Anticode bound to upper bound the left hand side of Inequality (5) to

$$\frac{q^v-1}{q^k-1}A_q(v-1,d;k-1)+1\le \frac{q^v-1}{q^k-1}\cdot q^{(v-k)(k-d/2)}\cdot \mu(k-1,d/2,q)+1,$$

where $\mu(a,b,q):=\prod_{i=b}^{a}\left(1-q^{-i}\right)^{-1}$. Thus, it suffices to verify

$$\frac{q^{k-d/2+1}}{q^k-1}\cdot \mu(k-1,d/2,q)+\frac{1}{f}\le 1, \tag{6}$$

where we have divided by

$$f:=\frac{q^v-1}{q^{v-k}}\cdot q^{(v-k-1)(k-d/2+1)}=\frac{q^v-1}{q}\cdot q^{(v-k-1)(k-d/2)}.$$

Since $d\ge 4$, we have $\mu(k-1,d/2,q)\le \prod_{i=2}^{\infty}\left(1-q^{-i}\right)^{-1}\le \prod_{i=2}^{\infty}\left(1-2^{-i}\right)^{-1}<1.74$.

Since $v\ge 4$ and $q\ge 2$, we have $\frac{1}{f}\le \frac{2}{15}$. Since $k\ge 2$, we have $\frac{q^{k-d/2+1}}{q^k-1}\le \frac{q}{q^2-1}$, which is at most $\frac{3}{8}$ for $q\ge 3$. Thus, Inequality (6) is valid for all $q\ge 3$.

If $d\ge 6$ and $q=2$, then $\mu(k-1,d/2,q)\le \prod_{i=3}^{\infty}\left(1-2^{-i}\right)^{-1}<1.31$ and $\frac{q^{k-d/2+1}}{q^k-1}\le \frac{1}{3}$, so that Inequality (6) is satisfied.

In the remaining part of the proof we assume $d=4$ and $q=2$. If $k=2$, then $\mu(k-1,d/2,q)=1$ and $\frac{q^{k-d/2+1}}{q^k-1}=\frac{2}{3}$. If $k=3$, then $\mu(k-1,d/2,q)=\frac{4}{3}$ and $\frac{q^{k-d/2+1}}{q^k-1}=\frac{4}{7}$. If $k\ge 4$, then $\frac{q^{k-d/2+1}}{q^k-1}\le \frac{8}{15}$, $\mu(k-1,d/2,q)\le 1.74$, and $\frac{1}{f}\le \frac{2}{255}$ due to $v\ge 2k\ge 8$. Thus, Inequality (6) is valid in all cases.   □

Knowing the optimal choice between Inequality (3) and Inequality (4), we can iteratively apply Theorem 7 in an ideal way initially assuming $k\le v/2$:

Corollary 2. (Implication of the Johnson type bound II)

$$A_q(v,d;k)\le \left\lfloor \frac{q^v-1}{q^k-1}\left\lfloor \frac{q^{v-1}-1}{q^{k-1}-1}\left\lfloor \cdots \left\lfloor \frac{q^{v-k+d/2+1}-1}{q^{d/2+1}-1}A_q(v-k+d/2,d;d/2)\right\rfloor \cdots \right\rfloor \right\rfloor \right\rfloor$$

We remark that this upper bound is commonly stated in an explicit version, where $A_q(v-k+d/2,d;d/2)\le \left\lfloor \frac{q^{v-k+d/2}-1}{q^{d/2}-1}\right\rfloor$ is inserted, see e.g. [17, Theorem 6], [29, Theorem 7], and [43, Corollary 3]. However, currently much better bounds for partial spreads are available.

It is shown in [43] that the Johnson bound of Theorem 7 improves on the Anticode bound in Theorem 5, see also [6]. To be more precise, removing the floors in the upper bound of Corollary 2 and replacing $A_q(v-k+d/2,d;d/2)$ by $\frac{q^{v-k+d/2}-1}{q^{d/2}-1}$ gives

$$\prod_{i=0}^{k-d/2}\frac{q^{v-i}-1}{q^{k-i}-1}=\frac{\prod_{i=0}^{k-1}\frac{q^{v-i}-1}{q^{k-i}-1}}{\prod_{i=k-d/2+1}^{k-1}\frac{q^{v-i}-1}{q^{k-i}-1}}=\frac{\left[\begin{smallmatrix}v\\k\end{smallmatrix}\right]_q}{\left[\begin{smallmatrix}v-k+d/2-1\\d/2-1\end{smallmatrix}\right]_q},$$

which is the right hand side of the Anticode bound for $k \leq v - k$. So, all upper bounds mentioned so far are (weakly) dominated by Corollary 2, if we additionally assume $k \leq v - k$. As a possible improvement [2, Theorem 3] was mentioned as [29, Theorem 8]. Here, we correct typos and give a slightly enlarged proof, thanks to a personal communication with Aydinian.

Theorem 8. *[2, Theorem 3] For integers $0 \leq t < r \leq k$, $k - t \leq m \leq v$, and $t \leq v - m$ we have*

$$A_q(v, 2r; k) \leq \frac{\left[{v \atop k} \right]_q A_q(m, 2r - 2t; k - t)}{\sum_{i=0}^t q^{i(m+i-k)} \left[{m \atop k-i} \right]_q \left[{v-m \atop i} \right]_q}.$$

Proof. Let W be a fixed subspace with $\dim(W) = m$ and define

$$\mathcal{B} = \left\{ U \in \left[{V \atop k} \right] \mid \dim(U \cap W) \geq k - t \right\},$$

so that $\#\mathcal{B}$ is given by Lemma 2. Consider a $(v, \#\mathcal{C}^*, d; k)$ code $\mathcal{C}^* \subseteq \mathcal{B}$ and take $\mathcal{C}' := \mathcal{C}^* \cap W$ noting that the latter has a minimum distance of at least $2r - 2t$. Two arbitrary codewords $U_1 \neq U_2 \in \mathcal{C}'$ have distance $d_s(U_1, U_2) \geq 2r - 2t + i + j$, where we write $\dim(U_1) = k - t + i$ and $\dim(U_2) = k - t + j$ for integers $0 \leq i, j \leq t$. Replacing each codeword of $\mathcal{C}'$ by an arbitrary $k - t$-dimensional subspace, we obtain a constant dimension code $\mathcal{C}$ with a minimum distance of at least $2r - 2t$. Since $t < r$ we have $\#\mathcal{C}^* = \#\mathcal{C}' = \#\mathcal{C}$, so that Corollary 1 gives the proposed upper bound.   □

As Theorem 8 has quite some degrees of freedom, we partially discuss the optimal choice of parameters. For $t = 0$ and $m \leq v - 1$, we obtain $A_q(v, d; k) \leq \left[{v \atop k} \right]_q / \left[{m \atop k} \right]_q \cdot A_q(m, d; k)$, which is the $(v - m)$-fold iteration of Inequality (4) of the Johnson bound (without rounding). Thus, $m = v - 1$ is the best choice for $t = 0$, yielding a bound that is equivalent to Inequality (4). For $t = 1$ and $m = v - 1$ the bound can be rewritten to $A_q(v, d; k) \leq A_q(v - 1, d - 2; k - 1)$, see the proof of Proposition 4. For $t > v - m$ the bound remains valid but is strictly weaker than for $t = v - m$. Choosing $m = v$ gives the trivial bound $A_q(v, 2r; k) \leq A_q(m, 2r - 2t; k - t)$. For the range of parameters $2 \leq q \leq 9$, $4 \leq v \leq 100$, limited facing numerical pitfalls, and $4 \leq d \leq 2k \leq v$, where q is of course a prime power and d is even, the situation is as follows. If $d \neq 2k$, there are no proper improvements with respect to Theorem 7. For the case $d = 2k$, i.e., partial spreads treated in the next subsection, we have some improvements compared to $\lfloor (q^v - 1)/(q^k - 1) \rfloor$ which is the most trivial bound for partial spreads. Within our numerical range, most of them are covered by the following proposition, where we apply Theorem 8 with $t = 1$ and $m = v - 1$ to $A_q(v, 2k; k)$. The other cases are due to the fact that Theorem 14 is tighter than Theorem 16 for larger values of z. In no case a proper improvement with respect to the tighter bounds from the next subsection emerged.

Proposition 4. *For $w \geq 1$ and $k \geq q^w + 3$ we have $A_q(2k + w, 2k; k) \leq$*

$$\left\lfloor \frac{\left[{2k+w \atop k} \right]_q A_q(2k + w - 1, 2k - 2; k - 1)}{\sum_{i=0}^1 q^{i(k+w-1+i)} \left[{2k+w-1 \atop k-i} \right]_q \left[{(2k+w)-(2k+w-1) \atop i} \right]_q} \right\rfloor < \left\lfloor \frac{q^{2k+w} - 1}{q^k - 1} \right\rfloor = q^{k+w} + q^w$$

Proof. Note that $k \geq q^w + 3$ implies $w < k$. The left hand side simplifies to

$$\frac{\left[\begin{smallmatrix} 2k+w \\ k \end{smallmatrix}\right]_q A_q(2k+w-1, 2k-2; k-1)}{\sum_{i=0}^{1} q^{i(k+w-1+i)} \left[\begin{smallmatrix} 2k+w-1 \\ k-i \end{smallmatrix}\right]_q \left[\begin{smallmatrix} (2k+w)-(2k+w-1) \\ i \end{smallmatrix}\right]_q} = A_q(2k+w-1, 2k-2; k-1).$$

Then we apply Theorem 16 with $t = 2$, $r = w + 1$, and $z = \left[\begin{smallmatrix} w \\ 1 \end{smallmatrix}\right]_q - 1$, which yields $A_q(2k+w-1, 2k-2; k-1) \leq q^{k+w}+1+q^w-q < q^{k+w}+q^w$ for $k-1 \geq q^w + 2$.   □

We remark that applying Theorems 14 and 16 directly is at least as good as the application of Theorem 8 with $t = 1$ and $m = v - 1$ for $d = 2k$.

The Delsarte linear programming bound for the q-Johnson scheme was obtained in [11]. However, numerical computations indicate that it is not better than the Anticode bound, see [6]. For $d \neq 2\min\{k, v - k\}$, i.e., the non-partial spread case, besides the stated bound only the following two specific bounds, based on extensive computer calculations, are known:

Theorem 9. *[26, Theorem 1]* $A_2(6, 4; 3) = 77$

Proposition 5. *[24]* $A_2(8, 6; 4) \leq 272$

As the authors of [24] have observed, the Johnson bound of Theorem 7 does not improve upon Corollary 2 when applied to Theorem 9 or Proposition 5.

If we additionally restrict ourselves to constant dimension codes, that contain a lifted MRD code, another upper bound is known:

Theorem 10. *[16, Theorem 10 and 11]* Let $\mathcal{C} \subseteq \left[\begin{smallmatrix} \mathbb{F}_q^v \\ k \end{smallmatrix}\right]$ *be a constant dimension code, with* $v \geq 2k$ *and minimum subspace distance* d, *that contains a lifted MRD code.*

- *If* $d = 2(k - 1)$ *and* $k \geq 3$, *then* $\#\mathcal{C} \leq q^{2(v-k)} + A_q(v - k, 2(k - 2); k - 1)$;
- *if* $d = k$, *where* k *is even, then* $\#\mathcal{C} \leq q^{(v-k)(k/2+1)} + \left[\begin{smallmatrix} v-k \\ k/2 \end{smallmatrix}\right]_q \frac{q^v-q^{v-k}}{q^k-q^{k/2}} + A_q(v - k, k; k).$

3.1   Upper Bounds for Partial Spreads

The case of constant dimension codes with maximum possible subspace distance $d = 2k$ is known under the name partial spreads. Counting points, i.e., 1-dimensional subspaces, in $\mathbb{F}_q^v$ and $\mathbb{F}_q^k$ gives the obvious upper bound $A_q(v, 2k; k) \leq \left[\begin{smallmatrix} v \\ 1 \end{smallmatrix}\right]_q / \left[\begin{smallmatrix} k \\ 1 \end{smallmatrix}\right]_q = (q^v - 1) / (q^k - 1)$. In the case of equality one speaks of spreads, for which a handy existence criterion is known from the work of Segre in 1964.

Theorem 11. *[37, Sect. 6]* $\mathbb{F}_q^v$ *contains a spread if and only if* k *is a divisor of* v.

If k is not a divisor of v, far better bounds are known including some recent improvements, which we will briefly summarize. For a more detailed treatment we refer to e.g. [27]. The best known parametric construction was given by Beutelspacher in 1975:

Theorem 12. *[7] For positive integers v, k satisfying $v = tk + r$, $t \geq 2$ and $1 \leq r \leq k - 1$ we have $A_q(v, 2k; k) \geq 1 + \sum_{i=1}^{t-1} q^{ik+r} = \frac{q^v - q^{k+r} + q^k - 1}{q^k - 1}$ with equality for $r = 1$.*

The determination of $A_2(v, 6; 3)$ for $v \equiv 2 \pmod 3$ was achieved more than 30 years later in [14] and continued to $A_2(v, 2k; k)$ for $v \equiv 2 \pmod k$ and arbitrary k in [33]. Besides the parameters of $A_2(8 + 3l, 6; 3)$, for $l \geq 0$, see [14] for an example showing $A_2(8, 6; 3) \geq 34$, no partial spreads exceeding the lower bound from Theorem 12 are known.

For a long time the best known upper bound on $A_q(v, 2k; k)$ was the one obtained by Drake and Freeman in 1979:

Theorem 13. *[13, Corollary 8] If $v = kt + r$ with $0 < r < k$, then*

$$A_q(v, 2k; k) \leq \sum_{i=0}^{t-1} q^{ik+r} - \lfloor \theta \rfloor - 1 = q^r \cdot \frac{q^{kt} - 1}{q^k - 1} - \lfloor \theta \rfloor - 1,$$

where $2\theta = \sqrt{1 + 4q^k(q^k - q^r)} - (2q^k - 2q^r + 1)$.

Quite recently this bound has been generalized to:

Theorem 14. *[34, Theorem 2.10] For integers $r \geq 1$, $t \geq 2$, $y \geq \max\{r, 2\}$, $z \geq 0$ with $\lambda = q^y$, $y \leq k$, $k = \begin{bmatrix} r \\ 1 \end{bmatrix}_q + 1 - z > r$, $v = kt + r$, and $l = \frac{q^{v-k} - q^r}{q^k - 1}$, we have $A_q(v, 2k; k) \leq lq^k + \left\lceil \lambda - \frac{1}{2} - \frac{1}{2}\sqrt{1 + 4\lambda(\lambda - (z + y - 1)(q - 1) - 1)} \right\rceil$.*

The construction of Theorem 12 is asymptotically optimal for $k \gg r = v \bmod k$, as recently shown by Năstase and Sissokho:

Theorem 15. *[36, Theorem 5] Suppose $v = tk + r$ with $t \geq 1$ and $0 < r < k$. If $k > \begin{bmatrix} r \\ 1 \end{bmatrix}_q$ then $A_q(v, 2k; k) = 1 + \sum_{i=1}^{t-1} q^{ik+r} = \frac{q^v - q^{k+r} + q^k - 1}{q^k - 1}$.*

Applying similar techniques, the result was generalized to $k \leq \begin{bmatrix} r \\ 1 \end{bmatrix}_q$:

Theorem 16. *[34, Theorem 2.9] For integers $r \geq 1$, $t \geq 2$, $u \geq 0$, and $0 \leq z \leq \begin{bmatrix} r \\ 1 \end{bmatrix}_q / 2$ with $k = \begin{bmatrix} r \\ 1 \end{bmatrix}_q + 1 - z + u > r$ we have $A_q(v, 2k; k) \leq lq^k + 1 + z(q - 1)$, where $l = \frac{q^{v-k} - q^r}{q^k - 1}$ and $v = kt + r$.*

Using Theorem 14 the restriction $z \leq \begin{bmatrix} r \\ 1 \end{bmatrix}_q / 2$ can be removed from Theorem 16, see [27].

Currently, Theorems 11, 14, and 16 constitute the tightest parametric bounds for $A_q(v, 2k; k)$. The only known improvements, by exactly one in every case, are given by the 21 specific bounds stated in [34], which are based on the linear programming method applied to projective q^{k-1}-divisible linear error-correcting codes over $\mathbb{F}_q$ with respect to the Hamming distance, see [27]. As this connection seemed to be overlooked before, it may not be improbable that more sophisticated methods from classical coding theory can improve further values, which then imply improved upper bounds for constant dimension codes via the Johnson bound of Theorem 7.

4 The Linkage Construction Revisited

A very effective and widely applicable construction of constant dimension codes was stated by Gluesing-Luerssen and Troha:

Theorem 17. *[22, Theorem 2.3], cf. [38, Corollary 39]   Let C_i be a $(v_i, N_i, d_i; k)_q$ constant dimension code for $i = 1, 2$ and let C_r be a $(k \times v_2, N_r, d_r)_q$ linear rank metric code. Then*

$$\{\tau^{-1}(\tau(U) \mid M) : U \in C_1, M \in C_r\} \cup \{\tau^{-1}(0_{k \times v_1} \mid \tau(W)) : W \in C_2\}$$

is a $(v_1 + v_2, N_1 N_R + N_2, \min\{d_1, d_2, 2d_r\}; k)_q$ constant dimension code.

Here $A|B$ denotes the concatenation of two matrices with the same number of rows and $0_{m \times n}$ denotes the $m \times n$-matrix consisting entirely of zeros. The resulting code depends on the choice of the codes C_1, C_2, C_r and their representatives within isomorphism classes, so that one typically obtains many isomorphism classes of codes with the same parameters.

We remark that [38, Theorem 37] corresponds to the weakened version of Theorem 17 where the codewords from the constant dimension code C_2 are not taken into account, cf. [21, Theorem 5.1]. In [38, Corollary 39] Silberstein and (Horlemann-)Trautmann obtain the same lower bound, assuming $d_1 = d_2 = 2d_r$, which is indeed the optimal choice, and $3k \leq v$.[2]

The main idea behind Theorem 17 is to consider two sets of codewords with disjoint pivot vectors across the two sets and to utilize the interplay between the rank and the subspace distance for a product type construction. Using Lemma 1 the restriction of the disjointness of the pivot vectors can be weakened, which gives the following improvement:

Theorem 18. *Let C_i be a $(v_i, N_i, d_i; k)_q$ constant dimension code for $i = 1, 2$, $d \in 2\mathbb{N}_{\geq 0}$ and let C_r be a $(k \times (v_2 - k + d/2), N_r, d_r)_q$ linear rank metric code. Then*

$$\mathcal{C} = \{\tau^{-1}(\tau(U) \mid M) : U \in C_1, M \in C_r\} \cup \{\tau^{-1}(0_{k \times (v_1 - k + d/2)} \mid \tau(W)) : W \in C_2\}$$

is a $(v_1 + v_2 - k + d/2, N_1 N_R + N_2, \min\{d_1, d_2, 2d_r, d\}; k)_q$ constant dimension code.

---

[2] It can be verified that for $2k \leq v \leq 3k-1$ the optimal choice of Δ in [38, Corollary 39] is given by $\Delta = v - k$. In that case the construction is essentially the union of a lifted MRD code with an $(v - k, \#\mathcal{C}', d; k)_q$ code $\mathcal{C}'$. Note that for $v - k < \Delta \leq v$ the constructed code is an embedded $(\Delta, \#\mathcal{C}', d; k)_q$ code $\mathcal{C}'$.

Proof. The dimension of the ambient space and the codewords of $\mathcal{C}$ directly follow from the construction. Since the constructed matrices all are in rref and pairwise distinct, $\mathcal{C}$ is well defined and we have $\#\mathcal{C} = N_1 N_R + N_2$. It remains to lower bound the minimum subspace distance of $\mathcal{C}$.

Let $A, C \in C_1$ and $B, D \in C_r$. If $A \neq C$, we have

$$d_s(\tau^{-1}((\tau(A) \mid B)), \tau^{-1}((\tau(C) \mid D))) = 2\left(\mathrm{rk}\begin{pmatrix} \tau(A) & B \\ \tau(C) & D \end{pmatrix} - k\right)$$

$$\geq 2\left(\mathrm{rk}\begin{pmatrix} \tau(A) \\ \tau(C) \end{pmatrix} - k\right) = d_s(A, C) \geq d_1$$

using Eq. (1) in the first step. If $A = C$ but $B \neq D$, we have

$$d_s(\tau^{-1}((\tau(A) \mid B)), \tau^{-1}((\tau(C) \mid D))) = 2\left(\mathrm{rk}\begin{pmatrix} \tau(A) & B \\ \tau(C) & D \end{pmatrix} - k\right)$$

$$\geq 2\left(\mathrm{rk}\begin{pmatrix} \tau(A) & B \\ 0 & D - B \end{pmatrix} - k\right) = 2(k + \mathrm{rk}(D - B) - k) \geq 2d_r.$$

For $A' \neq C' \in C_2$ applying Eq. (1) gives

$$d_s(\tau^{-1}(0_{k \times (v_1 - k + d/2)} \mid \tau(A')), \tau^{-1}(0_{k \times (v_1 - k + d/2)} \mid \tau(C'))) = d_s(A', C') \geq d_2.$$

Finally, for two codewords $U \in \{\tau^{-1}(\tau(U) \mid M) \mid U \in C_1, M \in C_r\}$ and $W \in \{\tau^{-1}(0_{k \times (v_1 - k + d/2)} \mid \tau(W)) \mid W \in C_2\}$, we can use the shape of the pivot vectors and apply Lemma 1. The pivot vector $p(U)$ has its k ones in the first v_1 positions and the pivot vector $p(W)$ has its k ones not in the first $v_1 - k + d/2$ positions, so that the ones can coincide at most at the positions $\{v_1 - k + d/2 + 1, \ldots, v_1\}$. Thus, $d_h(p(U), p(W)) \geq k - (k - d/2) + k - (k - d/2) = d$. Lemma 1 then gives $d_s(U, W) \geq d$. $\square$

An example where Theorem 18 yields a larger code than Theorem 17 is e.g. given for the parameters $q = 2$, $v = 7$, $d = 4$, and $k = 3$. In order to apply Theorem 17 we have to choose $v_1 + v_2 = 7$, $3 \leq v_1 \leq 4$, and $3 \leq v_2 \leq 4$. For $v_1 = 3$ we obtain $\#C_1 \leq A_2(3, 4; 3) = 1$ and $\#C_2 \leq A_2(4, 4; 3) = 1$. Since the size of the rank metric code is bounded by $\lceil 2^{4(3-2+1)} \rceil = 2^8$, the constructed code has a size of at most $1 \cdot 2^8 + 1 = 257$. For $v_1 = 4$ the roles of C_1 and C_2 interchange. Since the size of the rank metric code is bounded by $\lceil 2^{3(3-2+1)} \rceil = 2^6$, the constructed code has a size of at most $1 \cdot 2^6 + 1 = 65$. In Theorem 18 we can choose $d = 4$, so that we can drop one column of the zero matrix preceding the matrices of the second set of codewords, i.e., $v_1 + v_2 = 7 + 1 = 8$. Choosing $v_1 = 3$ and $v_2 = 5$ we can achieve $\#C_1 = A_2(3, 4; 3) = 1$ and $\#C_2 = A_2(5, 4; 3) = 9$. Since the size of the rank metric code can attain $\lceil 2^{4(3-2+1)} \rceil = 2^8$ we can construct a code of size $1 \cdot 2^8 + 9 = 265$. While for these parameters sill larger codes are known, the situation significantly changes in general. Considering the range of parameters $2 \leq q \leq 9$, $4 \leq v \leq 19$, and $4 \leq d \leq 2k \leq v$, where q is of course a prime power and d is even, Theorem 17 provides the best known lower bound for $A_q(v, d; k)$ in 41.8% of the cases, while Theorem 18 provides the best known lower bound in

65.6% of the cases. Since the sizes of both constructions can coincide, the sum of both constructions gives more than 100%. In just 34.4% of the cases strictly superior constructions are known compared to Theorem 18, where most of them arose from the so-called Echelon-Ferrers construction or one of their variants, see [23] and the corresponding webpage.[3]

If one is interested in codes of large size, then one should choose the parameters d_1, d_2, d_r, and d, in Theorem 18, as small as possible in order to maximize the sizes N_1, N_2, and N_r, i.e., we can assume $d_1 = d_2 = 2d_r = d$. Moreover, the codes C_1, C_2, and C_r should have the maximum possible size with respect to their specified parameters. For C_r the maximum possible size is $M(q, k, v_2 + d/2, d)$ and for C_i the maximum possible size is $A_q(v_1, d; k)$, where $i = 1, 2$.

Corollary 3. *For positive integers $k \leq \min\{v_1, v_2\}$ and $d \equiv 0 \pmod{2}$ we have* $A_q(v_1 + v_2 - k + d/2, d; k) \geq A_q(v_1, d; k) \cdot M(q, k, v_2 + d/2, d) + A_q(v_2, d; k)$.

Instead of $A_q(v_1, d; k)$ or $A_q(v_2, d; k)$ we may also insert any lower bound of these commonly unknown values. By a variable transformation we obtain:

Corollary 4. *For positive integers $k \leq m \leq v - d/2$ and $d \equiv 0 \pmod{2}$ we have* $A_q(v, d; k) \geq A_q(m, d; k) \cdot M(q, k, v - m + k, d) + A_q(v - m + k - d/2, d; k)$.

For the parameters of spreads the optimal choice of the parameter m in Corollary 4 can be determined analytically:

Lemma 4. *If $d = 2k$ and k divides v, then Corollary 4 gives $A_q(v, d; k) \geq \frac{q^v - 1}{q^k - 1}$ for all $m = k, 2k, \ldots, v - k$ and smaller sizes otherwise.*

Proof. Using $A_q(v', 2k; k) = (q^{v'} - 1)/(q^k - 1)$ for all integers v' being divisible by k, we obtain

$$A_q(v, d; k) \geq A_q(m, d; k) \cdot M(q, k, v - m + k, 2k) + A_q(v - m, 2k; k)$$
$$= \frac{q^m - 1}{q^k - 1} \cdot q^{v-m} + \frac{q^{v-m} - 1}{q^k - 1} = \frac{q^v - 1}{q^k - 1}$$

if k divides m. Otherwise, $A_q(m, 2k; k) \leq (q^m - 1)/(q^k - 1) - 1$ gives a lower bound. □

We remark that the tightest implications of Corollary 4 can be evaluated by dynamic programming. To this end we consider fixed parameters q, d, k and use the abbreviations $a(n) := A_q(n, d; k)$ and $b(n) := M(q, k, n + k, d)$ for integers n, so that the inequality of Corollary 4 reads

$$a(v) \geq a(m) \cdot b(v - m) + a(v - m + k - d/2). \qquad (7)$$

For a given maximal value v we initialize the values $a(n)$ for $1 \leq n \leq v$ by the best known lower bounds for $A_q(n, d; k)$ from other constructions. Then we loop over n from k to v and eventually replace $a(n)$ by

$$\max\{a(m) \cdot b(n - m) + a(n - m + k - d/2) \mid k \leq m \leq n - d/2\}.$$

---

[3] Entries of type `improved_linkage(m)` correspond to Corollary 4 with m chosen as parameter.

By an arithmetic progression we can use (7) in order to obtain a lower bound for $a(v) = A_q(v, d; k)$ given just two initial $a(i)$-values.

Proposition 6. *For positive integers $k \leq v_0$, $2s \geq d$, and $l \geq 0$, we have*

$$a(v_0 + ls) \geq a(v_0) \cdot b(s)^l + a(s - d/2 + k) \left[{l \atop 1} \right]_{b(s)}.$$

If additionally, $v_0 \geq 2k - d/2$ and $k \geq d/2$, then we have

$$a(v_0 + ls) \geq a(s + k - d/2) \cdot (q^{k-d/2+1})^{n_0 - k + d/2} \left[{l \atop 1} \right]_{q^{s(k-d/2+1)}} + a(v_0).$$

Proof. Using Inequality (7) with $v = v_0 + ls$ and $m = v_0 + (l-1)s$ gives

$$a(v_0 + ls) \geq a(v_0 + (l-1)s) \cdot b(s) + a(s + k - d/2).$$

By induction, we obtain

$$a(v_0 + ls) \geq a(v_0 + (l-i)s) \cdot b(s)^i + a(s + k - d/2) \left[{i \atop 1} \right]_{b(s)}$$

for all $0 \leq i \leq l$.

For the second part, applying Inequality (7) with $v = v_0 + ls$ and $m = s + k - d/2$ gives

$$a(v_0 + ls) \geq a(s + k - d/2) \cdot b(v_0 + (l-1)s - k + d/2) + a(v_0 + (l-1)s).$$

By induction, we obtain

$$a(v_0 + ls) \geq a(s + k - d/2) \cdot \sum_{j=1}^{i} b(v_0 + (l-j)s - k + d/2) + a(v_0 + (l-i)s)$$

for all $0 \leq i \leq l$.

If $v_0 \geq 2k - d/2$ and $k \geq d/2$, then

$$b(v_0 + (l-j)s - k + d/2) = (q^{k-d/2+1})^{v_0 + (l-j)s - k + d/2},$$

so that

$$\sum_{j=1}^{l} b(v_0 + (l-j)s - k + d/2) = \sum_{j=1}^{l} (q^{k-d/2+1})^{v_0 + (l-j)s - k + d/2}$$

$$= (q^{k-d/2+1})^{v_0 - k + d/2} \sum_{r=0}^{l-1} (q^{s(k-d/2+1)})^r = (q^{k-d/2+1})^{v_0 - k + d/2} \left[{l \atop 1} \right]_{q^{s(k-d/2+1)}}.$$

$\square$

Example 1. Using $A_2(13, 4; 3) = 1597245$ [9] and $A_2(7, 4; 3) \geq 333$ [23], applying Proposition 6 with $s = 6$ gives

$$A_2(13 + 6l, 4; 3) \geq 4096^l \cdot 1597245 + 333 \cdot \frac{4096^l - 1}{4095}$$

and

$$A_2(13 + 6l, 4; 3) \geq 333 \cdot 16777216 \cdot \frac{4096^l - 1}{4095} + 1597245$$

for all $l \geq 0$.

In the next section we will see that the first lower bound almost meets the Anticode bound.

We remark that Theorem 18 can be easily generalized to a construction based on a union of $m \geq 2$ sets of codewords.

Corollary 5. *For positive integers k, m, and $i = 1, \ldots, m$ let*

- *C_i be an $(v_i, N_i, d_i; k)_q$ constant dimension code,*
- *$\delta_i \in \mathbb{N}_{\geq 0}$, $\delta_m = 0$,*
- *C_i^R be a $(k \times v_i^R, N_i^R, d_i^R)_q$ linear rank metric code, where $v_i^R = \sum_{j=1}^{i-1}(v_j - \delta_j)$ and $i \neq 1$,*
- *$C_1^R = \emptyset$, $v_1^R = 0$, $N_1^R = 1$, and $d_1^R = \infty$.*

Then

$$\bigcup_{i=1}^m \{\tau^{-1}(0_{k \times (v - v_i - v_i^R)} \mid \tau(U_i) \mid M_i) : U_i \in C_i, M_i \in C_i^R\}$$

is a $(v, N, d; k)_q$ constant dimension code with

- *$v = \sum_{i=1}^m (v_i - \delta_i)$,*
- *$N = \sum_{i=1}^m N_i \cdot N_i^R$, and*
- *$d = \min\{d_i, 2d_i^R, 2(k - \delta_i) \mid i = 1, \ldots, m\}$.*

Proof. We prove by inductively applying Theorem 18 $m - 1$ times. Denote

$$\tilde{C}_i := \{\tau^{-1}(0_{k \times (v - v_i - v_i^R)} \mid \tau(U_i) \mid M_i) : U_i \in C_i, M_i \in C_i^R\}$$

for $i = 1, \ldots, m$, i.e., $\tilde{C}_i$ is a padded $(v_i + v_i^R, N_i \cdot N_i^R, \min\{d_i, 2d_i^R\}; k)_q$ constant dimension code. Applying Theorem 18 for $\tilde{C}_1$ and $\tilde{C}_2$ with $d = 2(k - \delta_1)$ yields a $(v_1 + v_2 - \delta_1, N_1 + N_2 \cdot N_2^R, \min\{d_1, d_2, 2d_2^R, 2(k - \delta_1)\}; k)_q$ constant dimension code. If the first m' codes, $\tilde{C}_1, \ldots, \tilde{C}_{m'}$ yield an $(\sum_{i=1}^{m'}(v_i - \delta_i) + \delta_{m'}, \sum_{i=1}^{m'} N_i \cdot N_i^R, \min\{d_i, 2d_i^R, 2(k - \delta_i) \mid i = 1, \ldots, m'\}; k)_q$ constant dimension code $\tilde{C}_{1,\ldots,m'}$, then performing Theorem 18 for this code and $\tilde{C}_{m'+1}$ with $d = 2(k - \delta_{m'})$ yields an $(\sum_{i=1}^{m'}(v_i - \delta_i) + \delta_{m'} + n_2 - \delta_{m'}, \sum_{i=1}^{m'} N_i \cdot N_i^R + N_{m'+1} \cdot N_{m'+1}^R, \min\{d_i, 2d_i^R, 2(k - \delta_i) \mid i = 1, \ldots, m' + 1\}; k)_q$ constant dimension code. □

Since the proof uses multiple applications of Theorem 18 this code can be found by the dynamic programming approach based on Theorem 18, i.e., Corollary 5 is redundant. However, it can be used to prove:

Corollary 6 (cf. [22, Theorem 4.6]). *Let C^R be an $(k \times v_1 + v_2, d)_q$ linear MRD code, where $k \leq v_i$, for $i = 1, 2$ and let C_i be an $(v_{i-2}, N_i, 2d; k)_q$ constant dimension codes for $i = 3, 4$. Then*

$$\{\tau^{-1}(I_{k \times k} \mid A) \mid A \in C^R\}$$
$$\cup \{\tau^{-1}(0_{k \times k} \mid \tau(A) \mid 0_{k \times v_2}) \mid A \in C_3\}$$
$$\cup \{\tau^{-1}(0_{k \times k} \mid 0_{k \times v_1} \mid \tau(A)) \mid A \in C_4\}$$

is a $(v_1 + v_2 + k, q^{(v_1+v_2)(k-d+1)} + N_3 + N_4, 2\min\{d, k\}; k)_q$ constant dimension code. Note that $k < d$ implies $N_3, N_4 \leq 1$.

Proof. Applying Corollary 5 with

- $m = 3$
- $\bar{C}_1 = C_4$, $\bar{C}_2 = C_3$,
- $\bar{C}_3 = \{\tau^{-1}(I_{k \times k})\}$ (i.e., an $(k, 1, \infty; k)_q$ constant dimension code)
- $\delta_1 = \delta_2 = \delta_3 = 0$
- $\bar{C}_1^R = \emptyset$
- $\bar{C}_2^R = \{0_{k \times v_2}\}$ (i.e., an $(k \times v_2, 1, \infty)_q$ rank metric code)
- $\bar{C}_3^R$ an $(k \times (v_1 + v_2, d))_q$ MRD code

yields an $(v_1 + v_2 + k, N_4 + N_3 + q^{(v_1+v_2)(k-d+1)}, 2\min\{d, k\}; k)_q$ constant dimension code:

$$\{\tau^{-1}(I_{k \times k} \mid M_3) : M_3 \in \bar{C}_3^R\}$$
$$\cup \{\tau^{-1}(0_{k \times k} \mid \tau(U_2) \mid 0_{k \times v_2}) : U_2 \in C_3\}$$
$$\cup \{\tau^{-1}(0_{k \times (v_1+k)} \mid \tau(U_1)) : U_1 \in C_4\}$$

$\square$

We remark that $\{(A \mid B) : A \in C_1^R, B \in C_2^R\}$ is a $(k \times (v_1 + v_2), d)_q$ linear MRD code, since each codeword has $\mathrm{rk}(A \mid B) \geq \mathrm{rk}(A) \geq d$. The other direction is not necessarily true, e.g., $\left(\begin{smallmatrix} I_{k-1} \\ 0 \end{smallmatrix} \mid 0 \mid \ldots \mid 0 \mid w \right)$, where w is a non-zero column, cannot be split in two matrices $\left(\begin{smallmatrix} I_{k-1} \\ 0 \end{smallmatrix} \mid 0 \mid \ldots \mid 0 \right)$ and $(0 \mid \ldots \mid 0 \mid w)$ both having rank distance at least d for $d \geq 2$. Hence, this corollary constructs codes of the same size as Theorem 4.6 in [22] but these codes are not necessarily equal.

5 Asymptotic Bounds

Kötter and Kschischang have stated the bounds

$$1 < q^{-k(v-k)} \cdot \begin{bmatrix} v \\ k \end{bmatrix}_q < 4$$

for $0 < k < v$ in [32, Lemma 4] for the q-binomial coefficients. They used this result in order to prove that the lifted MRD codes, they spoke about linearized polynomials, have at least a size of a quarter of the Singleton bound if v tends to infinity. Actually, they have derived a more refined bound, which can be best expressed using the so called q-Pochhammer symbol $(a; q)_n := \prod_{i=0}^{n-1} (1 - aq^i)$ specializing to $(1/q; 1/q)_n = \prod_{i=1}^{n} (1 - 1/q^i)$:

$$1 \leq \frac{\left[\begin{smallmatrix} v \\ k \end{smallmatrix}\right]_q}{q^{k(v-k)}} \leq \frac{1}{(1/q; 1/q)_k} \leq \frac{1}{(1/q; 1/q)_\infty} \leq \frac{1}{(1/2; 1/2)_\infty} \approx 3.4627, \quad (8)$$

where $(1/q; 1/q)_\infty$ denotes $\lim_{n \to \infty} (1/q; 1/q)_n$, cf. the estimation for the Anti-code bound in the proof of Proposition 3. The sequence $(1/q; 1/q)_\infty$ is monotonically increasing with q and approaches $(q-1)/q$ for large q, see e.g. [29,32] for some numerical values.

Lemma 5. *For each $b \in \mathbb{N}_{\geq 0}$ we have $\lim_{a \to \infty} \frac{\left[\begin{smallmatrix} a+b \\ b \end{smallmatrix}\right]_q}{q^{ab}} = \frac{1}{(1/q; 1/q)_b}$.*

Proof. Plugging in the definition of the q-binomial coefficient, we obtain

$$\lim_{a \to \infty} \frac{\left[\begin{smallmatrix} a+b \\ b \end{smallmatrix}\right]_q}{q^{ab}} = \lim_{a \to \infty} \frac{\prod_{i=1}^{b} \frac{q^{a+i}-1}{q^i-1}}{q^{ab}} = \prod_{i=1}^{b} \frac{q^i}{q^i-1} = \prod_{i=1}^{b} \frac{1}{1-1/q^i} = \frac{1}{(1/q; 1/q)_b}.$$

$\square$

Using this asymptotic result we can compare the size of the lifted MRD codes to the Singleton and the Anticode bound.

Proposition 7. *For $k \leq v - k$ the ratio of the size of a lifted MRD code divided by the size of the Singleton bound converges for $v \to \infty$ monotonically decreasing to $(1/q; 1/q)_{k-d/2+1} \geq (1/2; 1/2)_\infty > 0.288788$.*

Proof. Setting $z = k - d/2 + 1$ and $s = v - k$ the ratio is given by $g(s) := \frac{q^{sz}}{\left[\begin{smallmatrix} s+z \\ z \end{smallmatrix}\right]_q}$, so that Lemma 5 gives the proposed limit. The sequence is monotonically decreasing, since we have $0 \leq z - 1 < z \leq s + z$ and

$$\frac{g(s)}{g(s+1)} = \frac{q^{sz} \left[\begin{smallmatrix} s+1+z \\ z \end{smallmatrix}\right]_q}{\left[\begin{smallmatrix} s+z \\ z \end{smallmatrix}\right]_q q^{(s+1)z}} = \frac{q^z \left[\begin{smallmatrix} s+z \\ z \end{smallmatrix}\right]_q + \left[\begin{smallmatrix} s+z \\ z-1 \end{smallmatrix}\right]_q}{\left[\begin{smallmatrix} s+z \\ z \end{smallmatrix}\right]_q q^z} = 1 + \frac{\left[\begin{smallmatrix} s+z \\ z-1 \end{smallmatrix}\right]_q}{\left[\begin{smallmatrix} s+z \\ z \end{smallmatrix}\right]_q q^z} > 1.$$

$\square$

Proposition 8. *For $k \leq v - k$ the ratio of the size of a lifted MRD code divided by the size of the Anticode bound converges for $v \to \infty$ monotonically decreasing to $\frac{(1/q; 1/q)_k}{(1/q; 1/q)_{d/2-1}} \geq \frac{q}{q-1} \cdot (1/q; 1/q)_k \geq 2 \cdot (1/2; 1/2)_\infty > 0.577576$.*

Proof. The lifted MRD code has cardinality $q^{(v-k)(k-d/2+1)}$ and the Anticode bound is $\left[\begin{smallmatrix} v \\ k \end{smallmatrix}\right]_q / \left[\begin{smallmatrix} v-k+d/2-1 \\ d/2-1 \end{smallmatrix}\right]_q$. From Lemma 5 we conclude

$$\lim_{v \to \infty} \frac{\left[\begin{smallmatrix} (v-k)+k \\ k \end{smallmatrix}\right]_q}{q^{(v-k)k}} = \frac{1}{(1/q; 1/q)_k} \quad \text{and} \quad \lim_{v \to \infty} \frac{\left[\begin{smallmatrix} (v-k)+(d/2-1) \\ d/2-1 \end{smallmatrix}\right]_q}{q^{(v-k)(d/2-1)}} = \frac{1}{(1/q; 1/q)_{d/2-1}},$$

so that the limit follows. The subsequent inequalities follow from $d \geq 4$, the monotonicity of $(1/q; 1/q)_n$, and $q \geq 2$.

It remains to show the monotonicity of the sequence

$$g(v) := \frac{q^{(v-k)(k-d/2+1)} \begin{bmatrix} v-k+d/2-1 \\ d/2-1 \end{bmatrix}_q}{\begin{bmatrix} v \\ k \end{bmatrix}_q}.$$

Using the abbreviation $s = v - k$ we define

$$f(x) := \frac{\begin{bmatrix} s+x \\ s+1 \end{bmatrix}_q}{q^x \begin{bmatrix} s+x \\ s \end{bmatrix}_q} = \frac{\prod_{i=1}^{s+1} \frac{q^{x-1+i}-1}{q^i-1}}{q^x \prod_{i=1}^{s} \frac{q^{x+i}-1}{q^i-1}} = \frac{\frac{q^x-1}{q^{s+1}-1}}{q^x} = \frac{1-q^{-x}}{q^{s+1}-1}$$

and observe that f is strictly monotonically increasing in x, so that $f(k) > f(d/2 - 1)$. Using routine manipulations of q-binomial coefficients we compute

$$\frac{g(v)}{g(v+1)} = (1 + f(k)) \cdot (1 + f(d/2 - 1))^{-1} > 1.$$

$\square$

In other words the ratio between the best known lower bound and the best known upper bound for constant dimension codes is strictly greater than 0.577576 for all parameters and the most challenging parameters are given by $q = 2$, $d = 4$, and $k = \lfloor v/2 \rfloor$.

Replacing the Anticode bound by the Johnson bound of Theorem 2 does not change the limit behavior of Proposition 8 when v tends to infinity. As stated above, we obtain the Anticode bound if we remove the floors in Corollary 2 and replace $A_q(v - k + d/2, d; d/2)$ by $\frac{q^{v-k+d/2}-1}{q^{d/2}-1}$. First we consider the latter weakening. Applying the lower bound of Theorem 12 for $A_q(v', 2k'; k')$, where $v' = tk' + r$ with $1 \leq r \leq k' - 1$, we consider

$$\frac{q^{v'} - q^{k'+r} + q^{k'} - 1}{q^{k'} - 1} \Big/ \frac{q^{v'} - 1}{q^{k'} - 1} = 1 - \frac{q^{k'} \cdot (q^r - 1)}{q^{v'} - 1}.$$

If $v' \geq 3k'$, then the subtrahend on the right hand side is at most $\frac{q^{v'/3} \cdot (q^{v'/3}-1)}{q^{v'}-1}$. Otherwise we have $2k' \leq v' < 3k'$, so that $v' = 2k' + r$. Since $q^{k'} \cdot (q^r - 1) \cdot (q^{k'} + 1) = q^{2k'+r} - q^{2k'} + q^{k'+r} - q^{k'} \leq q^{2k'+r} - 1$ the subtrahend on the right hand side is at most $1/\left(q^{v'/3} + 1\right)$. Thus, the ratio between the lower and the upper bound for partial spreads tends to 1 if $v' = v - k + d/2$ tends to infinity. Since

$$\left. \left. \left\lfloor \frac{q^v-1}{q^k-1} \left\lfloor \frac{q^{v-1}-1}{q^{k-1}-1} \left\lfloor \cdots \left\lfloor \frac{q^{v-k+d/2+1}-1}{q^{d/2+1}-1} \left\lfloor \frac{q^{v-k+d/2}-1}{q^{d/2}-1} \right\rfloor \right\rfloor \cdots \right\rfloor \right\rfloor \right\rfloor \right.$$

$$\geq \left(\frac{q^v-1}{q^k-1} \left(\frac{q^{v-1}-1}{q^{k-1}-1} \left(\cdots \left(\frac{q^{v-k+d/2}-1}{q^{d/2}-1} - 1 \right) \cdots \right) - 1 \right) - 1 \right) - 1$$

$$\geq \frac{\left[\begin{smallmatrix} v \\ k \end{smallmatrix} \right]_q}{\left[\begin{smallmatrix} v-k+d/2-1 \\ d/2-1 \end{smallmatrix} \right]_q} - (k - d/2 + 1) \cdot \frac{\left[\begin{smallmatrix} v-1 \\ k-1 \end{smallmatrix} \right]_q}{\left[\begin{smallmatrix} v-k+d/2-1 \\ d/2-1 \end{smallmatrix} \right]_q}$$

$$\geq \frac{\left[\begin{smallmatrix} v \\ k \end{smallmatrix} \right]_q}{\left[\begin{smallmatrix} v-k+d/2-1 \\ d/2-1 \end{smallmatrix} \right]_q} \cdot \left(1 - \frac{4(k - d/2 + 1)}{q^{v-k}} \right)$$

the ratio between Corollary 2 and the Anticode bound tends to 1 as v tends to infinity.

Next, we consider the ratio between the lower bound from the first construction of Proposition 6 and the Anticode bound when l tends to infinity.

Proposition 9. *For integers satisfying the conditions of Proposition 6, $k \leq s$ and $d \leq 2k$, we have*

$$\lim_{l \to \infty} \left(b(s)^l a(v_0) + a(s - d/2 + k) \left[\begin{smallmatrix} l \\ 1 \end{smallmatrix} \right]_{b(s)} \right) \Big/ \frac{\left[\begin{smallmatrix} v_0+ls \\ k \end{smallmatrix} \right]_q}{\left[\begin{smallmatrix} v_0+ls-k+d/2-1 \\ d/2-1 \end{smallmatrix} \right]_q}$$

$$= \frac{a(v_0) + \frac{a(s-d/2+k)}{q^{s(k-d/2+1)}-1}}{q^{(v_0-k)(k-d/2+1)}} \cdot \prod_{i=d/2}^{k} \left(1 - \frac{1}{q^i} \right).$$

Proof. For $k \leq s$ and $k - d/2 + 1 \geq 1$ we have $b(s) \neq 1$, so that

$$b(s)^l a(v_0) + a(s') \left[\begin{smallmatrix} l \\ 1 \end{smallmatrix} \right]_{b(s)} = q^{lsk'} a(v_0) + a(s') \frac{q^{lsk'}-1}{q^{sk'}-1}$$

$$= q^{lsk'} \left(a(v_0) + \frac{a(s')}{q^{sk'}-1} \right) - \frac{a(s')}{q^{sk'}-1},$$

where we abbreviate $s' = s - d/2 + k$ and $k' = k - d/2 + 1$. Thus,

$$\lim_{l \to \infty} \left(b(s)^l a(v_0) + a(s') \left[\begin{smallmatrix} l \\ 1 \end{smallmatrix} \right]_{b(s)} \right) \Big/ q^{lsk'} = a(v_0) + \frac{a(s')}{q^{sk'}-1}.$$

Plugging in the definition of the q-binomial coefficients gives

$$\frac{\left[\begin{smallmatrix} v_0+ls \\ k \end{smallmatrix} \right]_q}{\left[\begin{smallmatrix} v_0+ls-k+d/2-1 \\ d/2-1 \end{smallmatrix} \right]_q} = \frac{\prod_{i=1}^{k} \frac{q^{v_0+ls-k+i}-1}{q^i-1}}{\prod_{i=1}^{d/2-1} \frac{q^{v_0+ls-k+i}-1}{q^i-1}} = \prod_{i=d/2}^{k} \frac{q^{v_0+ls-k+i}-1}{q^i-1},$$

so that

$$\lim_{l \to \infty} \frac{\left[\begin{smallmatrix} v_0+ls \\ k \end{smallmatrix} \right]_q}{\left[\begin{smallmatrix} v_0+ls-k+d/2-1 \\ d/2-1 \end{smallmatrix} \right]_q} \Big/ q^{lsk'} = \prod_{i=d/2}^{k} \frac{q^{v_0-k+i}}{q^i-1} = q^{(v_0-k)k'} \cdot \prod_{i=d/2}^{k} \frac{1}{1 - \frac{1}{q^i}}.$$

Dividing both derived limits gives the proposed result.   □

For Example 1 with $d = 4$ and $k = 3$, we obtain a ratio of

$$\left(1597245 + \frac{A_2(7,4;3)}{4095}\right) \cdot 21/2^{25} \in [0.99963386, 0.99963388]$$

for $v = 13 + 6l$ with $l \to \infty$ using $333 \leq A_2(7,4;3) \leq 381$, i.e., the Anticode bound is almost met by the underlying improved linkage construction.

6   Codes Better Than the MRD Bound

For constant dimension codes that contain a lifted MRD code, Theorem 10 gives an upper bound which is tighter than the Johnson bound of Theorem 7. In [5] two infinite series of constructions have been given where the code sizes exceed the MRD bound of Theorem 10 for $q = 2$, $d = 4$, and $k = 3$. Given the data available from [23] we mention that, besides $d = 4$, $k = 3$, the only other case where the MRD bound was superseded is $A_2(8,4;4) \geq 4801 > 4797$, see [10]. Next, we show that for $d = 4$ and $k = 3$ the MRD bound can be superseded for all field sizes q if v is large enough. For the limit of the achievable ratio we obtain:

Proposition 10. *For $q \geq 3$ we have* $\lim\limits_{v \to \infty} \frac{A_q(v,4;3)}{q^{2v-6} + \left[\begin{smallmatrix} v-3 \\ 2 \end{smallmatrix}\right]_q} \geq 1 + \frac{1}{2q^3}$.

Proof. For $q \geq 2$, [25, Theorem 4] gives

$$A_q(7,4;3) \geq q^8 + q^5 + q^4 + q^2 - q \geq q^8 + q^5 + q^4.$$

With this, we conclude

$$A_q(v_0,4;3) \geq A_q(7,4;3) \cdot q^{2v_0-14} + A_q(v_0-6,4;3) \geq q^{2v_0-10} \cdot \left(q^4 + q + 1\right)$$

from Corollary 4 choosing $m = 7$. Applying Proposition 6 with $s = 3$ gives

$$A_q(v_0+3l,4;3) \geq q^{6l} A_q(v_0,4;3) + \frac{q^{6l}-1}{q^6-1} \geq q^{6l} A_q(v_0,4;3)$$

for $v_0 \in \{12,13,14\}$, so that $A_q(v,4;3) \geq q^{2v-10} \cdot \left(q^4 + q + 1\right)$ for all $v \geq 12$.
    From Lemma 5 we conclude

$$\lim_{v \to \infty} \frac{q^{2v-6} + \left[\begin{smallmatrix} v-3 \\ 2 \end{smallmatrix}\right]_q}{q^{2v-10}} = q^4 + (1/q;1/q)_2 = \frac{q^3(q^4 - q^3 - q^2 + q + 1)}{(q-1)^2(q+1)}.$$

Since

$$\left(q^4 + q + 1\right) \Big/ \frac{q^3(q^4 - q^3 - q^2 + q + 1)}{(q-1)^2(q+1)} = 1 + \frac{1}{q^3} - \frac{q+1}{q^2(q^4 - q^3 - q^2 + q + 1)},$$

the statement follows for $q \geq 3$.   □

For $q = 2$ the estimations of the proof of Proposition 10 are too crude in order to obtain a factor larger than one. However, for the binary case better codes with moderate dimensions of the ambient space have been found by computer searches – with the prescription of automorphisms as the main ingredient in order to reduce the computational complexity, see e.g. [31].

Proposition 11. *For $v \geq 19$ we have* $\dfrac{A_2(v,4;3)}{2^{2v-6} + \begin{bmatrix} v-3 \\ 2 \end{bmatrix}_2} \geq 1.3056.$

Proof. Applying Proposition 6 with $s = 3$ and using $A_2(4,4;3) \geq 0$ gives $A_2(v_0 + 3l, 4; 3) \geq A_2(v_0, 4; 3) \cdot 2^{6l}$ for all $v_0 \geq 6$ and $l \geq 0$, so that

$$\frac{A_2(v_0 + 3l, 4, 3)}{2^{2(v_0+3l)-6} + \begin{bmatrix} (v_0+3l)-3 \\ 2 \end{bmatrix}_2} \geq \frac{A_2(v_0, 4; 3)}{\frac{7}{3} \cdot 2^{2v_0-7}}. \tag{9}$$

Using $A_2(7, 4; 3) \geq 333$ [23], $A_2(8, 4; 3) \geq 1326$ [10], $A_2(9, 4; 3) \geq 5986$ [10], and $A_2(13, 4; 3) = 1597245$ [9] we apply Corollary 4 with $m = 13$ to obtain lower bounds for $A_2(v_0, 4; 3)$ with $19 \leq v_0 \leq 21$. For these values of v_0 the minimum of the right hand side of Inequality (9) is attained at $v_0 = 20$ with value 1.3056442377. □

Note that the application of Proposition 6 was used in a rather crude estimation in the proof of Proposition 11. Actually, we do not use the codewords generated by the codewords of the constant dimension code C_2 in Theorem 18, so that we might have applied [38, Theorem 37] directly for this part of the proof – similarly for Proposition 10, which then allows to consider just one instead of $s = 3$ *starters*. In the latter part of the proof of Proposition 11 the use of Corollary 4 is essential in order to obtain large codes for medium sized dimensions of the ambient space from $A_2(13, 4; 3) = 1597245$ and relatively good lower bounds for small dimensions. This is a relative typical behavior of Corollary 4 and Proposition 6, i.e., the first few applications yield a significant improvement which quickly bottoms out – in a certain sense. As column bklb of Table 3 suggests, we may slightly improve upon the value stated in Proposition 11 by some fine-tuning effecting the omitted less significant digits.

In Tables 1, 2, and 3 we compare the sizes of different constructions with the lifted MRD and the best known upper bound. Here bklb and bkub stand for best known lower and upper bound respectively. The values of Theorem 10 are given in column mrdb. Applying Theorems 17 and 18 to the best known codes give the columns lold and lnew, respectively. The results obtained in [5] are stated in column ea. The achieved ratio between the mentioned constructions and the MRD bound can be found in Table 3. Since differences partially are beyond the given accuracy, we give absolute numbers in Table 1. Note that the values in column bklb of Table 3 show that Proposition 11 is also valid for $v \geq 16$, while we have a smaller ratio for $v < 16$. The relative advantage over lifted MRD codes is displayed in Table 2.

To conclude this section, we remark that an application of Corollary 4 with $2k \leq m \leq v - k$ using a lifted MRD in the constant dimension code C_1 cannot generate a code that exceeds the MRD bound of Theorem 10.

Table 1. Lower and upper bounds for $A_2(v, 4; 3)$.

v	bklb	mrdb	bkub	lold	lnew	ea
6	77	71	77	65	65	
7	333	291	381	257	265	301
8	1326	1179	1493	1033	1101	1117
9	5986	4747	6205	4929	4929	4852
10	23870	19051	24698	21313	21313	18924
11	97526	76331	99718	85249	85257	79306
12	385515	305579	398385	383105	383105	309667
13	1597245	1222827	1597245	1532417	1532425	1287958
14	6241665	4892331	6387029	6241665	6241665	4970117
15	24966665	19571371	25562941	24966657	24966665	20560924
16	102223681	78289579	102243962	102223681	102223681	79608330
17	408894729	313166507	409035142	408894721	408894729	
18	1635578957	1252682411	1636109361	1635578889	1635578957	
19	6542315853	5010762411	6544674621	6542315597	6542315853	5200895489

Table 2. Lower and upper bounds for $A_2(v, 4; 3)$ divided by the size of a corresponding lifted MRD code.

v	bklb	mrdb	bkub	lold	lnew	ea
6	1.203125	1.109375	1.203125	1.015625	1.015625	
7	1.300781	1.136719	1.488281	1.003906	1.035156	1.175781
8	1.294922	1.151367	1.458008	1.008789	1.075195	1.090820
9	1.461426	1.158936	1.514893	1.203369	1.203369	1.184570
10	1.456909	1.162781	1.507446	1.300842	1.300842	1.155029
11	1.488129	1.164719	1.521576	1.300797	1.300919	1.210114
12	1.470623	1.165691	1.519718	1.461430	1.461430	1.181286
13	1.523252	1.166179	1.523252	1.461427	1.461434	1.228292
14	1.488129	1.166423	1.522786	1.488129	1.488129	1.184968
15	1.488129	1.166545	1.52367	1.488129	1.488129	1.225527
16	1.523252	1.166606	1.523554	1.523252	1.523252	1.186257
17	1.523252	1.166636	1.523775	1.523252	1.523252	
18	1.523252	1.166651	1.523746	1.523252	1.523252	
19	1.523252	1.166659	1.523801	1.523252	1.523252	1.210928

Lemma 6. *Using the notation of Theorem 18, let $k \leq \min\{v_1 - k, v_2 - k + d/2\}$, C_r a linear MRD code, $d_r = d_1/2$, and C_1 contains a lifted MRD code (in $\begin{bmatrix} \mathbb{F}_q^{v_1} \\ k \end{bmatrix}$). Then, the codes constructed in Theorem 18 contain a lifted MRD code (in $\begin{bmatrix} \mathbb{F}_q^{v_1+v_2-k+d/2} \\ k \end{bmatrix}$).*

Table 3. Lower and upper bounds for $A_2(v, 4; 3)$ divided by the corresponding MRD bound.

v	bklb	mrdb	bkub	lold	lnew	ea
6	1.084507	1.0	1.084507	0.915493	0.915493	
7	1.144330	1.0	1.309278	0.883162	0.910653	1.034364
8	1.124682	1.0	1.266327	0.876166	0.933842	0.947413
9	1.261007	1.0	1.307141	1.038340	1.038340	1.022119
10	1.252953	1.0	1.296415	1.118734	1.118734	0.993334
11	1.277672	1.0	1.306389	1.116833	1.116938	1.038975
12	1.261589	1.0	1.303705	1.253702	1.253702	1.013378
13	1.306190	1.0	1.306190	1.253176	1.253182	1.053263
14	1.275806	1.0	1.305519	1.275806	1.275806	1.015900
15	1.275673	1.0	1.306140	1.275672	1.275673	1.050561
16	1.305712	1.0	1.305972	1.305712	1.305712	1.016845
17	1.305678	1.0	1.306127	1.305678	1.305678	
18	1.305661	1.0	1.306085	1.305661	1.305661	
19	1.305653	1.0	1.306124	1.305653	1.305653	1.037945

Proof. Let $\{\tau^{-1}(I_{k\times k} \mid M) : M \in R\} \subseteq C_1$ be the lifted MRD code in C_1. Since R is a $(k \times (v_1 - k), d_1/2)_q$ MRD code, we have $\#R = q^{(v_1-k)(k-d_1/2+1)}$. The first set of the construction contains

$$\{\tau^{-1}(I_{k\times k} \mid M \mid A) : M \in R, A \in C_r\}$$

in which $\{(M \mid A) : M \in R, A \in C_r\}$ forms a $(k \times (v_1 + v_2 - 2k + d/2), N, d_r)_q$ rank metric code of size $N = q^{(v_1+v_2-2k+d/2)(k-d_r+1)}$, hence it is a maximum rank metric code.   □

7   Conclusion

In this paper we have considered the maximal sizes of constant dimension codes. With respect to constructive lower bounds we have improved the so-called linkage construction, which then yields the best known codes for many parameters, see Footnote 2. With respect to upper bounds there is a rather clear picture. The explicit Corollary 2, which refers back to bounds for partial spreads, is the best known parametric bound in the case of $d \neq 2\min\{k, v - k\}$, while Theorem 8 or the linear programming method may possibly yield improvements. Since Theorem 8 implies the Johnson bound and so Corollary 2, it would be worthwhile to study whether it can be strictly sharper than Theorem 7 for $d \neq 2\min\{k, v - k\}$ at all. Compared to Corollary 2, the only two known improvements are given for the specific parameters from Theorem 9 and Proposition 5. In the case of partial spreads we have reported the current state-of-the-art mentioning that further improvements are far from being unlikely.

In general we have shown that the ratio between the best-known lower and upper bound is strictly larger than 0.577576 for all parameters. The bottleneck is formed by the parameters $q = 2$, $d = 4$, and $k = \lfloor v/2 \rfloor$, where no known method can properly improve that factor, see Footnote 2 for the linkage construction. For $d = 4$, $k = 3$ and general field sizes q we have applied the improved linkage construction in order to show that $A_q(v, d; k)$ is by a factor, depending on q, larger than the MRD bound for sufficiently large dimensions v.

Acknowledgement. The authors would like to thank Harout Aydinian for providing an enlarged proof of Theorem 8, Natalia Silberstein for explaining the restriction $3k \leq v$ in [38, Corollary 39], Heide Gluesing-Luerssen for clarifying the independent origin of the linkage construction, and Alfred Wassermann for discussions about the asymptotic results of Frankl and Rödl. We thank the reviewers for their comments that helped us to improve the presentation of the paper.

References

1. Agrell, E., Vardy, A., Zeger, K.: Upper bounds for constant-weight codes. IEEE Trans. Inform. Theory **46**(7), 2373–2395 (2000)
2. Ahlswede, R., Aydinian, H.: On error control codes for random network coding. In: Workshop on Network Coding, Theory, and Applications, NetCod 2009, pp. 68–73. IEEE (2009)
3. Ahlswede, R., Aydinian, H.K., Khachatrian, L.H.: On perfect codes and related concepts. Des. Codes Crypt. **22**(3), 221–237 (2001)
4. Ahlswede, R., Cai, N., Li, S.Y.R., Yeung, R.W.: Network information flow. IEEE Trans. Inf. Theory **46**(4), 1204–1216 (2000)
5. Ai, J., Honold, T., Liu, H.: The expurgation-augmentation method for constructing good plane subspace codes. arXiv preprint arXiv:1601.01502 (2016)
6. Bachoc, C., Passuello, A., Vallentin, F.: Bounds for projective codes from semidefinite programming. Adv. Math. Commun. **7**(2), 127–145 (2013)
7. Beutelspacher, A.: Partial spreads in finite projective spaces and partial designs. Math. Z. **145**(3), 211–229 (1975)
8. Blackburn, S.R., Etzion, T.: The asymptotic behavior of grassmannian codes. IEEE Trans. Inform. Theory **58**(10), 6605–6609 (2012)
9. Braun, M., Etzion, T., Östergård, P.R.J., Vardy, A., Wassermann, A.: Existence of q-analogs of steiner systems. Forum Math. Pi **4**, 1–14 (2016)
10. Braun, M., Östergård, P.R.J., Wassermann, A.: New lower bounds for binary constant-dimension subspace codes. Exp. Math., 1–5. doi:10.1080/10586458.2016.1239145
11. Delsarte, P.: Hahn polynomials, discrete harmonics, and t-designs. SIAM J. Appl. Math. **34**(1), 157–166 (1978)
12. Delsarte, P.: An algebraic approach to the association schemes of coding theory. Ph.D. thesis, Philips Research Laboratories (1973)
13. Drake, D., Freeman, J.: Partial t-spreads and group constructible (s, r, μ)-nets. J. Geom. **13**(2), 210–216 (1979)
14. El-Zanati, S., Jordon, H., Seelinger, G., Sissokho, P., Spence, L.: The maximum size of a partial 3-spread in a finite vector space over $GF(2)$. Des. Codes Crypt. **54**(2), 101–107 (2010)

15. Etzion, T., Silberstein, N.: Error-correcting codes in projective spaces via rank-metric codes and Ferrers diagrams. IEEE Trans. Inform. Theory **55**(7), 2909–2919 (2009)
16. Etzion, T., Silberstein, N.: Codes and designs related to lifted MRD codes. IEEE Trans. Inform. Theory **59**(2), 1004–1017 (2013)
17. Etzion, T., Vardy, A.: Error-correcting codes in projective space. IEEE Trans. Inform. Theory **57**(2), 1165–1173 (2011)
18. Frankl, P., Rödl, V.: Near perfect coverings in graphs and hypergraphs. Eur. J. Comb. **6**(4), 317–326 (1985)
19. Frankl, P., Wilson, R.M.: The Erdős-Ko-Rado theorem for vector spaces. J. Comb. Theory, Ser. A **43**(2), 228–236 (1986)
20. Gabidulin, E.: Theory of codes with maximum rank distance. Problemy Peredachi Informatsii **21**(1), 3–16 (1985)
21. Gluesing-Luerssen, H., Morrison, K., Troha, C.: Cyclic orbit codes and stabilizer subfields. Adv. Math. Commun. **9**(2), 177–197 (2015)
22. Gluesing-Luerssen, H., Troha, C.: Construction of subspace codes through linkage. Adv. Math. Commun. **10**(3), 525–540 (2016)
23. Heinlein, D., Kiermaier, M., Kurz, S., Wassermann, A.: Tables of subspace codes. arXiv preprint arXiv:601.02864 (2016)
24. Heinlein, D., Kurz, S.: A new upper bound for subspace codes. arXiv preprint arXiv:1703.08712 (2017)
25. Honold, T., Kiermaier, M.: On putative q-analogues of the Fano plane and related combinatorial structures. In: Dynamical Systems, Number Theory and Applications, pp. 141–175. World Sci. Publ, Hackensack, NJ (2016)
26. Honold, T., Kiermaier, M., Kurz, S.: Optimal binary subspace codes of length 6, constant dimension 3 and minimum subspace distance 4. In: Topics in finite fields, Contemp. Math., v ol. 632, pp. 157–176. Amer. Math. Soc., Providence, RI (2015)
27. Honold, T., Kiermaier, M., Kurz, S.: Partial spreads and vector space partitions. arXiv preprint arXiv:1611.06328 (2016)
28. Johnson, S.: A new upper bound for error-correcting codes. IRE Trans. Inform. Theory **8**(3), 203–207 (1962)
29. Khaleghi, A., Silva, D., Kschischang, F.R.: Subspace codes. In: Parker, M.G. (ed.) IMACC 2009. LNCS, vol. 5921, pp. 1–21. Springer, Heidelberg (2009). doi:10.1007/978-3-642-10868-6_1
30. Kiermaier, M., Kurz, S., Wassermann, A.: The order of the automorphism group of a binary q-analog of the fano plane is at most two. Des. Codes Crypt. (2017). doi:10.1007/s10623-017-0360-6
31. Kohnert, A., Kurz, S.: Construction of large constant dimension codes with a prescribed minimum distance. In: Calmet, J., Geiselmann, W., Müller-Quade, J. (eds.) Mathematical Methods in Computer Science. LNCS, vol. 5393, pp. 31–42. Springer, Heidelberg (2008). doi:10.1007/978-3-540-89994-5_4
32. Kötter, R., Kschischang, F.R.: Coding for errors and erasures in random network coding. IEEE Trans. Inform. Theory **54**(8), 3579–3591 (2008)
33. Kurz, S.: Improved upper bounds for partial spreads. Des. Codes Crypt. **85**(1), 97–106 (2017). doi:10.1007/s10623-016-0290-8
34. Kurz, S.: Packing vector spaces into vector spaces. Australas. J. Comb. **68**(1), 122–130 (2017)
35. MacWilliams, F.J., Sloane, N.J.A.: The theory of error-correcting codes. II. North-Holland Publishing Co., Amsterdam-New York-Oxford, North-Holland Mathematical Library, vol. 16 (1977)

36. Năstase, E., Sissokho, P.: The maximum size of a partial spread in a finite projective space. arXiv preprint arXiv:1605.04824 (2016)
37. Segre, B.: Teoria di galois, fibrazioni proiettive e geometrie non desarguesiane. Annali di Matematica **64**(1), 1–76 (1964)
38. Silberstein, N., Trautmann, A.L.: Subspace codes based on graph matchings, ferrers diagrams, and pending blocks. IEEE Trans. Inform. Theory **61**(7), 3937–3953 (2015)
39. Silva, D., Kschischang, F., Kötter, R.: A rank-metric approach to error control in random network coding. IEEE Trans. Inform. Theory **54**(9), 3951–3967 (2008)
40. Silva, D., Kschischang, F.R.: On metrics for error correction in network coding. IEEE Trans. Inform. Theory **55**(12), 5479–5490 (2009)
41. Tonchev, V.D.: Codes and designs. In: Handbook of Coding Theory, vol. 2, pp. 1229–1267 (1998)
42. Wang, H., Xing, C., Safavi-Naini, R.: Linear authentication codes: bounds and constructions. IEEE Trans. Inform. Theory **49**(4), 866–872 (2003)
43. Xia, S.T., Fu, F.W.: Johnson type bounds on constant dimension codes. Des. Codes Crypt. **50**(2), 163–172 (2009)

On Quasi-Abelian Complementary Dual Codes

Somphong Jitman[1], Herbert S. Palines[2,3](✉), and Romar B. dela Cruz[3]

[1] Department of Mathematics, Faculty of Science, Silpakorn University,
Nakhon Pathom 73000, Thailand
sjitman@gmail.com
[2] Institute of Mathematical Sciences and Physics,
University of the Philippines Los Baños, College, 4031 Laguna, Philippines
hspalines@up.edu.ph
[3] Institute of Mathematics, College of Science, University of the Philippines Diliman,
1101 Quezon City, Philippines
rbdelacruz@math.upd.edu.ph

Abstract. Linear codes that meet their dual trivially are also known as linear complementary dual codes. Quasi-abelian complementary dual codes are characterized using a known decomposition of a semisimple group algebra. Consequently, enumeration of such codes are obtained. More explicit formulas are given for the number of quasi-abelian complementary dual codes of index 2 with respect to Euclidean and Hermitian inner products. A sequence of asymptotically good binary quasi-abelian complementary dual codes of index 3 is constructed from an existing sequence of asymptotically good binary self-dual quasi-abelian codes of index 2.

Keywords: Quasi-abelian codes · Linear complementary dual codes · Asymptotically good codes

1 Introduction

Linear complementary dual (LCD) codes, introduced by Massey [21], are linear codes that have trivial intersection with their dual. It was reported in the same paper that the class of LCD codes is asymptotically good and this class can offer an optimum solution for the two-user binary adder channel. In [27], it was shown that LCD codes meet the Gilbert-Varshamov bound (GV-bound). LCD codes are also useful in information protection from side-channel attacks and hardware trojan horses (see [3,4,9,12,22,23]).

Another interesting class of linear codes are quasi-cyclic codes because of its rich mathematical theory and various practical applications (see [16–19,24,26], and references therein). This class contains cyclic codes which are well-studied for a relatively long period of time to date. The authors in [28] combined the concepts of LCD codes and cyclic codes and presented a necessary and sufficient condition for a cyclic code to be an LCD. This idea has been extended to quasi-cyclic codes with complementary duals (see [8,11]). It is interesting to note

© Springer International Publishing AG 2017
A.I. Barbero et al. (Eds.): ICMCTA 2017, LNCS 10495, pp. 192–206, 2017.
DOI: 10.1007/978-3-319-66278-7_16

that the two said papers used different techniques to characterize such codes. In [8], the authors offered a necessary and sufficient condition for a quasi-cyclic code to be an LCD code using properties of polynomial generators. On the other hand, [11] used the concatenated structure of quasi-cyclic codes in their characterization. In addition, asymptotic results are also derived. A technique of characterization similar to that of [11] is used in this work.

Quasi-abelian codes constitute an extensive class of codes which contains cyclic, quasi-cyclic and abelian codes. The readers are referred to [6,14,30] for a good background on quasi-abelian codes. Notably, there exist numerous asymptotically good quasi-abelian codes attaining the GV-bound [10]. This work is intended to characterize and study quasi-abelian codes with complementary duals.

A well-known decomposition of semisimple algebras (see [7,14]) is used and hence, quasi-abelian codes are treated as Cartesian products of linear codes of same length over some finite extension fields (possibly of different orders). This leads to a nice characterization and enumeration of quasi-abelian codes with complementary duals with respect to Euclidean and Hermitian inner products. More explicit formulas are derived for quasi-abelian complementary dual codes of index 2. Using an existing class of asymptotically good binary self-dual quasi-abelian codes of index 2 given in [14], a sequence of asymptotically good quasi-abelian complementary dual codes of index 3 is constructed.

The material is organized as follows. Section 2 recalls definitions, some basic properties of quasi-abelian codes and a well-known decomposition of quasi-abelian codes. In Sect. 3, characterization and enumeration of quasi-abelian codes with complementary duals are given. Construction of asymptotically good binary quasi-abelian complementary dual codes is presented in Sect. 4.

2   Preliminaries

Definitions and some basic properties of quasi-abelian codes are recalled in this section. Moreover, we also discuss two important decompositions of quasi-abelian codes which lead to convenient forms of Euclidean dual and Hermitian dual of such codes. These decompositions serve as a main tool in Sect. 3.

2.1   Notations

Let R denote a finite commutative ring with unity and let n be a positive integer. A *linear code* of length n over R is defined to be an R-submodule of R^n. A linear code B is said to be a *linear complementary dual (LCD) code* if $B \cap B^\perp = \{0\}$. In particular, if the given inner product is of Euclidean (resp., Hermitian) type, B is said to be a *Euclidean* (resp., *Hermitian*) *complementary dual code* or simply *ECD* (resp., *HCD*) code. Please see [11,21], respectively, for these definitions.

Denote by $\mathbb{F}_q$ the finite field of order q, where q is a prime power. Consider a finite abelian group G of order n, written additively. Let $\mathbb{F}_q[G]$ denote the *group algebra of G over $\mathbb{F}_q$*. The elements in $\mathbb{F}_q[G]$ will be written as $\sum_{g \in G} \alpha_g Y^g$, where $\alpha_g \in \mathbb{F}_q$. The addition and the multiplication in $\mathbb{F}_q[G]$ are given as in the usual polynomial rings over $\mathbb{F}_q$ with the indeterminate Y, where the indices are computed additively in G. As convention, Y^0 is treated as the multiplicative identity of $\mathbb{F}_q[G]$, where 0 is the identity of G. A (*linear*) *code* C in $\mathbb{F}_q[G]$ is an $\mathbb{F}_q$-subspace of $\mathbb{F}_q[G]$. This can be regarded as a linear code of length n over $\mathbb{F}_q$ by indexing each n-tuple codeword in the code by the elements of G.

Let H be a subgroup of G and $l := [G : H]$. A code C in $\mathbb{F}_q[G]$ is called an *H-quasi-abelian code* (specifically, an *H-quasi-abelian code of index l*) if C is an $\mathbb{F}_q[H]$-module, i.e., C is closed under addition and multiplication by the elements in $\mathbb{F}_q[H]$. Note that since $\mathbb{F}_q$ is naturally embedded in $\mathbb{F}_q[H]$, C can be viewed also as an $\mathbb{F}_q$-vector space. If H is a non-cyclic subgroup of G, then we say that C is a *strictly H-quasi-abelian code*. It is interesting to note the following specific cases of quasi abelian codes: C is a quasi-cyclic code if H is a proper cyclic subgroup, C is an abelian code if $H = G$, and C is a cyclic code if $H = G$ is a cyclic group. If it is clear in the context or if H is not specified, such a code will be called simply a *quasi-abelian code*.

Let C be a code in $\mathbb{F}_q[G]$. The *Hamming weight* of a codeword $\boldsymbol{u} = \sum_{g \in G} \alpha_g Y^g$ in C, denoted by $\mathrm{wt}(\boldsymbol{u})$, is defined to be the number of nonzero α_g in $\boldsymbol{u}$. The minimum Hamming distance of C is defined by $\mathrm{d}(C) := \min\{\mathrm{wt}(\boldsymbol{u}) \mid \boldsymbol{u} \in C, \boldsymbol{u} \neq 0\}$. Denote by $\dim_{\mathbb{F}_q}(C)$ the dimension of a code C as an $\mathbb{F}_q$-vector space in $\mathbb{F}_q[G]$.

In $\mathbb{F}_q^n$, the *Euclidean inner product* of $\boldsymbol{u} = (u_1, u_2, \ldots, u_n)$ and $\boldsymbol{v} = (v_1, v_2, \ldots, v_n)$ is defined to be $\langle \boldsymbol{u}, \boldsymbol{v} \rangle_e := \sum_{i=1}^n u_i v_i$. In a similar manner, the *Euclidean inner product* of $\boldsymbol{u} = \sum_{g \in G} \alpha_g Y^g$ and $\boldsymbol{v} = \sum_{g \in G} \beta_g Y^g$ in $\mathbb{F}_q[G]$ is defined as

$$\langle \boldsymbol{u}, \boldsymbol{v} \rangle_E := \sum_{g \in G} \alpha_g \beta_g.$$

If $q = q_0^2$, where q_0 is a prime power, then the *Hermitian inner product* of $\boldsymbol{u}$ and $\boldsymbol{v}$ in $\mathbb{F}_q^n$ can be defined as $\langle \boldsymbol{u}, \boldsymbol{v} \rangle_h := \sum_{i=1}^n u_i \overline{v_i}$, where $^-$ is the automorphism on $\mathbb{F}_q$ defined by $\alpha \mapsto \alpha^{q_0}$ for all $\alpha \in \mathbb{F}_q$. At the same time, the *Hermitian inner product* in $\mathbb{F}_q[G]$ is defined by

$$\langle \boldsymbol{u}, \boldsymbol{v} \rangle_H := \sum_{g \in G} \alpha_g \overline{\beta_g}$$

for all $\boldsymbol{u} = \sum_{g \in G} \alpha_g Y^g$ and $\boldsymbol{v} = \sum_{g \in G} \beta_g Y^g$ in $\mathbb{F}_q[G]$. The *Euclidean dual* and *Hermitian dual* of a code $C \subseteq \mathbb{F}_q[G]$ is given by

$$C^{\perp_E} := \{\boldsymbol{u} \in \mathbb{F}_q[G] \mid \langle \boldsymbol{u}, \boldsymbol{v} \rangle_E = 0 \text{ for all } \boldsymbol{v} \in C\}$$

and

$$C^{\perp_H} := \{\boldsymbol{u} \in \mathbb{F}_q[G] \mid \langle \boldsymbol{u}, \boldsymbol{v} \rangle_H = 0 \text{ for all } \boldsymbol{v} \in C\},$$

respectively. For a linear code B in $\mathbb{F}_q^n$, let $B^{\perp_e}$ and $B^{\perp_h}$ denote its Euclidean dual and Hermitian dual, respectively.

An H-quasi-abelian code C is said to be an *H-quasi-abelian Euclidean complementary dual (H-QAECD) code* if $C \cap C^{\perp_E} = \{0\}$. The *H-quasi-abelian Hermitian complementary dual (H-QAHCD) code* C is defined in the same fashion, i.e. if $C \cap C^{\perp_H} = \{0\}$. If it is clear in the context, we simply use QACD to indicate a quasi-abelian Euclidean or Hermitian complementary dual code.

Consider a fixed set of representatives of cosets of H in G given by $\{\mathfrak{g}_1, \mathfrak{g}_2, \dots, \mathfrak{g}_l\}$. Let $R := \mathbb{F}_q[H]$. Define $\Phi : \mathbb{F}_q[G] \to R^l$ by

$$\Phi\left(\sum_{h \in H}\sum_{i=1}^{l} \alpha_{h+\mathfrak{g}_i} Y^{h+\mathfrak{g}_i}\right) = (\alpha_1(Y), \alpha_2(Y), \dots, \alpha_l(Y)), \tag{1}$$

where $\alpha_i(Y) = \sum_{h \in H} \alpha_{h+\mathfrak{g}_i} Y^h \in R$, for all $i = 1, 2, \dots, l$. It is well known that Φ is an R-module isomorphism interpreted as follows.

Lemma 1. *The map Φ induces a one-to-one correspondence between H-quasi-abelian codes in $\mathbb{F}_q[G]$ and linear codes of length l over R.*

Define an involution * on R to be the $\mathbb{F}_q$-linear map that fixes $\mathbb{F}_q$ and sends Y^h to Y^{-h}. Following [14, Remark 2.1], the dual of a code D in R^l is given by

$$D^{\perp_*} = \{\boldsymbol{x} \in R^l \mid [\boldsymbol{x}, \boldsymbol{y}]_* = 0, \text{ for all } \boldsymbol{y} \in D\},$$

where the map $[\cdot, \cdot]_* : R^l \times R^l \to R$ is given by

$$[\boldsymbol{x}, \boldsymbol{y}]_* := \sum_{i=1}^{l} x_i y_i^*$$

for all $\boldsymbol{x} = (x_1, x_2, \dots, x_l), \boldsymbol{y} = (y_1, y_2, \dots, y_l) \in R^l$. This map resembles a Hermitian form in the sense that it is R-linear in the first component and behaves as a Hermitian form under conjugation.

2.2 Decompositions

The discussion in this section is found mainly in [13–15]. The concepts of q-cyclotomic classes and primitive idempotents are instrumental in this work thus, the readers are referred to [13, Sect. 2] and [15, Sect. 2] for details.

Given coprime positive integers i and j, the *multiplicative order of j modulo i*, denoted by $\mathrm{ord}_i(j)$, is defined to be the smallest positive integer s such that i divides $j^s - 1$. For each $a \in H$, denote by $\mathrm{ord}(a)$ the *additive order* of a in H.

Assume $\gcd(|H|, q) = 1$. A *q-cyclotomic class* of H containing $a \in H$, denoted by $S_q(a)$, is defined to be the set

$$S_q(a) := \{q^i \cdot a \mid i = 0, 1, \dots\} = \{q^i \cdot a \mid 0 \le i < \mathrm{ord}_{\mathrm{ord}(a)}(q)\},$$

where $q^i \cdot a := \sum_{j=1}^{q^i} a$ in H. For a positive integer r and $a \in H$, denote by $-r \cdot a$ the element $r \cdot (-a) \in H$.

We now describe the types of q-cyclotomic classes that play an important role in the decomposition of quasi-abelian codes with Euclidean duals. For $a \in H$, the q-cyclotomic class $S_q(a)$ is of one of the following three types: *type I_E* if $a = -a$ (i.e. $S_q(a) = S_q(-a)$), *type II_E* if $S_q(a) = S_q(-a)$ and $a \neq -a$, or *type III_E* if $S_q(a) \neq S_q(-a)$. For more details, see [13, Sect. 2].

An *idempotent* in a ring is a non-zero element e such that $e^2 = e$, and it is called a *primitive idempotent* if, for every other idempotent f, either $ef = e$ or $ef = 0$. The primitive idempotents in $R := \mathbb{F}_q[H]$ are induced by the q-cyclotomic classes of H [7, Proposition II.4].

Assume that H contains t q-cyclotomic classes. Let $\{a_1 = 0, a_2, \ldots, a_t\}$ be a complete set of representatives of the q-cyclotomic classes of H. Let $\{e_1, e_2, \ldots, e_t\}$ be the set of primitive idempotents of R induced by $\{S_q(a_i) \mid i = 1, 2, \ldots, t\}$, respectively.

In [25], $R := \mathbb{F}_q[H]$ is decomposed in terms of e_i's. Later, the components in the decomposition of R are rearranged in [13] and the following is obtained:

$$R = \bigoplus_{i=1}^{t} Re_i \cong \left(\prod_{i=1}^{r_{I_E}} \mathbb{F}_q \right) \times \left(\prod_{j=1}^{r_{II_E}} \mathbb{E}_j \right) \times \left(\prod_{r=1}^{r_{III_E}} (\mathbb{K}_r \times \mathbb{K}'_r) \right), \qquad (2)$$

where $\mathbb{F}_q \cong Re_i$ for $i = 1, 2, \ldots, r_{I_E}$, $\mathbb{E}_j \cong Re_{r_{I_E}+j}$, and $\mathbb{K}_r \cong \mathbb{K}'_r \cong Re_{r_{I_E}+r_{II_E}+r}$ are finite extension fields of $\mathbb{F}_q$ for all $j = 1, 2, \ldots, r_{II_E}$ and $r = 1, 2, \ldots, r_{III_E}$; r_{I_E}, r_{II_E} and $2r_{III_E}$ correspond to the number of types I_E, II_E and III_E q-cyclotomic classes of H, respectively; and $t = r_{I_E} + r_{II_E} + 2r_{III_E}$. Note that $\oplus$ denotes internal direct sum in R while $\prod$ and $\times$ denote Cartesian products.

Remark 2. The field extensions in the right-hand side of (2) are given as follows: $\mathbb{E}_j \cong \mathbb{F}_{q^{s_j}}$, $\mathbb{K}_r \cong \mathbb{F}_{q^{t_r}}$ and $\mathbb{K}'_r \cong \mathbb{F}_{q^{t'_r}}$, where $s_j := |S_q(a_{r_{I_E}+j})|$, $t_r := |S_q(a_{r_{I_E}+r_{II_E}+r})|$, and $t'_r := |S_q(a_{r_{I_E}+r_{II_E}+r_{III_E}+r})|$, such that $a_{r_{I_E}+r_{II_E}+r_{III_E}+r} = -a_{r_{I_E}+r_{II_E}+r}$, for $j = 1, 2, \ldots, r_{II_E}$ and $r = 1, 2, \ldots, r_{III_E}$. In this particular order of q-cyclotomic classes of type III_E, it will follow that $t_r = t'_r$ for each $r = 1, 2, \ldots, r_{III_E}$ and thus, $\mathbb{K}_r \cong \mathbb{K}'_r$ for each $r = 1, 2, \ldots, r_{III_E}$.

From (1), (2) and Remark 2,

$$\mathbb{F}_q[G] \cong R^l \cong \left(\prod_{i=1}^{r_{I_E}} \mathbb{F}_q^l \right) \times \left(\prod_{j=1}^{r_{II_E}} \mathbb{F}_{q^{s_j}}^l \right) \times \left(\prod_{r=1}^{r_{III_E}} \left(\mathbb{F}_{q^{t_r}}^l \times \mathbb{F}_{q^{t'_r}}^l \right) \right), \qquad (3)$$

where the isomorphisms can be viewed as $\mathbb{F}_q$-module isomorphisms. Consequently, an H-quasi-abelian code C in $\mathbb{F}_q[G]$ can be represented as

$$C \cong \left(\prod_{i=1}^{r_{I_E}} B_i \right) \times \left(\prod_{j=1}^{r_{II_E}} C_j \right) \times \left(\prod_{r=1}^{r_{III_E}} (D_r \times D'_r) \right), \qquad (4)$$

where B_i, C_j, D_r and D'_r are linear codes, each of length l, over finite fields $\mathbb{F}_q$, $\mathbb{F}_{q^{s_j}}$, $\mathbb{F}_{q^{t_r}}$ and $\mathbb{F}_{q^{t'_r}}$, respectively, for $i = 1, 2, \ldots, r_{I_E}$, $j = 1, 2, \ldots, r_{II_E}$ and $r = 1, 2, \ldots, r_{III_E}$. From [14, Proposition 4.1], the Euclidean dual of C in (4) is given by

$$C^{\perp_E} \cong \left(\prod_{i=1}^{r_{I_E}} B_i^{\perp_e} \right) \times \left(\prod_{j=1}^{r_{II_E}} C_j^{\perp_h} \right) \times \left(\prod_{r=1}^{r_{III_E}} ((D'_r)^{\perp_e} \times D_r^{\perp_e}) \right). \quad (5)$$

Let $q = q_0^2$. In the Hermitian case, the following types of q-cyclotomic classes are considered: *type* I_H if $S_q(a) = S_q(-q_0 \cdot a)$ or *type* II_H if $S_q(a) \neq S_q(-q_0 \cdot a)$, for $a \in H$. In a similar manner, let $\{a_1 = 0, a_2, \ldots, a_t\}$ be the set of representatives of q-cyclotomic classes H and let $\{e_1, e_2, \ldots, e_t\}$ be the set of primitive idempotents in $R := \mathbb{F}_q[H]$ induced by $\{S_q(a_i) \mid i = 1, 2, \ldots, t\}$, respectively.

Let $s_i = |S_q(a_i)|$, $t_j = |S_q(a_{r_{I_H}+j})|$ and $t'_j = |S_q(a_{r_{I_H}+r_{II_H}+j})|$, such that $a_{r_{I_H}+r_{II_H}+j} = -a_{r_{I_H}+j}$ for all $i = 1, 2, \ldots, r_{I_H}$ and $j = 1, 2, \ldots, r_{II_H}$. Following the above discussion for the Euclidean case,

$$R = \bigoplus_{i=1}^{t} Re_i \cong \left(\prod_{i=1}^{r_{I_H}} \mathbb{E}_i \right) \times \left(\prod_{j=1}^{r_{II_H}} \mathbb{K}_j \times \mathbb{K}'_j \right), \quad (6)$$

and

$$\mathbb{F}_q[G] \cong R^l \cong \left(\prod_{i=1}^{r_{I_H}} \mathbb{F}_{q^{s_i}}^l \right) \times \left(\prod_{j=1}^{r_{II_H}} (\mathbb{F}_{t_j}^l \times \mathbb{F}_{t'_j}^l) \right), \quad (7)$$

where $\mathbb{E}_i \cong Re_i \cong \mathbb{F}_{q^{s_i}}$, $\mathbb{K}_j \cong Re_{r_{I_H}+j} \cong \mathbb{F}_{q^{t_j}}$ and $\mathbb{K}'_j \cong Re_{r_{I_H}+r_{II_H}+j} \cong \mathbb{F}_{q^{t'_j}}$, for $i = 1, 2, \ldots r_{I_H}$ and $j = 1, 2, \ldots, r_{II_H}$; r_{I_H} and $2r_{II_H}$ correspond to the number of q-cyclotomic classes of H of types I_H and II_H, respectively; and $t = r_{I_H} + 2r_{II_H}$. Note that $t_j = t'_j$ and hence, $\mathbb{K}_j \cong \mathbb{K}'_j$ for all j. As a result, a quasi-abelian code C in $\mathbb{F}_q[G]$ can be also represented as

$$C \cong \left(\prod_{i=1}^{r_{I_H}} C_i \right) \times \left(\prod_{j=1}^{r_{II_H}} (D_j \times D'_j) \right), \quad (8)$$

where C_i, D_j and D'_j are linear codes, each of length l, over $\mathbb{F}_{q^{s_i}}$, $\mathbb{F}_{q^{t_j}}$ and $\mathbb{F}_{q^{t'_j}}$, respectively.

Following similar arguments as in the proofs of [15, Proposition 2.7] and [14, Proposition 4.1], it can be deduced that the Hermitian dual of C in (8) is of the form

$$C^{\perp_H} \cong \left(\prod_{i=1}^{r_{I_H}} C_i^{\perp_h} \right) \times \left(\prod_{j=1}^{r_{II_H}} ((D'_j)^{\perp_e} \times D_j^{\perp_e}) \right). \quad (9)$$

3   Characterization and Enumeration of QACD Codes

In this section, QACD codes are characterized using decompositions (4), (5), (8) and (9), accordingly. Enumeration of QACD codes naturally follows from the said characterization.

From the definition of a QACD code, it is clear that a code C given in (4) is an H-QAECD code if every factor in its decomposition meet each corresponding factor in $C^{\perp_{\mathrm{E}}}$ in (5) trivially. Adding the fact that $D'_r \cap (D_r)^{\perp_{\mathrm{e}}} = \{0\}$ is equivalent to $D_r \oplus (D'_r)^{\perp_{\mathrm{e}}} = \mathbb{F}_{q^{t_r}}^l$, for each $r = 1, 2, \ldots, r_{III_{\mathrm{E}}}$, the following proposition is obtained.

Proposition 3. *Let $H \leq G$ be finite abelian groups such that $\gcd(|H|, q) = 1$ and $l = [G : H]$. Assume that $\mathbb{F}_q[H]$ contains $r_{I_{\mathrm{E}}}$, $r_{II_{\mathrm{E}}}$ and $2r_{III_{\mathrm{E}}}$ primitive idempotents of types I_{E}, II_{E} and III_{E}, respectively. Assume further that the primitive idempotents of types I_{E}, II_{E} and III_{E} are induced by q-cyclotomic classes of sizes equal to $s'_i = 1$, s_j, $t_r = t'_r$, respectively, for all $i = 1, 2, \ldots, r_{I_{\mathrm{E}}}$, $j = 1, 2, \ldots, r_{II_{\mathrm{E}}}$ and $r = 1, 2, \ldots, r_{III_{\mathrm{E}}}$. Then a quasi-abelian code C of index l given in (4) is an H-QAECD code if and only if the following conditions hold:*

(i) B_i is an ECD code, for each $i = 1, 2, \ldots r_{I_{\mathrm{E}}}$,
(ii) C_j is an HCD code, for each $j = 1, 2, \ldots r_{II_{\mathrm{E}}}$, and
(iii) $D_r \oplus (D'_r)^{\perp_{\mathrm{e}}} = \mathbb{F}_{q^{t_r}}^l$ for each $r = 1, 2, \ldots, r_{III_{\mathrm{E}}}$.

A similar result for the Hermitian case is given below.

Proposition 4. *Let $H \leq G$ be finite abelian groups such that $\gcd(|H|, q) = 1$, $q = q_0^2$ and $l = [G : H]$. Assume further that $\mathbb{F}_q[H]$ contains $r_{I_{\mathrm{H}}}$ and $2r_{II_{\mathrm{H}}}$ primitive idempotents of types I_{H} and II_{H}, respectively. Assume further that the primitive idempotents of types I_{H} and II_{H} are induced by q-cyclotomic classes of sizes equal to s_i, $t_j = t'_j$, respectively, for all $i = 1, 2, \ldots, r_{I_{\mathrm{H}}}$ and $j = 1, 2, \ldots, r_{II_{\mathrm{H}}}$. Then a quasi-abelian code C of index l given in (8) is an H-QAHCD code if and only if the two conditions hold:*

(i) C_i is an HCD code, for each $i = 1, 2, \ldots r_{I_{\mathrm{H}}}$, and
(ii) $D_j \oplus (D'_j)^{\perp_{\mathrm{e}}} = \mathbb{F}_{q^{t_j}}^l$ for each $j = 1, 2, \ldots, r_{II_{\mathrm{H}}}$.

Remark 5. An $[n, k]$ linear code over a finite field, generated by a $k \times n$ matrix A, is a Euclidean or Hermitian LCD code if and only if AA^T is invertible (see [21, Proposition 1] and [11, Proposition 3.5]). Note that from conditions (iii) and (ii) of Propositions 3 and 4, respectively, it is necessary that D_r and D'_r (resp., D_j and D'_j) have the same dimensions for each $r = 1, 2, \ldots, r_{III_{\mathrm{E}}}$ (resp., $j = 1, 2, \ldots, r_{II_{\mathrm{H}}}$) for C to be an H-QAECD (resp., H-QAHCD) code. Indeed, if D and D' satisfy condition (iii) of Proposition 3 or condition (ii) of Proposition 4, then $\dim(D) = l - \dim((D')^\perp) = l - (l - \dim(D'))$.

Propositions 3 and 4 provide a convenient way to count $QACD$ codes. The general counting strategy can be summarized into three parts: (1) counting ECD codes, (2) counting HCD codes and (3) counting complementary linear codes D

and $(D')^{\perp_e}$ such that $D \oplus (D')^{\perp_e} = F^l$, where D and D' are linear codes of length l over some finite extension field F of $\mathbb{F}_q$.

Let $N_{ECD}(q, l)$ (resp., $N_{HCD}(q, l)$) be the number of distinct ECD (resp., HCD) codes over $\mathbb{F}_q$ of length l. Moreover, consider the Gaussian binomial coefficient, denoted by $\begin{bmatrix} n \\ k \end{bmatrix}_q$, which gives the number of distinct k-dimensional subspaces of $\mathbb{F}_q^n$.

Lemma 6. *Suppose D is an $[l, k]$ linear code over $\mathbb{F}_q$. Let $N_\oplus(q, k, l)$ denotes the number of $[l, l - k]$ linear codes B over $\mathbb{F}_q$ such that $D \oplus B = \mathbb{F}_q^l$. Then*

$$N_\oplus(q, k, l) = (q^k)^{l-k}.$$

Proof. Suppose $T = \{v_1, v_2, \ldots, v_k\} \subseteq \mathbb{F}_q^l$ is a basis of D. The number of ways that T can be extended to the set $\{v_1, v_2, \ldots, v_k, v_{k+1}, \ldots, v_l\}$ as a basis of $\mathbb{F}_q^l$ is

$$\frac{1}{(l-k)!} \prod_{i=k}^{l-1} (q^l - q^i).$$

By taking $\{v_{k+1}, v_{k+2}, \ldots, v_l\}$ as a basis for B and considering the fact that the number of different bases for B is given by $\frac{1}{(l-k)!} \prod_{i=0}^{l-k-1} (q^{l-k} - q^i)$ [20, Theorem 4.1.15], then it follows that

$$N_\oplus(q, k, l) = \frac{\prod_{i=k}^{l-1} (q^l - q^i)}{\prod_{i=0}^{l-k-1} (q^{l-k} - q^i)} = \prod_{i=k}^{l-1} \frac{q^k (q^{l-k} - q^{i-k})}{(q^{l-k} - q^{i-k})} = (q^k)^{l-k}.$$

$\square$

Lemma 6 is useful to the following propositions.

Proposition 7. *Let $H \leq G$ be finite abelian groups such that $\gcd(|H|, q) = 1$ and $l = [G : H]$. Assume that $\mathbb{F}_q[H]$ contains r_{I_E}, r_{II_E} and $2r_{III_E}$ primitive idempotents of types I_E, II_E and III_E, respectively. Assume further that the primitive idempotents of types I_E, II_E and III_E are induced by q-cyclotomic classes of sizes equal to $s_i' = 1$, s_j, $t_r = t_r'$, respectively, for all $i = 1, 2, \ldots, r_{I_E}$, $j = 1, 2, \ldots, r_{II_E}$ and $r = 1, 2, \ldots, r_{III_E}$. Then the number of H-QAECD codes C in (4), is given by*

$$\left(\prod_{i=1}^{r_{I_E}} N_{ECD}(q, l) \right) \left(\prod_{j=1}^{r_{II_E}} N_{HCD}(q^{s_j}, l) \right) \left(\prod_{r=1}^{r_{III_E}} \left(2 + \sum_{k=1}^{l-1} \left(\begin{bmatrix} l \\ k \end{bmatrix}_{q^{t_r}} \cdot N_\oplus(q^{t_r}, k, l) \right) \right) \right). \quad (10)$$

Proof. It is clear from (4) and Proposition 3 that we need to count the number of ECD codes B_i and HCD codes C_j, for each $i = 1, 2, \ldots, r_{I_E}$ and $j = 1, 2, \ldots, r_{II_E}$, to obtain the first and second factors in (10). The third factors are obtained by getting the number of codes (D_r, D_r') such that $D_r \oplus (D_r')^{\perp_E} = \mathbb{F}_{q^{t_r}}^l$ for each $r = 1, 2, \ldots, r_{III_E}$ and for all possible dimensions k of D_r, where $0 \leq k \leq l$. Apply Lemma 6 and consider the fact that there are $\begin{bmatrix} l \\ k \end{bmatrix}_{q^{t_r}}$ distinct k-dimensional subspaces in $\mathbb{F}_{q^{t_r}}^l$, $N_\oplus(q^{t_r}, 0, l) = 1 = N_\oplus(q^{t_r}, l, l)$, and $\begin{bmatrix} l \\ 0 \end{bmatrix}_{q^{t_r}} = 1 = \begin{bmatrix} l \\ l \end{bmatrix}_{q^{t_r}}$, for each r and k. $\square$

The following result is for the Hermitian case where its proof follows from that of the Euclidean case.

Proposition 8. *Let $H \leq G$ be finite abelian groups such that $\gcd(|H|, q) = 1$, $q = q_0^2$ and $l = [G : H]$. Assume further that $\mathbb{F}_q[H]$ contains r_{I_H} and $2r_{II_H}$ primitive idempotents of types I_H and II_H, respectively. Assume further that the primitive idempotents of types I_H and II_H are induced by q-cyclotomic classes of sizes equal to s_i, $t_j = t'_j$, respectively, for all $i = 1, 2, \ldots, r_{I_H}$ and $j = 1, 2, \ldots, r_{II_H}$. Then the number of H-QAHCD codes C in (8), is given by*

$$\left(\prod_{i=1}^{r_{I_H}} N_{HCD}(q^{s_i}, l)\right) \left(\prod_{j=1}^{r_{II_H}} \left(2 + \sum_{k=1}^{l-1} \left(\begin{bmatrix} l \\ k \end{bmatrix}_{q^{t_j}} \cdot N_{\oplus}(q^{t_j}, k, l)\right)\right)\right). \tag{11}$$

Remark 9. It should be noted that the requirement $q = q_0^2$ in Proposition 8 is needed to make sense of the Hermitian dual of C. However, in Proposition 7, q can be a non-square even if there are Hermitian dual codes in the factors of $C^{\perp_E}$ in (5). Since these factors correspond to q-cyclotomic classes of type II_E, wherein s_j is even (see [13, Remark 2.6]) for each $j = 1, 2, \ldots, r_{II_E}$, then q^{s_j} is always a square and hence, set $q_0 = q^{s_j/2}$.

Considering a relatively small fixed value of the index l serves some theoretical and practical use (see for instance [5, 17, 19]). In this work, fixing the index l, say $l = 2$, gives rise to specific formulas for $N_{ECD}(q, 2)$, $N_{HCD}(q, 2)$ and $N_{\oplus}(q, k, 2)$, which can be seen in the following results.

In the lemma below, some properties of the square map on $\mathbb{F}_q^* := \mathbb{F}_q \setminus \{0\}$ are recalled. This can be found also in [29, Theorem 6.18].

Lemma 10. *Let $f : \mathbb{F}_q^* \to \mathbb{F}_q^*$, such that $f(x) = x^2$.*

(i) *The mapping f is a group homomorphism.*
(ii) *If $q = 2^r$, for some $r > 0$, then f is an isomorphism. Consequently, every element in $\mathbb{F}_q^*$ has a unique square root. That is, $x^2 = a$ has a unique solution in $\mathbb{F}_q^*$ for each $a \in \mathbb{F}_q^*$.*
(iii) *If q is odd, then every square element in $\mathbb{F}_q^*$ has two distinct square roots in $\mathbb{F}_q^*$.*
(iv) *Let q be odd. An element $\alpha \in \mathbb{F}_q^*$ is a square if and only if $\alpha^{\frac{q-1}{2}} = 1$. Consequently, $-1 \in \mathbb{F}_q^*$ is a square if and only if $q \equiv 1 \pmod 4$.*

Lemma 11. *Let q be a prime power. Then*

(i)

$$N_{ECD}(q, 2) = \begin{cases} q + 2 & \text{if } q \text{ is even,} \\ q + 1 & \text{if } q \equiv 1 \pmod 4 \\ q + 3 & \text{if } q \equiv 3 \pmod 4, \end{cases}$$

(ii) if $q = q_0^2$,

$$N_{HCD}(q, 2) = q - q_0 + 2$$

Proof.

(i) Let C be a linear code over $\mathbb{F}_q$ of length 2. If $C = \{0\}$ or $C = \mathbb{F}_q^2$, then C is trivially an ECD code. Suppose C is of dimension 1. Then, C has the following possible distinct generator matrices: $[\alpha, 0]$, $[0, \alpha]$ and $[1, \alpha]$ where $\alpha \in \mathbb{F}_q^*$. It is clear that the first two generator matrices will generate ECD codes since α^2 is invertible in $\mathbb{F}_q$. We only need to count the number of distinct generator matrices of the form $[1, \alpha]$ such that $1 + \alpha^2 \neq 0$. Define the map $f : \mathbb{F}_q^* \to \mathbb{F}_q^*$ such that $f(\alpha) = \alpha^2$ for all $\alpha \in \mathbb{F}_q^*$. Let $O := \{\alpha \in \mathbb{F}_q^* \mid f(\alpha) = \alpha^2 = -1\}$. Note $|O| = 1$ if q is even (Lemma 10 (ii)) or $|O| = 2$ if $q \equiv 1 \pmod 4$ (Lemma 10 (iii) and (iv)) or $|O| = 0$ if $q \equiv 3 \pmod 4$ (Lemma 10 (iv)). So, if q is even, there are $(q - 1) - |O| = q - 2$ elements α in $\mathbb{F}_q^*$ such that $1 + \alpha^2$ is nonzero, if $q \equiv 1 \pmod 4$ or $q \equiv 3 \pmod 4$, there are $(q - 1) - 2 = q - 3$ or $q - 1$ such elements, respectively. Sum up the number of LCD codes of dimensions 0, 1 and 2 for each case of q.

(ii) For the Hermitian case, note that $q = q_0^2$. This proof is similar to that of the Euclidean case. However, the norm function $Nrm : \mathbb{F}_q^* \to \mathbb{F}_{q_0}^*$, given by $Nrm(\alpha) = \alpha\overline{\alpha}$ for all $\alpha \in \mathbb{F}_q^*$, is needed to derive the formula. Note that Nrm is surjective. Let $O' := \{\alpha \in \mathbb{F}_q^* \mid Nrm(\alpha) = \alpha\overline{\alpha} = -1\}$. It is known that $|Nrm^{-1}(u)| = \frac{|\mathbb{F}_q^*|}{|\mathbb{F}_{q_0}^*|} = |Nrm^{-1}(v)|$ for any $u, v \in \mathbb{F}_{q_0}^*$. Thus, $|O'| = |Nrm^{-1}(-1)| = \frac{q-1}{q_0-1} = q_0 + 1$. Thus, there are $(q - 1) - |O'| = q - q_0 - 2$ elements $\alpha \in \mathbb{F}_q^*$ such that $[1, \alpha]$ is a generator of an HCD code, i.e. $1 + \alpha\overline{\alpha} \neq 0$. Combining all the numbers we got from HCD codes of dimensions 0, 1, and 2, we have $1 + (2 + q - q_0 - 2) + 1 = q - q_0 + 2$.

$\square$

The following corollary follows directly from Propositions 7 and 8 and Lemma 11.

Corollary 12. *Using the same assumptions given in Propositions 7 and 8 and letting $l = 2$, the following statements hold.*

(i) The number of H-QAECD codes C of index 2 in (4), is given by

$$\begin{cases} \left(\prod_{i=1}^{r_{I_E}}(q+2)\right) \left(\prod_{j=1}^{r_{II_E}}(q^{s_j} - q^{s_j/2} + 2)\right) \left(\prod_{r=1}^{r_{III_E}}(q^{2t_r} + q^{t_r} + 2)\right) & \text{if } q \text{ is even,} \\[2ex] \left(\prod_{i=1}^{r_{I_E}}(q+1)\right) \left(\prod_{j=1}^{r_{II_E}}(q^{s_j} - q^{s_j/2} + 2)\right) \left(\prod_{r=1}^{r_{III_E}}(q^{2t_r} + q^{t_r} + 2)\right) & \text{if } q \equiv 1 \pmod 4, \\[2ex] \left(\prod_{i=1}^{r_{I_E}}(q+3)\right) \left(\prod_{j=1}^{r_{II_E}}(q^{s_j} - q^{s_j/2} + 2)\right) \left(\prod_{r=1}^{r_{III_E}}(q^{2t_r} + q^{t_r} + 2)\right) & \text{if } q \equiv 3 \pmod 4. \end{cases}$$

(ii) *The number of H-QAHCD codes C of index 2 in (8), is given by*

$$\left(\prod_{i=1}^{r_{I_H}}(q^{s_i} - q^{s_i/2} + 2)\right)\left(\prod_{j=1}^{r_{II_H}}(q^{2t_j} + q^{t_j} + 2)\right).$$

Proof. (i) Use Proposition 7 and Lemma 11 (i) correspondingly for each case of q.

(ii) Use Proposition 8 and Lemma 11 (ii) to obtain the result.   □

Example 13.

1. Consider $q = 2$ and $H = (\mathbb{Z}_3)^s$, for $s = 1, 2$. Let $\mathbb{Z}_3 := \{0, 1, 2\}$ and $(\mathbb{Z}_3)^2 := \{(x, y) \mid x, y \in \mathbb{Z}_3\}$. The 2-cyclotomic classes of $\mathbb{Z}_3$ are given by $S_2(0) = \{0\}$ and $S_2(1) = \{1, 2\}$ which are of type I_E and type II_E, respectively. For $(\mathbb{Z}_3)^2$, its 2-cyclotomic classes consist of $S_2((0,0))$ which is of type I_E and all the other four are of type II_E with cardinality 2 each. It follows that $\mathbb{F}_2[\mathbb{Z}_3] \cong \mathbb{F}_2 \times \mathbb{F}_{2^2}$ and $\mathbb{F}_2[(\mathbb{Z}_3)^2] \cong \mathbb{F}_2 \times \left(\prod_{j=1}^4 \mathbb{F}_{2^2}\right)$. This information is summarized in Table 1 which also shows the number of corresponding H-QAECD codes of index 2 in the last column. Let $A := \prod_{i=1}^{r_{I_E}} N_{ECD}(q, 2)$ and $B := \prod_{j=1}^{r_{II_E}} N_{HCD}(q^{s_j}, 2)$.

2. For the Hermitian case, let $q = 4$ and $H = (\mathbb{Z}_3)^s$, for $s = 1, 2$. In this case, the 4-cyclotomic classes of $\mathbb{Z}_3$ and $(\mathbb{Z}_3)^2$ are all of type I_H with cardinality 1 each. Thus, we can write $\mathbb{F}_4[\mathbb{Z}_3] \cong \mathbb{F}_4 \times \mathbb{F}_4 \times \mathbb{F}_4$ and $\mathbb{F}_4[(\mathbb{Z}_3)^2] \cong \prod_{i=1}^9 \mathbb{F}_4$. The number of corresponding H-QAHCD codes of index 2 are given in the last column of Table 2.

Table 1. Number of H-QAECD codes of index 2 in $\mathbb{F}_2[\mathbb{Z}_2 \times H]$, $H = (\mathbb{Z}_3)^s$

q	s	H	r_{I_E}	r_{II_E}	r_{III_E}	s_j	A	B	$A \cdot B$
2	1	$\mathbb{Z}_3$	1	1	-	$s_1 = 2$	4	4	16
	2	$(\mathbb{Z}_3)^2$	1	4	-	$s_j = 2$ for all j	4	4^4	1024

Table 2. Number of H-QAHCD codes of index 2 in $\mathbb{F}_4[\mathbb{Z}_2 \times H]$, $H = (\mathbb{Z}_3)^s$

q	s	H	r_{I_H}	r_{II_H}	s_i	$\prod_{i=1}^{r_{I_H}} N_{HCD}(q^{s_i}, 2)$
4	1	$\mathbb{Z}_3$	3	-	$s_i = 1$ for all i	64
	2	$(\mathbb{Z}_3)^2$	9	-	$s_i = 1$ for all i	4^9

4   Asymptotically Good Binary QAECD Codes of Index 3

In this section, a sequence of asymptotically good binary QAECD codes of index 3 is constructed through an existing sequence of asymptotically good binary self-dual quasi-abelian codes of index 2 presented in [14]. Recall that $R := \mathbb{F}_q[H]$ where H is a subgroup of a finite abelian group G, written additively. Moreover from Lemma 1, the map $\Phi : \mathbb{F}_q[G] \to R^l$ induces a one-to-one correspondence between linear codes of length $l = [G : H]$ over R and H-quasi abelian codes in $\mathbb{F}_q[G]$. Thus, quasi-abelian codes can be studied as linear codes over R of length l.

If C is an H-quasi abelian code in $\mathbb{F}_q[G]$, then $\Phi(C)^{\perp *} = \Phi(C^{\perp_E})$ [14, Corollary 2.1]. This means that $C \cap C^{\perp_E} = \{0\}$ if and only if $\Phi(C) \cap \Phi(C)^{\perp *} = \{0\}$. Hence, the map Φ also induces a one-to-one correspondence between LCD codes in $\mathbb{F}_q[G]$ and R^l.

Consider linear codes $C_{(a,b)} := \{(fa, fb) \mid f \in R\} \subseteq R^2$ and $C_{(a,b,1)} := \{(fa, fb, f) \mid f \in R\} \subseteq R^3$. Note that $C_{(a,b)}$ and $C_{(a,b,1)}$ correspond to quasi-abelian codes in $\mathbb{F}_q[G]$ of index 2 and index 3, respectively.

Starting from this point, assume that $R := \mathbb{F}_2[H]$ ($\gcd(|H|, 2) = 1$ is no longer required). From [14, Lemma 7.1], $C_{(a,b)}$ is self-dual if and only if $aa^* = bb^* \in \mathcal{U}(R)$, where $\mathcal{U}(R)$ denotes the group of units in R. Since $\mathrm{char}(R) = 2$, the given condition is equivalent to $aa^* + bb^* = 0$ and $a, b \in \mathcal{U}(R)$. It follows that if $C_{(a,b)}$ is self-dual, then $C_{(a,b,1)}$ is an ECD code from the fact that $aa^* + bb^* + 1 = 1$. Indeed, it can be easily shown that if $(u, v, w) \in C_{(a,b,1)} \cap C_{(a,b,1)}^{\perp *}$, then $(u, v, w) = (0, 0, 0)$. In general, we have the following proposition.

Proposition 14. *Let $H \leq G$ be finite abelian groups and let $R := \mathbb{F}_q[H]$. Suppose C is a 1-generator quasi-abelian code in $\mathbb{F}_q[G]$ generated by $g \in R$. If $[g, g]_* \in \mathcal{U}(R)$, then C is an H-QAECD code.*

We are now ready to show that there exists a sequence of asymptotically good binary QAECD codes of index 3. First, consider the main tool for that purpose. Let m be an odd integer and let $s(m)$ denote the multiplicative order of 2 modulo m. Consequently, let $l(m) := \min\{s(p) \mid p \text{ is a prime divisor of } m\}$ which is the smallest dimension of a nontrivial $\mathbb{F}_2$-representation of H (or equivalently, the smallest size of nonzero 2-classes of H) [1, Lemma 2.5].

Theorem 15 [14, Theorem 7.1]. *Let (H_{m_k}) be a sequence of abelian groups of odd order m_k, where (m_k) is a strictly increasing sequence, $s(m_k)$ is odd and $l(m_k) \gg \log m_k$. Let $A_k = H_{m_k} \oplus \mathbb{Z}_4$, $R_k = \mathbb{F}_2[A_k]$ and*

$$\Omega = \{(a, b) \in \mathcal{U}(R_k) \times \mathcal{U}(R_k) \mid \mathrm{wt}(a) + \mathrm{wt}(b) \equiv 0 \ (\mathrm{mod}\ 4)\}.$$

For each k, assume that $(a, b) \in \Omega$ is chosen as random and let $C_{(a,b)}^{(k)} := \{(fa, fb) \mid f \in R_k\}$. Then

(i) $C_{(a,b)}^{(k)}$ is a self-dual doubly even quasi-abelian code, and

(ii) there exists $\delta > 0$ such that for k large enough, the codes $C_{(a,b)}^{(k)}$ have relative minimum distance

$$\frac{\mathrm{d}\left(C_{(a,b)}^{(k)}\right)}{8m_k} \geq \delta. \tag{12}$$

Theorem 15 gives the assurance that there is a sequence of self-dual codes $\left(C_{(a,b)}^{(k)}\right)$ over R_k which is asymptotically good. Equivalently, there exists a sequence of asymptotically good binary self-dual quasi-abelian codes of index 2. This sequence of codes is used to construct a sequence of asymptotically good binary QAECD codes of index 3, as presented in the following proposition.

Proposition 16. Let (H_{m_k}) be a sequence of abelian groups of odd order m_k, where (m_k) is a strictly increasing sequence, $s(m_k)$ is odd and $l(m_k) \gg \log m_k$. Let $A_k = H_{m_k} \oplus \mathbb{Z}_4$, $R_k = \mathbb{F}_2[A_k]$ and

$$\Omega = \{(a,b) \in \mathcal{U}(R_k) \times \mathcal{U}(R_k) \mid \mathrm{wt}(a) + \mathrm{wt}(b) \equiv 0 \,(\mathrm{mod}\ 4)\}.$$

For each k, assume that $(a,b) \in \Omega$ is chosen as random and let $C_{(a,b)}^{(k)} := \{(fa, fb) \mid f \in R_k\}$. Define a sequence of codes given by $\left(C_{(a,b,1)}^{(k)}\right)$ where $C_{(a,b,1)}^{(k)} := \{(fa, fb, f) \mid f \in R_k\}$. Then

(i) $C_{(a,b,1)}^{(k)}$ is a QAECD code for each k,
(ii) the codes $C_{(a,b,1)}^{(k)}$ have relative minimum distance

$$\frac{\mathrm{d}\left(C_{(a,b,1)}^{(k)}\right)}{12m_k} > 0,$$

and have a positive constant relative rate.

Proof.

(i) Follows directly from Theorem 15 (i) and from the discussion above.
(ii) From Theorem 15 (ii), it follows that the relative minimum distance of the codes $C_{(a,b,1)}^{(k)}$ is given by

$$\frac{\mathrm{d}\left(C_{(a,b,1)}^{(k)}\right)}{12m_k} \geq \frac{\mathrm{d}\left(C_{(a,b)}^{(k)}\right)}{12m_k} = \frac{8}{12} \cdot \frac{\mathrm{d}\left(C_{(a,b)}^{(k)}\right)}{8m_k} \geq \frac{2}{3} \cdot \delta > 0,$$

for some $\delta > 0$ and for k large enough.

The relative rate of the codes $C_{(a,b,1)}^{(k)}$ is given by $\dfrac{\dim_{\mathbb{F}_2}\left(C_{(a,b,1)}^{(k)}\right)}{12m_k} = \dfrac{4m_k}{12m_k} = \dfrac{1}{3}$ for all k.

□

Note that in [11], it was shown that there exists an asymptotically good sequence of quasi-cyclic complementary dual codes. We have to emphasize that Proposition 16 also considers the case of strictly quasi-abelian codes since A_k can be non-cyclic for each k.

Summary

Through a known decomposition of semisimple group algebras, quasi-abelian complementary dual codes are characterized and enumerated. Explicit formulas for the number of quasi-abelian Euclidean and Hermitian complementary dual codes of index 2 were given. A class of asymptotically good binary index 3 quasi-abelian Euclidean complementary dual codes were constructed from an existing class of asymptotically good binary self-dual quasi-abelian codes of index 2.

Recall that if $H \leq G$ are finite abelian groups and $R := \mathbb{F}_q[H]$, then $\mathbb{F}_q[G] \cong R^l$ as R-modules, where $l := [G : H]$. For non-semisimple case (i.e. $\gcd(|H|, q) \neq 1$), one might study the algebraic structure of quasi-abelian complementary dual codes purely as linear codes over R or as linear codes in $\mathbb{F}_q[G]$. The asymptotic goodness of QACD codes of index 2 is still a viable problem for semisimple and non-semisimple cases.

Acknowledgment. S. Jitman was supported by the Thailand Research Fund under Research Grant MRG6080012. H. S. Palines would like to extend his sincerest gratitude to the following institutions: University of the Philippines Los Baños, University of the Phillipines System, Department of Science and Technology-Science Education Institute (DOST-SEI) of the Philippines, and Mathematics Department, Faculty of Science, Silpakorn University, Nakhon Pathom, Thailand.

References

1. Bazzi, L.M.J., Mitter, S.K.: Some randomized code constructions from group actions. IEEE Trans. Inf. Theory **52**, 3210–3219 (2006)
2. Bosma, W., Cannon, J., Playoust, C.: The Magma algebra system I: the user language. J. Symb. Comput. **24**, 235–265 (1997)
3. Carlet, C., Guilley, S.: Complementary dual codes for countermeasures to side-channel attacks. Coding Theor. Appl. **3**, 97–105 (2015)
4. Carlet, C., Daif, A., Danger, J.L., Guilley, S., Najm, Z., Ngo, X.T., Portebouef, T., Tavernier, C.: Optimized linear complementary codes implementation for hardware trojan prevention. In: Proceedings of European Conference on Circuit Theory and Design, 24–26 August 2015, Trondheim, Norway. IEEE, Piscataway (2015)
5. Dey, B.K.: On existence of good self-dual quasi-cyclic codes. IEEE Trans. Inform. Theory **50**, 1794–1798 (2004)
6. Dey, B.K., Rajan, B.S.: Codes closed under arbitrary abelian group of permutations. SIAM J. Discrete Math. **18**, 1–18 (2004)
7. Ding, C., Kohel, D.R., Ling, S.: Split group codes. IEEE Trans. Inform. Theory **46**, 485–495 (2000)
8. Esmaeili, M., Yari, S.: On complementary-dual quasi-cylic codes. Finite Fields Their Appl. **15**, 375–386 (2009)
9. Etesami, J., Hu, F., Henkel, W.: LCD codes and iterative decoding by projections, a first step towards an intuitive description of iterative decoding. In: Proceedings of IEEE Globecom, 5–9 December 2011, Texas, USA. IEEE, Piscataway (2011)
10. Fan, Y., Lin, L.: Thresholds of random quasi-abelian codes. IEEE Trans. Inform. Theory **61**, 82–90 (2015)

11. Guneri, C., Ozkaya, B., Solé, P.: Quasi-cylic complementary dual codes. Finite Fields Their Appl. **42**, 67–80 (2016)
12. Ishai, Y., Sahai, A., Wagner, D.: Private circuits: securing hardware against probing attacks. In: Boneh, D. (ed.) CRYPTO 2003. LNCS, vol. 2729, pp. 463–481. Springer, Heidelberg (2003). doi:10.1007/978-3-540-45146-4_27
13. Jitman, S., Ling, S., Liu, H., Xie, X.: Abelian codes in principal ideal group algebras. IEEE Trans. Inform. Theory **59**, 3046–3058 (2013)
14. Jitman, S., Ling, S.: Quasi-abelian codes. Des. Codes Crypt. **74**, 511–531 (2015)
15. Jitman, S., Ling, S., Solé, P.: Hermitian self-dual abelian codes. IEEE Trans. Inform. Theory **60**, 1496–1507 (2014)
16. Lally, K., Fitzpatrick, P.: Algebraic structure of quasicyclic codes. Discrete Appl. Math. **111**, 157–175 (2001)
17. Ling, S., Solé, P.: On the algebraic structure of quasi-cyclic codes I: finite fields. IEEE Trans. Inform. Theory **47**, 2751–2760 (2001)
18. Ling, S., Solé, P.: Good self-dual quasi-cyclic codes exist. IEEE Trans. Inform. Theory **49**, 1052–1053 (2003)
19. Ling, S., Solé, P.: On the algebraic structure of quasi-cyclic codes III: generator theory. IEEE Trans. Inform. Theory **51**, 2692–2700 (2005)
20. Ling, S., Xing, C.: Coding Theory, A First Course. Cambridge University Press, New York (2004)
21. Massey, J.L.: Linear codes with complementary duals. Discrete Math. **106**(107), 337–342 (1992)
22. Ngo, X.T., Guilley, S., Bhasin, S., Danger, J.L., Najm, Z.: Encoding the state of integrated circuits: a proactive and reactive protection against hardware trojans horses. In: Proceedings of WESS 2014, 12–17 October 2014, New Delhi, India. ACM, New York (2014)
23. Ngo, X.T., Bhasin, S., Danger, J.L., Guilley, S., Najm, Z.: Linear complementary dual code improvement to strengthen encoded cirucit against Hardware Trojan Horses. In: Proceedings of IEEE International Symposium on Hardware Oriented Security and Trust (HOST): 2015 May 2015, Washington DC Metropolitan Area, USA. IEEE, Piscataway (2015)
24. Pei, J., Zhang, X.: 1-generator quasi-cyclic codes. J. Syst. Sci. Complex. **20**, 554–561 (2007)
25. Rajan, B.S., Siddiqi, M.U.: Transform domain characterization of abelian codes. IEEE Trans. Inform. Theory **38**, 1817–1821 (1992)
26. Séguin, G.: A class of 1-generator quasi-cyclic codes. IEEE Trans. Inform. Theory **50**, 1745–1753 (2004)
27. Sendrier, N.: Linear codes with complementary duals meet the Gilber-Varshamov bound. Discrete Math. **285**, 345–347 (2004)
28. Yang, X., Massey, J.L.: The condition for a cyclic code to have a complementary dual. Discrete Math. **126**, 391–393 (1994)
29. Wan, Z.X.: Finite Fields and Galois Rings. World Scientific Pub. Co. Pte. Ltd., Singapore (2012)
30. Wasan, S.K.: Quasi abelian codes. Publ. Inst. Math. **35**, 201–206 (1977)

Relative Generalized Hamming Weights
and Extended Weight Polynomials
of Almost Affine Codes

Trygve Johnsen[(✉)] and Hugues Verdure

UiT - The Arctic University of Norway, 9037 Tromsø, Norway
{Trygve.Johnsen,Hugues.Verdure}@uit.no

Abstract. This paper is devoted to giving a generalization from linear codes to the larger class of almost affine codes of two different results. One such result is how one can express the relative generalized Hamming weights of a pair of codes in terms of intersection properties between the smallest of these codes and subcodes of the largest code. The other result tells how one can find the extended weight polynomials, expressing the number of codewords of each possible weight, for each code in an infinite hierarchy of extensions of a code over a given alphabet. Our tools will be demi-matroids and matroids.

Keywords: Pairs of almost affine codes · Relative generalized hamming weights · Extended weight polynomials

1 Introduction

We will focus on almost affine codes as defined in [14], that is: $C \subset F^n$ for some finite alphabet F, and the projection C_X has cardinality $|F|^s$ for a non-negative integer s for each $X \subset \{1, \cdots, n\}$. It is well known that this is a class of codes, which contain linear codes over fields F as a proper subclass. The intermediate class of affine codes are translates of linear codes. Another intermediate class is that of multilinear codes. It is also well known ([14]) that C defines a matroid M_C through the rank function

$$r(X) = \log_{|F|} |C_X|.$$

Such codes were studied in connection with access structures over $E = \{1, 2, \cdots, n\}$ and are strongly related to ideal perfect secret sharing schemes for such access structures. See e.g. [14].

In this note we will demonstrate how two different results for linear codes can be generalized to find analogous results for almost affine codes in general.

First, we will recall some known results and terminology for almost affine codes in Sect. 2. Then, in Sect. 3, we will study generalized Hamming weight (RLDP, in the sense of [2]) of pairs $C_2 \subset C_1$ of almost affine codes, and we will investigate to which extent it is possible to generalize the results in [12,15],

© Springer International Publishing AG 2017
A.I. Barbero et al. (Eds.): ICMCTA 2017, LNCS 10495, pp. 207–216, 2017.
DOI: 10.1007/978-3-319-66278-7_17

where one only treats linear codes. There one expresses these relative generalized weights as the minimum weights of subcodes of C_1 of various dimensions, intersecting C_2 only in the zero element. In one of our two main results, Theorem 2, we show an analogue of this result for almost affine codes. In Remark 3 we show, however, that the situation is not completely like in the case of linear codes.

There are many applications of relative generalized Hamming weights, as referred to in [12,15], for pairs of linear codes. In addition, relating to secret sharing schemes, which is a particularly natural topic, when working with almost affine codes, we would like to mention the significance of relative generalized Hamming weights described in [3,11].

In the last chapter, Sect. 4, we study another aspect of the relationship between almost affine codes and matroids. In [6], and in [5, p. 323], one points out that for linear block codes of length n over a finite field $\mathbb{F}_q$, one can produce an infinite series of codes by extending the alphabet to $\mathbb{F}_{q^s}$, for $s = 1, 2, \cdots$, and nevertheless find polynomials $A_0, \cdots, A_n$, such that $A_j(q^s)$ computes the number of codewords of weight j, for all s simultaneously, for each of $j = 0, \cdots, n$. We will show that a corresponding result holds for almost affine codes, and we use the arguments in [7, Sect. 3] as a stepping stone to find weight polynomials for a similar infinite series of almost affine codes C^s, all of the same block length, but over growing alphabets F^s as s grows. A main point in the linear case is that the polynomials A_j are only dependent on the associated matroid of C, and that we have matroids that play a completely analogous role, and that are equally simple to handle in the general case of almost affine codes.

2   Matroids, Demi-Matroids and Almost Affine Codes

In this section, we essentially recall relevant material that will be needed in the sequel, and we do not claim to have any new results here. We refer to [13] for the theory of matroids, to [1] for an introduction on demi-matroids and to [14] for an introduction on almost affine codes, and we will use their notation.

2.1   Matroids and Demi-Matroids

A matroid is a combinatorial structure that extend the notion of linear (in)dependency. There are many equivalent definitions, but we will give just one here.

Definition 1. *A matroid is a pair $M = (E, r)$ where E is a finite set, and r a function on the power set of E into $\mathbb{N}$ satisfying the following axioms:*

(R1) $r(\emptyset) = 0$,
(R2) *for every subset $X \subset E$ and $x \in E$, $r(X) \leqslant r(X \cup \{x\}) \leqslant r(X) + 1$,*
(R3) *for every $X \subset E$ and $x, y \in E$, if $r(X) = r(X \cup \{x\}) = r(X \cup \{y\})$, then $r(X \cup \{x, y\}) = r(X)$.*

Demi-matroids were introduced in [1]. They are a generalization of matroids in the following way:

Definition 2. *A demi-matroid is a pair $M = (E, r)$ where E is a finite set, and r a function on the power set of E into $\mathbb{N}$ satisfying axioms (R1) and (R2) above. The rank of M is $r(E)$.*

Matroids and demi-matroids have duals defined in the following way:

Proposition 1. *Let $M = (E, r)$ be a matroid (respectively a demi-matroid). Then $M^* = (E, r^*)$ with r^* defined as*

$$r^*(X) = |X| + r(E \backslash X) - r(E)$$

is a matroid (respectively a demi-matroid). Moreover, $(M^*)^* = M$.

The matroid (respectively demi-matroid) M^* is called the dual (respectively the dual or first dual) of M. It has rank $|E| - r(M)$. Demi-matroids have another dual, called the supplement dual or second dual. See [1, Theorem 4]:

Proposition 2. *Let $M = (E, r)$ be a demi-matroid. Then $\overline{M} = (E, \overline{r})$ with $\overline{r}$ defined as*

$$\overline{r}(X) = r(E) - r(E \backslash X)$$

is a demi-matroid. Moreover, we have $\overline{\overline{M}} = M$ and $\overline{M^*} = \overline{M}^*$.

2.2 Almost Affine Codes

Almost affine codes were first introduced in [14], and are a combinatorial generalization of affine codes.

Definition 3. *An almost affine code over a finite alphabet F, of length n and dimension k, is a subset $C \subset F^n$ such that $|C| = |F|^k$ and such that for every subset $X \subset E = \{1, \cdots, n\}$,*

$$\log_{|F|} |C_X| \in \mathbb{N},$$

where C_X is the puncturing of C with respect to $E \backslash X$.
An almost affine subcode of C is a subset $D \subset C$ which is itself an almost affine code over the same alphabet.

Remark 1. Any linear or affine code is obviously an almost affine code.

To any almost affine code C of length n and dimension k on the alphabet F, we can associate a matroid M_C on the ground set $E = \{1, \cdots, n\}$ and with rank function

$$r(X) = \log_{|F|} |C_X|,$$

for $X \subset E$.

Definition 4. *Let C be a block code of length n, and let $\mathbf{c} \in C$ be fixed. The $\mathbf{c}$-support of any codeword $\mathbf{w}$ is*

$$Supp(\mathbf{w}, \mathbf{c}) = \{i, c_i \neq w_i\}.$$

The $\mathbf{c}$-support of C is

$$Supp(C, \mathbf{c}) = \bigcup_{\mathbf{w} \in C} Supp(\mathbf{w}, \mathbf{c}).$$

Note that the $\mathbf{c}$-support of an almost affine code is independent of the choice of $\mathbf{c} \in C$ (see [8, Lemma 1]), and it will therefore be denoted by $Supp(C)$ without reference to any codeword. This observation gives rise to:

Definition 5. *The weight of an almost code C is $w(C) = |Supp(C)|$.*

Definition 6. *Let C be an almost affine code of length n, and let $\mathbf{c} \in F^n$ be fixed. Then*

$$C(X, \mathbf{c}) = \{\mathbf{w} \in C, \mathbf{w}_X = \mathbf{c}_X\},$$

where $\mathbf{c}_X$ is the projection of $\mathbf{w}$ to X. Such a subcode of C is called a standard subcode.

This might be empty, or not be an almost affine code, but when we take $\mathbf{c} \in C$, we get the following ([14, Corollary 1]):

Proposition 3. *Let C be an almost affine code of length n and dimension k over the alphabet F. Let $\mathbf{c} \in C$. Let $X \subset \{1, \cdots, n\}$. Then $C(X, \mathbf{c})$ is an almost affine subcode of C. Its asscociated matroid $M_{C(X,\mathbf{c})}$ is the contracted matroid M_C/X with rank function ρ given by*

$$\rho(Y) = r(X \cup Y) - r(X)$$

where r is the rank function of the matroid M_C. In particular,

$$|C(X, \mathbf{c})| = |F|^{k - r(X)}.$$

Remark 2. Not all subcodes of C are of the form $C(X, \mathbf{c})$, i.e. not all subcodes are standard subcodes.

Corollary 1. *Every almost affine code C of dimension k has almost affine subcodes of dimension $0 \leqslant i \leqslant k$.*

2.3   Generalized Hamming Weights

For a demi-matroid $D = (E, r)$ of rank $n - k$ we define:

Definition 7. *The generalized Hamming weights for a demi-matroid of dimension k are*

$$m_i(D) = \min\{|X|, \ n(X) = |X| - r(X) = i\}$$

for $1 \leqslant i \leqslant k$.

Definition 8. *The generalized Hamming weights for an almost affine code C of dimension k are*

$$d_i(C) = m_i(M_C^*) = \min\{|X|,\ |X| - r^*(X) = \bar{r}(X) = i\}$$

for $1 \leqslant i \leqslant k$, where r^* is the rank function of M_C^*, and $\bar{r}$ is the rank function of $\overline{M_C}$.

In fact the following was proved in [9]:

Theorem 1. *Let C be an almost affine code of length n and dimension k on an alphabet F of cardinality q, and let $\mathbf{c} \in C$. Then the generalized Hamming weights for C are*

$$
\begin{aligned}
d_i(C) &= \min\{|Supp(D)|,\ D \text{ is an almost affine subcode of dim. } i \text{ of } C\} \\
&= \min\{|Supp(D)|,\ D \text{ is a standard subcode of dim. } i \text{ of } C\} \\
&= n - \max\{|X|,\ |C(X, \mathbf{c})| = q^j\},
\end{aligned}
$$

for $1 \leqslant i \leqslant k$.

3   Equivalent Formulations of Some Hamming Weights of Pairs of Codes

From [10] we have:

Proposition 4. *Let $C_2 \subset C_1$ be two almost affine codes with rank functions r_2 and r_1. Then the pair (E, ρ) is a demi-matroid, for $\rho = r_1 - r_2$.*

Definition 9. *For $0 \leqslant i \leqslant \dim C_1 - \dim C_2$, we define the RLDP (Relative Length/Dimension Profile), or relative generalized Hamming weight, of the pair (C_1, C_2) as follows:*

$$m_i = \min\{|X|,\ \bar{\rho}(X) = i\}.$$

We observe that if $C_2 = 0$, this is the d_i associated to C_1.

For linear codes the most usual way in the literature (see e.g. [12]) to express the m_i is perhaps as

$$\min\{|X|\,|\, \dim(C_1(E\backslash X, \underline{0}) - \dim(C_2(E\backslash X, \underline{0}) \geqslant i\}.$$

It is easy to see that our Definition 9 above of the m_i gives the same values for linear codes. In [12, Lemma 1], and [15, Proposition 2], however, one gives an alternative, and less trivial reformulation of the m_i for linear codes. This is as:

$$\min\{w(D),\ |D \cap C_2| = 1,\ D \subset C_1 \text{ is a linear subcode with } \dim D = i\}.$$

Another, similar variant is given in [15, Prop.4]. We will now investigate the possibility of a reformulation of the m_i in analogous ways, not only for linear codes, but for almost affine codes in full generality. Our result is given in the theorem below:

Theorem 2. *Let $C_2 \subset C_1$ be a pair of almost affine codes with associated demi-matroid (E, ρ). Then for $0 \leqslant i \leqslant \dim C_1 - \dim C_2$,*

$$m_i = \min\{w(D),\ |D \cap C_2| = 1,\ D \subset C_1 \text{ is a standard subcode with } \dim D = i\}.$$

Proof. Let b_i be the right hand side of the above equality. We fix $v \in C_2$. All the standard subcodes, and all supports, considered in this proof, are with respect to v, and we omit its reference in the rest of the proof. For simplicity, denote $k_j = \dim C_j$ and $r_j = r_{C_j}$ for $j = 1, 2$.

Let $X \subset E$ be such that $|X| = m_i$ and $\overline{\rho}(X) = i$, that is

$$k_1 - r_1(E \backslash X) - k_2 + r_2(E \backslash X) = i$$

or equivalently

$$\dim C_1(E \backslash X) - \dim C_2(E \backslash X) = i.$$

We have the obvious inclusions

$$Supp(C_2(E \backslash X)) \subset Supp(C_1(E \backslash X)) \subset X.$$

We claim that the second inclusion is actually an equality. Indeed, if not, let $y \in X \backslash Supp(C_1(E \backslash X))$ and consider $Y = X \backslash \{y\}$. Then, for $j = 1, 2$, if $w \in C_j(E \backslash X)$, $w_y = v_y$ since y is not in the support, and in turn, the natural inclusions

$$C_j(E \backslash Y) \subset C_j(E \backslash X)$$

are equalities. This contradicts the minimality of X since Y also satisfies $\overline{\rho}(Y) = \overline{\rho}(X) = i$.

Let $Z \subset E \backslash X$ be a maximal independent subset of $E \backslash X$ for the matroid M_{C_2}, that is

$$|Z| = r_2(Z) = r_2(E \backslash X).$$

Let $Z' \subset X$ be such that $Z \cup Z'$ is a basis of M_{C_2}. Obviously, we have

$$r_2(Z') = |Z'| = k_2 - |Z| = k_2 - r_2(E \backslash X).$$

Let $W = X - Z$. Note that $Z \cup Z' \subset E \backslash W$. Then

$$C_2 \cap C_1(E \backslash W) = \{v\}.$$

Namely,

$$v \in C_2 \cap C_1(E \backslash W) = C_2(E \backslash W)$$

and

$$\dim C_2(E \backslash W) = k_2 - r_2(E \backslash W) = 0.$$

Moreover, we have

$$\begin{aligned}
r_1(E \backslash W) &\geqslant r_1(E \backslash X) + |Z'| \\
&\leqslant k_1 - k_2 + r_2(E \backslash X) - i + k_2 - r_2(E \backslash X) \\
&\leqslant k_1 - i
\end{aligned}$$

that is,

$$\dim C_1(E\backslash W) \geq i.$$

Take now any standard subcode of $C_1(E\backslash W)$ of dimension i. Then of course we have

$$v \in D \cap C_2 \subset C_1(E\backslash W) \cap C_2 = \{v\}$$

and

$$Supp(D) \subset Supp(C_1(E\backslash W)) \subset Supp(C_1(E\backslash X)) = X,$$

which implies that

$$b_i \leq Supp(D) \leq |X| = m_i.$$

For the converse, let $Y \subset E$ be such that $|C_1(Y) \cap C_2| = 1$. Then $C_1(Y) \cap C_2 = \{v\}$. Assume that $w(C_1(Y)) = b_i$ and $\dim C_1(Y) = i$. Let $Y' = E \backslash Supp(C_1(Y))$. Obviously, $Y \subset Y'$. Let $w \in C_1(Y)$. For any $y \in Y'$, $y \notin Supp(C_1(Y))$ so that $w_y = v_y$, and in turn $w \in C_1(Y')$. Hence, the natural inclusion $C_1(Y') \subset C_1(Y)$ is actually an equality.

Let $X = E - Y'$. Then we have

$$|X| = |E\backslash Y'| = |E\backslash(E\backslash Supp(C_1(Y))| = |Supp(C_1(Y))| = b_i$$

and

$$C_2(E\backslash X) = C_2(Y') = C_2 \cap C_1(Y') = C_2 \cap C_1(Y) = \{v\},$$

which implies that

$$\dim C_1(E\backslash X) - \dim C_2(E\backslash X) = \dim C_1(Y') - 0 = \dim C_1(Y) = i$$

and finally

$$m_i \leq |X| = b_i.$$

3.1 An Open Question Concerning Subcodes

Remark 3. Let

$$b_i' = \min\{w(D),\ |D \cap C_2| = 1,\ D \subset C_1 \text{ is a subcode with } \dim D = i\},$$

that is we allow D to be any subcode, not only a standard subcode. Obviously, for $1 \leq i \leq k_1 - k_2$, we have

$$m_i = b_i \geq b_i'.$$

It is an open question whether the last inequality is an equality, and will be the topic of further research. For linear codes, [15, Proposition 2] gives an analogous statement with equality. On the other hand, while b_i is defined just for $0 \leq i \leq k_1 - k_2$ (for $i > k_1 - k_2$ it is not difficult to show that any standard subcode of C_1 will have a non-trivial intersection with C_2), b_i' might be defined for $i > k_1 - k_2$. Consider namely the following codes: let $F = \{0, 1, 2, 3\}$ and $C_1 = F^3$. Let C_2 and D be the subcodes

$$\{000, 012, 023, 031, 103, 110, 121, 132, 201, 213, 222, 230, 302, 311, 320, 333\}$$

and

$$\{000, 011, 022, 033, 102, 113, 120, 131, 203, 210, 221, 232, 301, 312, 323, 330\}$$

respectively. Both subcodes have dimension 2, while C_1 has dimension 3. But we have $C_2 \cap D = \{000\}$ and $\dim D > \dim C_1 - \dim C_2 = 1$. Hence b_2' is defined (and is at most 3), while m_2 and b_2 could be said to be ∞ if one insists on defining them.

4   Extended Weight Polynomials of Almost Affine Codes

In [6], and in [5, p. 323], one points out that for linear block codes of length n over a finite field $\mathbb{F}_q$, one can produce an infinite series of codes by extending the alphabet to $\mathbb{F}_{q^s}$, for $s = 1, 2, \cdots$, and nevertheless find polynomials $A_0, \cdots, A_n$, such that $A_j(q^s)$ computes the number of codewords of weight j, for all s simultaneously, for each of $j = 0, \cdots, n$. Hence knowledge of a finite number of coefficients of the A_j compute an infinite number of weights. We will show that a corresponding result holds for almost affine codes, and we will mimick the arguments in [7, Section 3] to find weight polynomials for an infinite series of almost affine codes C_s, which we will now define.

Let $q = |F|$, where F is the alphabet over which an almost affine code C of block length n is defined. Then C^s is a code of block length n over the alphabet F^s, if an element $((c_{1,1}, \cdots, c_{1,n}), \cdots, (c_{s,1}, \cdots, c_{s,n}))$ instead is interpreted as:

$$((c_{1,1}, \cdots, c_{s,1}), \cdots, (c_{1,n}, \cdots, c_{s,n})). \tag{1}$$

It is then automatic that $|(C^s)_X| = (q^s)^r$ if $|C_X| = q^r$, for some $X \subset E = \{1, 2, \cdots, n\}$, and natural number r. Hence C^s is an almost affine code over F^s, since C is an almost affine code over F. Moreover the matroid $M_{C^s} = M_C$ since the rank functions are the same. Call the rank function r. Put $k = r(E)$.

Let $U \subset E$, and let c_Q be a fixed codeword in C^s. Similarly as in [7] we define: $S_U(s)$ is the subset of C^s, viewed over F^s, with the same coordinates as c_Q in the positions corresponding to U, in other words $S_U(s) = C^s(U, c_Q)$. But, since C^s is an almost affine code we see that $|S_U(s)| = (q^s)^{k-r(U)}$. In the next definition, there is no explicit reference to the codeword c_Q, since this is independent of the word chosen.

Definition 10. *For each $j = 1, \cdots, n$ let $A_{C,j}(s)$ be the number of codewords of weight j in C^s.*

Using the exclusion/inclusion principle we obtain the same formula as in [7, Formula (9) p. 638]:

$$A_{C,n}(s) = (-1)^n \sum_{U \subseteq E} (-1)^{|U|} (q^s)^{n^*(U)}.$$

Here $n^*(Y) = |Y| - r^*(Y)$ is the nullity function associated to the dual matroid (E, r^*) of M_C.

For each $X \subset E$ let $a_{X,C}(s)$ be the number of codewords with support exactly X. We then obtain in a similar way:

Lemma 1.
$$a_{X,C}(s) = (-1)^{|X|} \sum_{U \subset X} (-1)^{|U|} (q^s)^{n_X^*(U)},$$

where n_X^* is the nullity function of the dual of the rank function associated to the code $C(E \backslash X, c_Q)$.

A refined study, using Proposition 3, also gives

Lemma 2. For any $U \subset X$ we have: $n_X^*(U) = n^*(U)$.

Combining Lemmas 1 and 2 we then obtain an analogous formula as in [7, p. 638], and obtain:
This gives:

Proposition 5. For each $j = 0, 1, \cdots, n$ there are polynomials

$$A_{C,j}(s) = (-1)^j \sum_{|X|=j} \sum_{Y \subset X} (-1)^{|Y|} (q^s)^{n^*(Y)}.$$

counting the number over codewords of weight j in C^s.

In [7, Sects. 4 and 5], one shows how this matroid expression can be expressed by $\mathbb{N}_0$-graded Betti numbers of the Stanley-Reisner rings of the matroid M_C^* and its elongations, viewed as simplicial complexes via their independence sets ([7, Theorem 5.1]). From the arguments above we now see that its concequence, [7, Corollary 5.1], formulated for linear codes in that corollary, carries over to almost affine codes, except that the matroid $M(H)$ appearing in [7, Corollary 5.1], must be replaced by the matroid dual M_C^*. See also [7, Proposition 4.1], which can be applied to determine the generalized Hamming weights for almost affine codes from the degrees of the polynomials $A_j(s)$.
We also observe:

Example 1. As one sees from Proposition 5 the formula for $A_{C,j}(s)$ is only dependent on the polynomial (in the variable Q)

$$P_j(Q) = (-1)^j \sum_{|X|=j} \sum_{Y \subset X} (-1)^{|Y|} (Q)^{n^*(Y)},$$

which is defined for any (demi-)matroid, since it only uses its dual nullity function.
Let C be the almost affine code in [14, Example 2]. This is a code of rank 3 over the alphabet $\mathbb{F}_3^2$ of cardinality 9. Its length is also 9, and its well known ([14, Example 2]) that its associated matroid M_C is the non-Pappus matroid. In [4, p. 102] one calculated the polynomials $P_j(Q)$, for $j = 0, \cdots, 9$ without relating them to any code, since one knew that this matroid is not linearly representable. The results, however, automatically carry over to determining the $A_{C,j}(s)$ for the non-linear almost affine code C, and we obtain from [4, p. 102], or from usual inclusion/exclusion methods:

$$A_{C,0}(s) = 1,$$

$$A_{C,1}(s) = A_{C,2}(s) = A_{C,3}(s) = A_{C,4}(s) = A_{C,5}(s) = 0,$$

$$A_{C,6}(s) = 8q^s - 8,$$

$$A_{C,7}(s) = 12q^s - 12,$$

$$A_{C,8}(s) = 3q^{2s} - 18q^s + 15,$$

$$A_{C,9}(s) = q^{3s} - 9q^{2s} + 28q^s - 20.$$

$q^{3s} = |C^s|.$

References

1. Britz, T., Johnsen, T., Mayhew, D., Shiromoto, K.: Wei-type duality theorems for matroids. Des. Codes Cryptogr. **62**(3), 331–341 (2012)
2. Forney, G.F.: Dimension/length profiles and trellis complexity of linear block codes. IEEE Trans. Inform. Theory **40**(6), 1741–1752 (1994)
3. Geil, O., Martin, S., Matsumoto, R., Ruano, D.: Relative generalized Hamming weights of one-point algebraic geometric codes. IEEE Trans. Inform. Theory **60**(10), 5938–49 (2014)
4. Huerga Represa, V.: Towers of Betti Numbers of Matroids and Weight Distribution of Linear Codes and their Duals, Master's thesis in Pure Mathematics, University of Tromsø - The Arctic University of Norway (2015). http://hdl.handle.net/10037/7736
5. Jurrius, R.P.M.J.: Weight enumeration of codes from finite spaces Des. Codes Crypt. **63**(3), 321–330 (2012)
6. Jurrius, R.P.M.J., Pellikaan, G.R.: Algebraic geometric modeling in information theory. In: Codes, arrangements and matroids. Seroes on Coding Theory and Cryptology. World Scientific Publishing, Hackensack (2001)
7. Johnsen, T., Roksvold, J., Verdure, H.: Generalized weight polynomials of matroids. Discrete Math. **339**(2), 632–645 (2016)
8. Johnsen, T., Verdure, H.: Hamming weights of linear codes and Betti numbers of Stanley-Reisner rings associated to matroids. Appl. Algebra Engrg. Comm. Comput. **24**(1), 73–93 (2013)
9. Johnsen, T., Verdure, H.: Generalized Hamming weights for almost affine codes. IEEE Trans. Inform. Theory **63**(4), 1941–1953 (2017)
10. Johnsen, T., Verdure, H.: Flags of almost affine codes, arXiv:1704.02819 (2017)
11. Kurihara, J., Uyematsu, T., Matsumoto, R.: Secret sharing schemes based on linear codes can be precisely characterized by the relative generalized Hamming weights. IEICE Trans. Fundam. Electron. Commun. Comput. Sci. **95**(11), 2067–75 (2012)
12. Liu, Z., Chen, W., Luo, Y.: The relative generalized Hamming weight of linear q-ary codes and their subcodes. Des. Codes Crypt. **48**(2), 111–123 (2008)
13. Oxley, J.G.: Matroid Theory. Oxford University Press, New York (1992)
14. Simonis, J., Ashikhmin, A.: Almost affine codes. Des. Codes Crypt. **14**(2), 179–197 (1998)
15. Zhuang, Z., Dai, B., Luo, Y., Han-Vinck, A.J.: On the relative profiles of a linear code and a subcode. Des. Codes Crypt. **72**(2), 219–247 (2014)

On the Performance of Block Woven Codes Constructions with Row-Wise Permutations

Alexey Kreshchuk$^{(\boxtimes)}$ (iD), Igor Zhilin (iD), and Victor Zyablov (iD)

Laboratory 3, Institute for Information Transmission Problems,
Russian Academy of Science, Moscow, Russia
{krsch,zhilin,zyablov}@iitp.ru

Abstract. In this paper we propose a woven block code construction based on two convolutional codes. We also propose a soft-input decoder that allows this construction to have better error correction performance than the turbo codes with a conventional decoder. Computer simulation has showed a 0.1 dB energy gain relative to the LTE turbo code. Asymptotically the proposed code has distance greater than the product of free distances of component codes.

1 Introduction

Evolution of communication systems' standards requires improving coding gain and achieving new goals along with keeping complexity of the proposed codecs low enough. One of the ways to achieve these goals is to carefully design concatenated codes. It allows one to use simpler codes like convolutional codes as constituent codes to reduce implementation cost and time.

Our main goal is to design a code construction that would have better error correction performance than the widely used LTE turbo code.

In this paper we propose a block woven code construction that is based on two convolutional codes. Like turbo codes woven codes are a particular case of the concatenated codes. Earlier works mostly considered normal woven codes [8] and direct product convolutional codes [10] which are both convolutional codes. This makes it hard to improve the error-correction performance by introducing interleavers like it was done in turbo codes. In this work we also propose a block woven code with row-wise permutations that has better error correction performance than a block woven code without additional permutations. A similar construction without row-wise permutations and its distance properties were studied in [11].

This work is organized as follows. In Sect. 2 we describe the proposed code construction and the encoding algorithm. Section 3 is devoted to describing the proposed decoding algorithm. In Sect. 4 we study distance properties of the proposed code. In Sect. 5 we present results of computer simulation of proposed code and its soft decoder.

This work has been supported by RScF, research project No. 14-50-00150. This work was carried out using high-performance computing resources of federal center for collective usage at NRC "Kurchatov Institute", http://computing.kiae.ru/.

A.I. Barbero et al. (Eds.): ICMCTA 2017, LNCS 10495, pp. 217–227, 2017.
DOI: 10.1007/978-3-319-66278-7_18

2   Block Woven Code Construction and Encoding

Woven convolutional codes were introduced in 1997 [8]. Their distance properties and encoder design are studied in [7] and their error rates and decoder design are studied in [9].

An important reason to consider woven codes is the existence of woven convolutional codes with distance growing linearly with the number of constituent codes [4] whereas the distance of turbo codes always grows sublinearly [2].

Two block code constructions similar to the one considered in this paper were described in [3,5]. Let us describe the main differences. In [3] Serially Concatenated Block Codes use a random interleaver whereas the code considered uses a combination of matrix transpose and a row-wise permutation. In [5] each row of the Woven Block Code (with outer warp) word was terminated separately and the code considered has all rows encoded as one word of the outer code. Overall the code we propose in this work can be considered as a combination of ideas behind these two papers.

We consider a *block* code that uses *terminated* convolutional codes as constituent codes. Let us first make simplified description of the proposed woven code that doesn't account for parity check symbols introduced by termination and row-wise permutations that generate an ensemble of woven block codes.

2.1   Simplified Deterministic Description

Let us describe a simplified construction. We consider this construction with two constituent codes that can be the same code. This construction was earlier described in [12]. It does not picture termination symbols and row-wise permutations.

Let us write information symbols of this code as matrix:

$$\mathbf{I} = \quad \left.\vphantom{\rule{0pt}{2em}}\right\} k_A$$
$$\underbrace{}_{k_B}$$

At first the information matrix is read in row-wise order and encoded by the outer convolutional coder, $\mathbf{I} =$

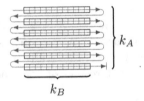

The resulting matrix is written in row-wise order too, $\mathbf{I}_A = Enc_B(\mathbf{I}) =$

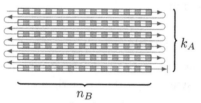

Information sequence is terminated in a usual way. White cells represent information symbols and grey cells represent parity-check symbols. It is worth noting that all this matrices are processed row-by-row by a single convolutional encoder. This differs, for example, from the work [6] where authors considered encoding all rows by several independent encoders.

Then $\mathbf{I}_A$ is read in column-wise order by inner convolutional code encoder, $\mathbf{I}_A =$

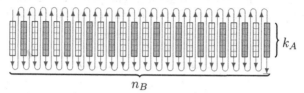

And written in the same column-wise order to a matrix that is a codeword, $\mathbf{C} = Enc_A(\mathbf{I}_A) = Enc_A(Enc_B(\mathbf{I})) =$

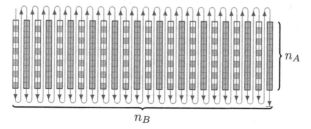

The result is a codeword:

$$\mathbf{C} = Enc_A(Enc_B(\mathbf{I})) =$$

Shaded cells correspond to parity-check symbols.

Note: a single encoder is used for encoding all rows. Then a single encoder is used for encoding all columns.

2.2   Proposed Code Ensemble

In a more precise description matrix $\mathbf{I}$ has a couple of termination symbols (they are in fact parity check symbols) in the lower right corner:

$$\mathbf{I} = \left.\rule{0pt}{16pt}\right\} k_A$$

$$\underbrace{\rule{80pt}{0pt}}_{k_B}$$

We introduce an ensemble of woven block codes by applying random permutations to each row of matrix $\mathbf{I}_A$ to get $\mathbf{I}'_A$. The permutation for different rows are selected independently and uniformly on the ensemble of all permutations.

$$\mathbf{I}'_A = \left.\rule{0pt}{16pt}\right\} k_A$$

$$\underbrace{\rule{120pt}{0pt}}_{n_B}$$

Note, that the last row of $\mathbf{I}_A$ has more parity check symbols than the others. Matrix $\mathbf{C}$ has some additional symbols introduced by termination that do not fit into the matrix.

$$\mathbf{C} = \left.\rule{0pt}{30pt}\right\} n_A$$

$$\underbrace{\rule{120pt}{0pt}}_{n_B}$$

3   Decoding Algorithm

The proposed decoding algorithm is iterative decoding algorithm based on sequential decoding of inner and outer codes by BCJR [1] soft-input soft-output algorithm. It has some resemblance of turbo codes' decoding algorithm in general, but has some significant differences due to specific features of the construction under consideration. The main difference is the fact that the inner and outer codes have different length, thus the decoder of the inner code can not get extrinsics for its parity-check symbols from the outer code.

We won't stop on the details of the BCJR decoding implementation since it is a very well-known decoder. Let us only note that BCJR decoder can output

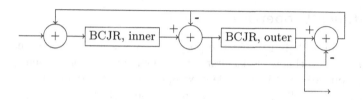

Fig. 1. Decoder scheme

LLRs for either each code symbol or each information symbol. Both modes will be used in the proposed decoder.

Let us describe decoding of the construction in general. Let us denote all decoder outputs by o and the inputs by i. The channel output is o^{channel}. The decoder is iterative and each iteration is split into these steps (schematically depicted on Fig. 1):

1. Compute input for the inner code. For information symbols it is:

$$i_n^{\text{inner,i}} = o^{\text{channel}} + o_{n-1}^{\text{outer}},$$

where o_{n-1}^{outer} is the extrinsics of the outer code on the previous iteration. For parity check symbols it is just:

$$i_n^{\text{inner,pc}} = o^{\text{channel}}.$$

2. Decode the inner code and get LLRs per information symbol:

$$o_n^{\text{inner}} = Dec_A(i_n^{\text{inner}}) - o_{n-1}^{\text{outer}}.$$

3. If this is not the last iteration, decode the outer code and get LLRs per code symbol:

$$o_n^{\text{outer}} = Dec_B(o_n^{\text{inner}}) - o_n^{\text{inner}}.$$

4. If this is the last iteration, decode the outer code and stop:

$$o = Dec_B(o_n^{\text{inner}}).$$

This algorithm has hard output and it always performs the same number of iterations. On the first iteration $o_0^{\text{outer}} = 0$.

There are two main differences between the proposed algorithm and the turbo decoder arising from the differences between their codes.

1. On step 2 the decoder does not output the extrinsics. That is a simple way to add n_A channel LLRs to the n_B outer code symbols.
2. On step 3 we get LLRs for each code symbols, not just for information symbols.

4   Distance Properties

Let us find the probability that the proposed code would have distance greater then $d_A d_B$. To do this we must first estimate the number of segments of length n_A taken from words of inner code of weight not greater than d_A.

There are generally three kinds of these code segments:

1. having zero syndrome both at the start and at the end,
2. having nonzero syndrome either at the start or at the end,
3. having nonzero syndrome both at the start and at the end.

The number of segments belonging to the last two kinds is limited by some constant c_0 due to increasing active distance of the inner code. The number of segments belonging to the first kind is less than $n_A c_1$, where c_1 is the number of words of minimal weight. For convolutional code defined by its generator polynomials $(7, 5)$, $c_1 = 1$.

Next let us calculate the probability that a random permutation will map a word of weight w and length n to a specified word. The number of such permutations is $w!(n - w)!$, and the number of all permutations is $n!$. Therefore this probability is $p_p = \binom{n}{w}^{-1}$.

Every word of the proposed code of weight $d_A d_B$ has a dense submatrix of size $d_B \times d_A$ (with all elements equal to 1). The corresponding matrix $\mathbf{I}'_A$ has a dense submatrix of size $w_B \times d_A$, where w_B is the weight of the information sequence generating a outer code sequence of minimal weight. For code $(7, 5)$, $w_B = 3$. The proposed code will contain no such words if all matrices $\mathbf{I}_A$ with w_B rows of weight d_A are permuted so that no dense submatrix. The number of these matrices is:

$$N_{\mathbf{I}_A} \leq \binom{n_B}{w_B}(c_0 + c_1 n_A).$$

The probability that all nonzero rows of a row-wise permutation of matrix $\mathbf{I}_A$ will have ones at the same positions is:

$$p_1 = p_p^{w_B - 1} = \binom{n_A}{d_A}^{-(w_B - 1)}$$

The probability the all matrices $\mathbf{I}_A$ with w_B rows of weight d_A are permuted so that no dense submatrix of size $d_A \times d_B$ exist is:

$$p_g \geq 1 - N_{\mathbf{I}_A} p_1 \geq 1 - \binom{n_B}{w_B}(c_0 + c_1 n_A)\binom{n_A}{d_A}^{-(w_B - 1)}$$

So if $\lim_{n_A, n_B \to \infty} N_{\mathbf{I}_A} p_1 = 0$ then most of the codes don't have codewords of weight $d_A d_B$. Let's compute this limit for $n_B = \alpha n_A$:

$$N_{\mathbf{I}_A} p_1 \leq \binom{n_B}{w_B}(c_0 + c_1 n_A)\binom{n_A}{d_A}^{-(w_B - 1)} \leq \frac{n_B^{w_B}}{w_B!}c_1 n_A\left(\frac{n_A}{d_A}\right)^{-(w_B - 1)d_A}$$

$$= \frac{c_1 d_A^{(w_B - 1)d_A} \alpha^{w_B}}{w_B!} n_A^{w_B + 1 - (w_B - 1)d_A}$$

For $w_B \geq 2$ and $d_A \geq 3$ this expression tends to zero with $n_A \to \infty$. We can state this fact as a theorem:

Theorem 1. *If $w_B \geq 2$ and $d_A \geq 3$ then for $n_A \to \infty$, $n_B \to \infty$, $\frac{n_A}{n_B} \to$ const the probability of the proposed code having distance $d_A d_B$ or less tends to zero.*

Proof. Described earlier in this section.

5   Numerical results

We implemented the decoding algorithm described above and measured its performance for selected codes.

The codes were simulated in an additive white gaussian noise (AWGN) channel with quadrature phase shift keying (QPSK).

The following code parameters were chosen: $k_A = 15$, $n_A = 30$, $k_B = 24$, $n_B = 48$. Woven code has dimension $K = k_A k_B - \nu_A = 358$, length $N = n_A n_B + 2\nu_B = 1444$ and rate $R = 0.248$, where $\nu = 2$ denotes the delay of convolutional code. Both inner and outer codes have generator polynomials $(7, 5)$.

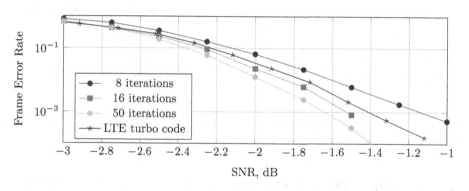

Fig. 2. Comparison with the LTE turbo code. Component codes are $(7, 5)$ convolutional codes.

Figure 2 compares frame error rates between the proposed code and the LTE turbo code. LTE turbo code has $k = 352$, $n = 1440$ and rate 0.244. Its decoder used 8 iterations. Its constituent codes' generator polynomials are $(13, 15)$ therefore their trellises have twice as many states as the one for $(7, 5)$. That's why it is reasonable to compare the LTE turbo codes performance to the proposed woven code decoded with 16 iterations. Then the energy gain of the proposed code relative to the LTE is about 0.1 dB. Increasing the number of iterations allows for greater energy gains.

Figure 3 shows the corresponding bit error rates. The constituent codes are $(7, 5)$ convolutional codes as in previous figure. Its main point is the intersection of decoder BER with the channel BER. From its location we can conclude that the decoder starts improving error rates at SNR -3.25 dB.

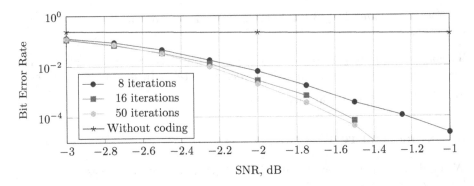

Fig. 3. Bit error rate depending on the number of iterations. Component codes are $(7, 5)$ convolutional codes.

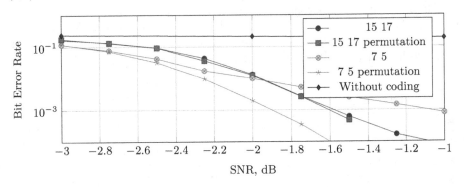

Fig. 4. Bit error rate depending on the constituent code with and without row-wise permutations

Figure 4 displays the effect of the row-wise permutations in the code construction: the difference in slope. The construction with $(7, 5)$ constituent codes and row-wise permutations has much steeper slope than the one without permutations. Supposedly, it is due to an increase in the code distance. Simulation showed no difference in slope for the construction with $(15, 17)$ constituent codes, but it is possible that this difference will appear at higher SNR. The permutations were selected at random, without any heuristics or picking.

Let us look at Fig. 4 to compare the error correction performance for different constituent codes. The construction with $(7, 5)$ constituent code starts correcting errors 0.25 earlier than the one with $(15, 17)$, but without row-wise permutations it has less steep slope. These permutations allow $(7, 5)$ construction to have the same slope as the $(15, 17)$ construction. Let us study the error correction performance of the constituent codes separately.

The difference between error rates of $(7, 5)$ and $(15, 17)$ convolutional codes might seem very small on Fig. 5, but as we have seen earlier it has huge effect on the proposed construction error rates. The $(15, 17)$ code has better error

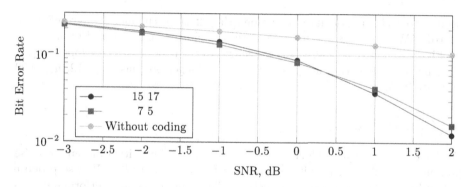

Fig. 5. Component codes bit error rates

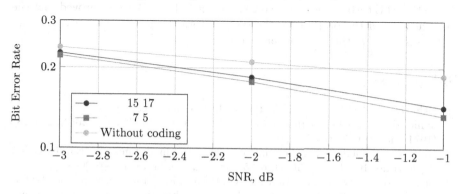

Fig. 6. Component codes bit error rates, enlarged

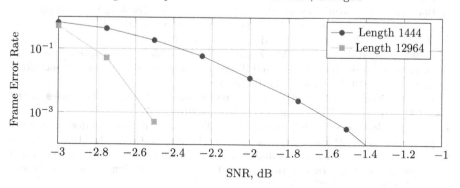

Fig. 7. Error rates depending on code length.

correction performance for SNR above 0.25 dB and we are confident that the proposed code with $(15, 17)$ constituent code would have lower error rates in this SNR range. Unfortunately it is very hard to test with computer simulation as these rates seem to be very small. The fact that without row-wise permutations $(15, 17)$ construction has lower BER than $(7, 5)$ construction above -2 dB might be due to fact that on later iterations the effective channel is "better" than 0.25 dB AWGN. The same bit error rates are plotted enlarged on Fig. 6.

Figure 7 shows that increasing the length of the code makes the slope steeper. The shorter code is the same as in Fig. 2, the longer one has $k_A = 72$, $n_B = 90$ and rate 0.25.

6   Conclusion

We have presented new block woven code construction with random row-wise permutations. We consider it a competitor to the turbo codes. We also propose a decoder for the proposed construction. The error correction performance was studied through a computer simulation. It has shown a 0.1 dB energy gain relative to the LTE turbo code at rate 0.24 and length 1440. We have proved that the code distance of the proposed code is asymptotically greater than the product of free distances of the component codes.

References

1. Bahl, L., Cocke, J., Jelinek, F., Raviv, J.: Optimal decoding of linear codes for minimizing symbol error rate (corresp.). IEEE Trans. Inf. Theory **20**(2), 284–287 (1974). doi:10.1109/tit.1974.1055186
2. Bazzi, L., Mahdian, M., Spielman, D.A.: The minimum distance of turbo-like codes. IEEE Trans. Inf. Theory **55**(1), 6–15 (2009). doi:10.1109/tit.2008.2008114
3. Benedetto, S., Divsalar, D., Montorsi, G., Pollara, F.: Serial concatenation of interleaved codes: performance analysis, design, and iterative decoding. IEEE Trans. Inf. Theory **44**(3), 909–926 (1998). doi:10.1109/18.669119
4. Bocharova, I.E., Kudryashov, B.D., Johannesson, R., Zyablov, V.V.: Asymptotically good woven codes with fixed constituent convolutional codes. In: Proceeding of ISIT 2007. IEEE, June 2007. doi:10.1109/isit.2007.4557169
5. Freudenberger, J., Bossert, M., Zyablov, V., Shavgulidze, S.: Woven codes with outer warp: variations, design, and distance properties. IEEE J. Sel. Areas Commun. **19**(5), 813–824 (2001). doi:10.1109/49.924865
6. Gazi, O., Yilmaz, A.O.: Turbo product codes based on convolutional codes. ETRI J. **28**(4), 453–460 (2006). doi:10.4218/etrij.06.0105.0187
7. Höst, S., Johannesson, R., Zyablov, V.: Woven convolutional codes. I. encoder properties. IEEE Trans. Inf. Theory **48**(1), 149–161 (2002). doi:10.1109/18.971745
8. Höst, S., Johannesson, R., Zyablov, V.V.: A first encounter with binary woven convolutional codes. In: 4th International Symposium on Communication Theory and Applications (1997)
9. Jordan, R., Höst, S., Johannesson, R., Bossert, M., Zyablov, V.: Woven convolutional codes II: decoding aspects. IEEE Trans. Inf. Theory **50**(10), 2522–2529 (2004). doi:10.1109/tit.2004.834790
10. Sidorenko, V., Bossert, M., Vatta, F.: Properties and encoding aspects of direct product convolutional codes. In: Proceeding ISIT 2012, pp. 2351–2355. IEEE, July 2012. doi:10.1109/isit.2012.6283934
11. Zhilin, I., Kreshchuk, A., Zyablov, V.: On the code distance of a woven block code construction. In: 2017 IEEE International Symposium on Information Theory (ISIT 2017), Aachen, Germany, pp. 16–20, June 2017

12. Zhilin, I., Zyablov, V., Zigangirov, D.: A binary block concatenated code based on two convolutional codes. In: Fifteenth International Workshop on Algebraic and Combinatorial Coding Theory ACCT2016, pp. 307–312, June 2016

New Lower Bounds on Error-Correcting Ternary, Quaternary and Quinary Codes

Antti Laaksonen[⊠] and Patric R.J. Östergård

Department of Communications and Networking, School of Electrical Engineering, Aalto University, P.O. Box 15400, 00076 Aalto, Finland
{antti.2.laaksonen,patric.ostergard}@aalto.fi

Abstract. Let $A_q(n, d)$ denote the maximum size of a q-ary code with size n and minimum distance d. For most values of n and d, only lower and upper bounds on $A_q(n, d)$ are known. In this paper we present 19 new lower bounds where $q \in \{3, 4, 5\}$. The bounds are based on codes whose automorphisms are prescribed by transitive permutation groups. An exhaustive computer search was carried out to find the new codes.

Keywords: Bounds on codes · Error-correcting codes · Transitive groups

1 Introduction

A q-ary code C of length n is a subset of Z_q^n where $Z_q = \{0, 1, \ldots, q-1\}$. Each element $c \in C$ is called a *codeword*, and Z_q is called the *alphabet* of C. The *size* of C is $|C|$, and the *minimum distance* of C is $\min_{a,b \in C, a \neq b} d_H(a, b)$ where d_H denotes the Hamming distance. A q-ary code with length n, size M and minimum distance d is called an $(n, M, d)_q$ code.

Let $A_q(n, d)$ denote the maximum size of an $(n, M, d)_q$ code. As it is difficult to determine exact values of $A_q(n, d)$, an important problem in coding theory is to find lower and upper bounds on the function. While binary codes have received the most attention [1,4,17], also ternary [8], quaternary [5] and quinary [7] codes have been studied.

Lower bounds on $A_q(n, d)$ can be found by discovering codes: if there is an $(n, M, d)_q$ code, then $A_q(n, d) \geq M$. Computer search techniques are often used to find such codes. However, as the search space is typically very large, assumptions about the structure of the code are usually needed to make the search efficient enough.

One way to limit the search space is to assume that there are *symmetries* in the code and only consider codes with prescribed automorphisms [12,14,19]. Such automorphisms may permute coordinates and coordinate values of codewords. In [16], new binary codes were found by focusing on codes whose groups of automorphisms are transitive permutation groups. In this paper, we extend this approach to ternary, quaternary and quinary codes.

© Springer International Publishing AG 2017
Á.I. Barbero et al. (Eds.): ICMCTA 2017, LNCS 10495, pp. 228–237, 2017.
DOI: 10.1007/978-3-319-66278-7_19

We carry out computer searches to systematically go through transitive permutation groups and search for codes with automorphisms prescribed by those groups. For a fixed group of automorphisms, the problem of finding a large code can be transformed into a graph problem where each vertex of the graph consists of an orbit of codewords and each clique in the graph corresponds to a code with the given automorphisms.

It turns out that several lower bounds on the maximum size of ternary, quaternary and quinary error-correcting codes can be improved by creating codes whose symmetries are prescribed by transitive permutation groups. We present 17 new codes, each of which yields a new lower bound on $A_q(n, d)$ where $q \in \{3, 4, 5\}$. In addition, two more lower bounds can be derived from the new codes.

The structure of the rest of the paper is as follows: In Sect. 2, we discuss the method using which we construct codes with prescribed automorphisms. In Sect. 3, we describe the computer searches we used to find the new codes. Finally, in Sect. 4, we present the new lower bounds on $A_q(n, d)$.

2   Code Construction

An *automorphism* of a code is a mapping from the code to itself that may permute coordinates and coordinate values of codewords. The general idea in our work is to search for codes with prescribed groups of automorphisms, i.e., to focus on codes that have certain symmetries.

In this context, it is convenient to represent codewords as sets of integers as follows: Let $[n] = \{1, 2, \ldots, n\}$. The representation of a q-ary codeword $c_1 c_2 \cdots c_n$ is a set

$$\{c_k n + k \mid k \in [n]\},$$

so that each codeword is an n-element subset of $[nq]$. The idea in using this representation is that we can permute both coordinates and coordinate values of codewords using permutations of $[nq]$.

Our general method to construct a q-ary code of length n is as follows: Let G be a permutation group of degree nq such that the group has a *block system* where each block is of the form

$$\{k, n + k, \ldots, (q - 1)n + k\}$$

where $k \in [n]$. Such a group corresponds to a group of automorphisms of a q-ary code of length n. The *orbit* of a codeword c is

$$\{gc \mid g \in G\}.$$

We construct the code as a union of orbits of codewords. Thus, for each orbit, we either include all words in the code or none of them.

Example: Let us construct a ternary code of length 4 whose group of automorphisms is

$$G = \langle (1\ 5\ 9) \rangle.$$

Here G determines that whenever we include a codeword $c_1c_2c_3c_4$ in the code, we also include all other codewords of the form $xc_2c_3c_4$ where x is any element. For example, we may create a code

$$C = \{0000, 1000, 2000, 0111, 1111, 2111, 0222, 1222, 2222\}$$

that consists of 3 orbits whose representatives are 0000, 1111 and 2222.

To prove that $A_q(n, d) \geq M$, it suffices to find a code with size M and minimum distance d. This corresponds to finding a clique of weight M in a graph that is generated as follows: Each vertex of the graph corresponds to an orbit where the minimum distance of any two words is at least d. The weight of such a vertex is the number of words in the orbit. There is an edge between two vertices if the minimum distance between any two words in the corresponding orbits is at least d.

Thus, to find a maximum-size code based on a permutation group, we should find a maximum-weight clique in a graph. Unfortunately, this is an NP-hard problem in general graphs. However, in many cases, it may be possible to find a clique whose weight is large by using, for example, backtracking or stochastic algorithms. The benefit in focusing on codes that consist of orbits of codewords is that the size of the resulting graph is moderate.

3   Computer Search

We carried out computer searches to find codes whose automorphisms are prescribed by *transitive* permutation groups. In [16], new binary codes were found using this approach; now we focus on ternary, quaternary and quinary codes. Transitive permutation groups have been classified [9,13] up to degree 47, so it is possible to systematically go through them for the parameters considered in this work.

Let T denote the collection of transitive permutation groups up to degree 47. We performed four separate searches over all groups in T. We describe the first search in detail, and the other searches are variations of it. Consider a group $G \in T$ whose degree is d. We first generate all block systems of G with block size $q \in \{3, 4, 5\}$. Then, for each such block system, we relabel the elements of $[d]$ so that the blocks are of the form

$$\{k, n + k, \ldots, (q - 1)n + k\}$$

where $n = d/q$ and $k \in [n]$. This yields a group of automorphisms for a q-ary code of length n. Finally, we select the orbits that produce the code by conducting a clique search in the corresponding graph.

In the second search, we searched for q-ary codes of length $n + 1$ such that G acts transitively on n coordinates (in the manner described above) and fixes one coordinate. In the third search, we searched for q-ary codes of length nk using k copies of G that act transitively and simultaneously on n coordinates each. Finally, in the fourth search, we searched for q-ary codes of length $nk + 1$ by combining the two previous techniques.

We used the *Cliquer* software [18] to find maximum-weight cliques in orbit graphs. We restricted ourselves to graphs that contain at most 5000 vertices, because processing larger graphs would have been too slow. Each clique search was run for at most 1000 s; in most cases the maximum-weight clique was found in a couple of seconds.

4   New Lower Bounds

We found several new codes that improve lower bounds on $A_q(n, d)$ where $q \in \{3, 4, 5\}$. The groups and orbits are given in the Appendix.

Table 1 summarizes the new lower bounds. As many as 17 new lower bounds follow directly from the new codes. In addition, using the facts

$$A_q(n, d) \geq A_q(n + 1, d)/q$$

and

$$A_q(n, d) \geq A_q(n + 1, d + 1)$$

we obtain two more lower bounds (on $A_3(14, 4)$ and $A_4(8, 5)$), resulting in a total of 19 new lower bounds.

Table 1. The new lower bounds

Old lower bound	New lower bound
$A_3(13, 4) \geq 8559$ [10]	$A_3(13, 4) \geq 13122$
$A_3(14, 4) \geq 24786$ [10]	$A_3(14, 4) \geq 27702$
$A_3(15, 4) \geq 72171$ [20]	$A_3(15, 4) \geq 83106$
$A_3(15, 5) \geq 6561$ [11]	$A_3(15, 5) \geq 7812$
$A_3(15, 6) \geq 2187$ [11]	$A_3(15, 6) \geq 3321$
$A_3(16, 7) \geq 729$ [2]	$A_3(16, 7) \geq 1026$
$A_3(16, 8) \geq 297$ [20]	$A_3(16, 8) \geq 387$
$A_4(8, 4) \geq 320$ [5]	$A_4(8, 4) \geq 352$
$A_4(8, 5) \geq 70$ [5]	$A_4(8, 5) \geq 76$
$A_4(9, 4) \geq 1024$ [3]	$A_4(9, 4) \geq 1152$
$A_4(9, 6) \geq 64$ [3]	$A_4(9, 6) \geq 76$
$A_4(10, 3) \geq 17408$ [5]	$A_4(10, 3) \geq 24576$
$A_4(10, 4) \geq 4096$ [3]	$A_4(10, 4) \geq 4192$
$A_4(11, 3) \geq 65536$ [15]	$A_4(11, 3) \geq 77056$
$A_5(8, 4) \geq 1125$ [7]	$A_5(8, 4) \geq 1225$
$A_5(8, 5) \geq 160$ [7]	$A_5(8, 5) \geq 165$
$A_5(9, 4) \geq 3750$ [7]	$A_5(9, 4) \geq 4375$
$A_5(9, 5) \geq 625$ [6]	$A_5(9, 5) \geq 725$
$A_5(10, 4) \geq 15625$ [3]	$A_5(10, 4) \geq 17500$

One of the codes in the Appendix gives the bound $A_3(15,5) \geq 7452$ when prescribing the given group G. By augmenting this code with 360 additional codewords, the final lower bound $A_3(15,5) \geq 7812$ is obtained. The additional codewords are presented via another group H—which is a subgroup of G—and orbits under the action of H. No codewords can be added to the other codes.

Acknowledgments. This work was supported in part by the Academy of Finland, Project #289002.

Appendix: Codes for the New Lower Bounds

Bound: $A_3(13,4) \geq 13122$
Generators of G:
(1 10 35 32 31 28)(2 14 36 22 19 5)(3 25 20 4 24 21)
(6 18 15 27 23 9)(7 17 37 34 29 12)(8 16 38 33 30 11),
(1 25 22 21 5 17)(2 24 32 3 23 33)(4 27 38 35 8 31)
(6 16 36 7 28 11)(9 34 18 30 14 12)(10 20 15 37 19 29)
Orbit representatives:
1112000000000, 2001001000000, 0110201000000, 0220021000000,
0202002000000, 1201000000001, 2111010000001, 2022110000001,
2200120000001, 0002000100001, 1011100000002, 2220010000002,
0212110000002, 1202011000002, 2200021100002

Bound: $A_3(15,4) \geq 83106$
Generators of G:
(1 41 6)(2 45 7 5 42 25)(3 4 8 9 28 44)(10 32 30 37 35 27)(11 21 31)
(12 40 17 15 22 20)(13 14 18 19 38 39)(16 26 36)(23 24 43 29 33 34),
(1 2 10 44)(3 33)(4 11 7 35)(5 19 26 37)(6 27 45 39)(8 43)(9 36 42 30)
(12 15 24 21)(13 38)(14 31 32 25)(16 17 40 29)(18)(20 34 41 22)(23 28)
Orbit representatives:
200101000000000, 111201000000000, 022202000000000, 120112000000000,
222210100000000, 012120100000000, 102200200000000, 021111010000000,
020022110000000, 220100210000000, 210021210000000, 201210020000000,
001022020000000, 222001001000000, 210201201000000, 011012011000000,
012021111000000, 211001121000000, 201222221000000, 222211012000000,
210101112000000, 211110222000000, 001102222000000, 000000110100000,
221021021100000, 010112112200000

Bound: $A_3(15,5) \geq 7812$
Generators of G:
(1 13 34 22)(2 6 23 39 17 21 38 9 32 36 8 24)
(3 14 27 26 33 44 42 41 18 29 12 11)(4 7 16 43)
(5 45 35 30 20 15)(19 37 31 28),
(1 45 41 25 21 5)(2 4 12 29 7 9)(3 23 28)
(6 35 16 30 26 10)(8 13 18)(11 40 36 20 31 15)

(14 22 24 17 34 27)(19 42 44 37 39 32)(33 38 43)
Orbit representatives:
212221100000000, 002122200000000, 210110001000000, 112100102000000,
121212212000000, 000002022010000, 010202102122000
Generators of H:
(16 4 7 10 43)(31 19 37 40 28)(2 35 38 26 29)(17 5 8 41 14)
(32 20 23 11 44)(3 36 39 42 15)(18 21 24 27 45)(33 6 9 12 30)
(34 22 25 13 1)
Orbit representatives:
000011021101212, 000022002020201, 000111111111010, 000112122210020,
000122202220010, 000201021012210, 000201122122112, 000210120202021,
000220012002020, 000220120210100, 001010010100011, 001011112001020,
001022001012200, 001110111110022, 001121201222012, 001122121212102,
001200121121111, 001201102011212, 002010011001022, 002021000011202,
002022120122210, 002121120211101, 002200101010211, 002200220120102,
010002012000221, 010012110022220, 010102212200000, 010200100220120,
010211001022200, 010211102102102, 010212110100100, 011000101100201,
011001011002220, 011101210120002, 011102101222122, 011211112021202,
011211201101120, 012001010021222, 012002100102200, 012010111020221,
012210111101101, 012210200100122, 020010022000211, 020110112200110,
020112221101020, 021012021002210, 021111221212022, 022010020011212,
022110220211021, 100001221212110, 100020212202220, 100102000212100,
100111211001212, 100112210200221, 100120202210101, 101000220211112,
101101002211102, 101110220000111, 102000001210111, 102002222010110,
102110011200222, 110011201222100, 110100212220121, 110112010222120,
111010200221102, 111101211202120, 111111012221122, 112012202220101,
112100210201122, 120112222210111, 121112221202110, 122111220201112

Bound: $A_3(15,6) \geq 3321$
Generators of G:
(1 33 20 7 24 11 13 45 17 34 36 8 40 27 44)
(2 4 6 23 10 42 29 31 18 5 22 39 26 28 15)
(3 35 37 9 41 43 30 32 19 21 38 25 12 14 16),
(1 36 11)(2 10 42 5 22 15)(3 29 43 39 38 4)
(6 26 31)(7 45 17 40 27 20)(8 19 33 14 13 9)
(12 35 37 30 32 25)(16 21 41)(18 44 28 24 23 34)
Orbit representatives:
022102200000000, 111000121000000, 222222012000000, 110012000100000,
101120020200000

Bound: $A_3(16,7) \geq 1026$
Generators of G:
(1 21 18 38)(2 22 17 37)(3 23 36 8)(4 24 35 7)(5 34 6 33)
(9 29 26 46)(10 30 25 45)(11 31 44 16)(12 32 43 15)
(13 42 14 41)(19 39 20 40)(27 47 28 48),

(1 35 5)(2 36 6)(3 37 33)(4 38 34)(9 43 13)(10 44 14)
(11 45 41)(12 46 42)(17 19 21)(18 20 22)(25 27 29)(26 28 30)
Orbit representatives:
2000000020000000, 0122222201000000, 2221220000100000,
2010010012212000, 0112121210101010, 2001010102101010,
0221120112201010, 1012022020102010, 2002122110201020

Bound: $A_3(16,8) \geq 387$
Generators of G:
(1 20 7 8 5 38)(2 3 18 35 34 19)(4 23 40 21 22 17)(6 33 36 39 24 37)
(9 28 15 16 13 46)(10 11 26 43 42 27)(12 31 48 29 30 25)(14 41 44 47 32 45),
(1 23 8 3 34 22 36)(2 38 4 33 7 24 35)(6 20 17 39 40 19 18)
(9 31 16 11 42 30 44)(10 46 12 41 15 32 43)(13)(14 28 25 47 48 27 26)
Orbit representatives:
2200210222100000, 0201000120210000, 0000221122221100,
0202211002022110, 2020120102022110, 1111002202022110

Bound: $A_4(8,4) \geq 352$
Generators of G:
(1 20)(2 19)(3 26 11 18 27 10)(4 25 12 17 28 9)
(5 24)(6 23)(7 30 15 22 31 14)(8 29 16 21 32 13),
(1 3)(2 4)(5 7)(6 8)(9 27)(10 28)(11 25)(12 26)(13 31)
(14 32)(15 29)(16 30)(17 19)(18 20)(21 23)(22 24),
(1 17 9)(2 18 10)(3 11 19)(4 12 20)(5 21 13)(6 22 14)(7 15 23)(8 16 24)
Orbit representatives:
30100000, 21320000, 22002200, 11112200, 10201010, 01021010, 21212010,
12122010, 33003010, 03322110, 30232110, 30321210, 03231210, 11220310

Bound: $A_4(9,4) \geq 1152$
Generators of G:
(1 8 28 35 10 17 19 26)(2 7 29 34 11 16 20 25)
(3 33 21 6 12 24 30 15)(4 32 22 5 13 23 31 14),
(1 32 30 34 10 23 21 25)(2 33 31 35 11 24 22 26)
(3 16 28 5 12 7 19 14)(4 17 29 6 13 8 20 15),
(1 13)(2 12)(3 11)(4 10)(5 24)(6 23)(7 35)(8 34)(14 33)
(15 32)(16 26)(17 25)(19 22)(20 21)(28 31)(29 30)
Orbit representatives:
210020000, 022220000, 203220100, 202030300, 332020001, 311030101,
123230101, 130220201, 111220002, 013020102, 320220102, 031230302,
112030003, 221230003, 121020103, 303020203

Bound: $A_4(9,6) \geq 76$
Generators of G:
(1 30 24 14)(2 35 25 36)(3 33 23 19)(4 13)(5 10 12 6)(7 9 11 8)
(15 32 28 21)(16 18 29 17)(20 26 34 27)(22 31),

(1 11 34)(2 7 28)(3 22 17)(4 8 30)(5 9 33)(6 32 27)(10 29 16)
(12 31 35)(13 26 21)(14 18 24)(15 23 36)(19 20 25)
Orbit representatives:
221012000, 000322200

Bound: $A_4(10,3) \geq 24576$
Generators of G:
(1 37 31 7)(2 26 12 36)(3 5)(4 14)(6 22 16 32)(8 40 18 30)(9 19)
(10 38 20 28)(11 27 21 17)(13 15)(23 35)(24 34)(25 33)(29 39),
(1 6 21 36)(2 35)(3 4 13 14)(5 22)(7 10)(8 39 18 29)(9 38 19 28)
(11 16 31 26)(12 25)(15 32)(17 20)(23 34 33 24)(27 30)(37 40)
Orbit representatives:
1310000000, 3120220000

Bound: $A_4(10,4) \geq 4192$
Generators of G:
(1 29 4 6 11 39 24 36)(2 30 3 5 12 40 23 35)(7 38 27 18)(8 37 28 17)
(9 14 16 21 19 34 26 31)(10 13 15 22 20 33 25 32),
(1 19)(2 20)(3 7 23 27 33 37)(4 8 24 28 34 38)(5 36 25 6 35 26)
(9 11 29 31 39 21)(10 12 30 32 40 22)(13 17)(14 18)(15 16)
Orbit representatives:
0000000000, 3311000000, 0110301000, 1001301000, 2332301000,
1032121000, 1210303010

Bound: $A_4(11,3) \geq 77056$
Generators of G:
(1 32 23 10)(2 31)(3 41 14 19)(4 7 37 40)(5 6 16 28)(8 25 30 36)
(9 35)(12 21 34 43)(13 42)(15 18 26 29)(17 27 39 38)(20 24),
(1 35)(2 34 24 23 13 12)(3 10 36 43 14 32)(4 9 26 31 37 42)
(5 30 27 19 38 8)(6 29 39 40 17 7)(15 20)(16 41)(18 28)(21 25)
Orbit representatives:
10120000000, 02130000000, 23330000000, 02312000000, 12000000001,
31220000001, 20130000001, 33000000002, 10200000002, 01120000002,
11100000003, 22300000003, 30002000003

Bound: $A_5(8,4) \geq 1225$
Generators of G:
(1 20 40 10 27 7 17 36 16 26 3 23 33 12 32 2 19 39 9 28 8 18 35 15 25 4 24 34 11
31) (5 22 37 14 29 6 21 38 13 30),
(1 28 5 26 3 30)(2 27 6 25 4 29)(7 32)(8 31)(9 20 13 18 11 22)
(10 19 14 17 12 21)(15 24)(16 23)(33 36 37 34 35 38)(39 40)
Orbit representatives:
00000000, 41131000, 02241000, 24332000, 43411010, 34212010,
01010110, 02020220

Bound: $A_5(8,5) \geq 165$
Generators of G:
(1 32 19 5 33 24 11 37 25 16 3 29 17 8 35 21 9 40 27 13)
(2 31 20 6 34 23 12 38 26 15 4 30 18 7 36 22 10 39 28 14),
(1 10 17 26 33 2 9 18 25 34)(3 12 19 28 35 4 11 20 27 36)
(5 14 21 30 37 6 13 22 29 38)(7 16 23 32 39 8 15 24 31 40)
Orbit representatives:
33330000, 40304100, 13013200, 12340210, 30134210

Bound: $A_5(9,4) \geq 4375$
Generators of G:
(1 32 10 41 19 5 28 14 37 23)(2 33 11 42 20 6 29 15 38 24)
(3 43 21 25 39 7 12 34 30 16)(4 44 22 26 40 8 13 35 31 17),
(1 3 20 13)(2 4 19 12)(5 16 6 17)(7 33 26 23)(8 32 25 24)
(10 30 11 31)(14 34 42 44)(15 35 41 43)(21 29 40 37)(22 28 39 38)
Orbit representatives:
200020000, 112040000, 104010100, 232030100, 311000001, 023010101,
342010201, 121020301, 433000002, 143040102, 231010003, 411020103,
244000004, 322020004, 402040304

Bound: $A_5(9,5) \geq 725$
Generators of G:
(1 44 37 26)(2 45 38 27)(3 43 39 25)(7 12 16 30)(8 10 17 28)(9 11 18 29)
(13 22 40 31)(14 23 41 32)(15 24 42 33)(19 35)(20 36)(21 34),
(1 9 13 19 27 31 37 45 4 10 18 22 28 36 40)
(2 7 15 20 25 33 38 43 6 11 16 24 29 34 42)
(3 8 14 21 26 32 39 44 5 12 17 23 30 35 41)
Orbit representatives:
444222000, 231140100, 312401100, 003121100, 123014100

Bound: $A_5(10,4) \geq 17500$
Generators of G:
(1 32 33 44 5)(2 3 14 35 21)(4 45 11 42 43)(6 37 38 49 10)
(7 8 19 40 26)(9 50 16 47 48)(12 13 24 25 31)
(15 41 22 23 34)(17 18 29 30 36)(20 46 27 28 39),
(1 42)(2 11)(3 5 13 45 23 35 33 25 43 15)(4 44 34 24 14)
(6 47)(7 16)(8 10 18 50 28 40 38 30 48 20)(9 49 39 29 19)
(12 21)(17 26)(22 31)(27 36)(32 41)(37 46),
(1 32 33 44 25)(2 3 14 5 21)(4 15 11 42 43)(6 37 38 49 30)
(7 8 19 10 26)(9 20 16 47 48)(12 13 24 45 31)
(17 18 29 50 36)(22 23 34 35 41)(27 28 39 40 46)
Orbit representatives:
0000000000, 1331000000, 4342110000, 4112220000, 0442030000,
3114040000, 2333240000, 1222340000

References

1. Agrell, E., Vardy, A., Zeger, K.: A table of upper bounds for binary codes. IEEE Trans. Inform. Theory **47**, 3004–3006 (2001)
2. Assmus, E.F., Mattson, H.F.: New 5-designs. J. Combin. Theory **6**, 122–151 (1969)
3. Assmus, E.F., Mattson, H.F.: On weights in quadratic-residue codes. Discrete Math. **3**, 1–20 (1972)
4. Best, M., Brouwer, A.E., MacWilliams, F.J., Odlyzko, A.M., Sloane, N.J.A.: Bounds for binary codes of length less than 25. IEEE Trans. Inform. Theory **24**, 81–93 (1978)
5. Bogdanova, G.T., Brouwer, A.E., Kapralov, S.N., Östergård, P.R.J.: Error-correcting codes over an alphabet of four elements. Des. Codes Cryptogr. **23**, 333–342 (2001)
6. Boukliev, I., Kapralov, S., Maruta, T., Fukui, M.: Optimal linear codes of dimension 4 over F_5. IEEE Trans. Inform. Theory **43**, 308–313 (1997)
7. Bogdanova, G.T., Östergård, P.R.J.: Bounds on codes over an alphabet of five elements. Discrete Math. **240**, 13–19 (2001)
8. Brouwer, A.E., Hämäläinen, H.O., Östergård, P.R.J., Sloane, N.J.A.: Bounds on mixed binary/ternary codes. IEEE Trans. Inform. Theory **44**, 140–161 (1998)
9. Cannon, J.J., Holt, D.F.: The transitive permutation groups of degree 32. Exp. Math. **17**, 307–314 (2008)
10. Código [pseud.], A discussion on Foros de Free1X2.com, cited by Andries Brouwer at https://www.win.tue.nl/~aeb/codes/ternary-1.html
11. Conway, J.H., Pless, V., Sloane, N.J.A.: Self-dual codes over GF(3) and GF(4) of length not exceeding 16. IEEE Trans. Inform. Theory **25**, 312–322 (1979)
12. Elssel, K., Zimmermann, K.-H.: Two new nonlinear binary codes. IEEE Trans. Inform. Theory **51**, 1189–1190 (2005)
13. Hulpke, A.: Constructing transitive permutation groups. J. Symbolic Comput. **39**, 1–30 (2005)
14. Kaikkonen, M.K.: Codes from affine permutation groups. Des. Codes Cryptogr. **15**, 183–186 (1998)
15. Kschischang, F.R., Subbarayan, P.: Some ternary and quaternary codes and associated sphere packings. IEEE Trans. Inform. Theory **38**, 227–246 (1992)
16. Laaksonen, A., Östergård, P.R.J.: Constructing error-correcting binary codes using transitive permutation groups (submitted). Preprint at. arXiv:1604.06022
17. MacWilliams, F.J., Sloane, N.J.A.: The Theory of Error-Correcting Codes, North-Holland, Amsterdam (1977)
18. Niskanen, S., Östergård, P.R.J.: Cliquer User's Guide, Version 1.0, Communications Laboratory, Helsinki University of Technology, Espoo, Finland, Technical report T48 (2003)
19. Östergård, P.R.J.: Two new four-error-correcting binary codes. Des. Codes Cryptogr. **36**, 327–329 (2005)
20. Plotkin, M.: Binary codes with specified minimum distance. IRE Trans. Inform. Theory **6**, 445–450 (1960)

A State Space Approach to Periodic Convolutional Codes

Diego Napp[1], Ricardo Pereira[1(✉)], and Paula Rocha[2]

[1] CIDMA - Center for Research and Development in Mathematics
and Applications, Department of Mathematics,
University of Aveiro, Aveiro, Portugal
{diego,ricardopereira}@ua.pt
[2] SYSTEC, Faculty of Engineering, University of Porto, Porto, Portugal
mprocha@fe.up.pt

Abstract. In this paper we study periodically time-varying convolutional codes by means of input-state-output representations. Using these representations we investigate under which conditions a given time-invariant convolutional code can be transformed into an equivalent periodic time-varying one. The relation between these two classes of convolutional codes is studied for period 2. We illustrate the ideas presented in this paper by constructing a periodic time-varying convolutional code from a time-invariant one. The resulting periodic code has larger free distance than any time-invariant convolutional code with equivalent parameters.

Keywords: Convolutional codes · Periodically time-varying codes · Input-state-output representations

1 Introduction

Convolutional codes [10] are an important type of error correcting codes that can be represented as a time-invariant discrete linear system over a finite field [20]. They are used to achieve reliable data transfer, for instance, in mobile communications, digital video and satellite communications [10,23].

Since the sixties it has been widely known that convolutional codes and linear systems defined over a finite field are essentially the same objects [20]. More recently, there has been a new and increased interest in this connection and many advances have been derived from using the system theoretical framework

This work was supported in part by the Portuguese Foundation for Science and Technology (FCT-Fundação para a Ciência e a Tecnologia), through CIDMA - Center for Research and Development in Mathematics and Applications, within project UID/MAT/04106/2013 and also by Project POCI-01-0145-FEDER-006933 - SYSTEC - Research Center for Systems and Technologies - funded by FEDER funds through COMPETE2020 - Programa Operacional Competitividade e Internacionalização (POCI) - and by national funds through FCT - Fundação para a Ciência e a Tecnologia.

© Springer International Publishing AG 2017
Á.I. Barbero et al. (Eds.): ICMCTA 2017, LNCS 10495, pp. 238–247, 2017.
DOI: 10.1007/978-3-319-66278-7_20

when dealing with convolutional codes, see [12,17]. Most of the large body of literature on convolutional codes and on the relation of these codes with linear systems has been devoted to the "time-invariant" case.

In this work we aim at studying time-varying convolutional codes from a system theoretical point of view. These codes have attracted much attention after Costello conjectured in [5] that nonsystematic time-varying convolutional codes can attain can larger free distance than the nonsystematic time-invariant ones. Since then, several researchers have investigated such codes [3,15,16,18]. Moreover, in combination with wavelets [6] time-varying convolutional codes yield unique trellis structures that resulted in fast and low computational complexity decoding algorithms. However, little is known on the relation of time-varying convolutional codes and time-varying linear systems and very few general constructions of time-varying convolutional codes with designed distances are known.

Here we deal with periodically time-varying convolutional codes (for short, periodic codes) using input-state-output representations, and investigate some of their special properties and structures. In particular, we associate periodic codes with suitably defined time-invariant convolutional codes. This allows us to derive constructions of (periodic) input-state-output representations for periodic codes from the better understood time-invariant class.

2 Preliminaries

In the sequel we shall follow the system theory notation and consider column vectors rather than row vectors.

2.1 Time-Invariant Convolutional Codes

A time-invariant convolutional code is a set of finite support sequences, called *codewords*, obtained as the image of a polynomial shift operator (the *encoder*) acting on finite support sequences that correspond to the original *information*. More precisely this can be defined as follows.

Definition 1. *Let $\mathbb{F}$ be a finite field and n, k be positive integers with $k < n$. A time-invariant convolutional code $\mathcal{C}$ of rate k/n is a set of finite support sequences described as*

$$\mathcal{C} = \left\{ v : v(\ell) = \left(G\left(\sigma^{-1}\right) u \right)(\ell); \ \ell \in \mathbb{N}_0, u \in \left[\left(\mathbb{F}^k\right)^{\mathbb{N}_0} \right]_{\mathsf{FS}} \right\}$$

where $G(z) \in \mathbb{F}^{n \times k}[z]$ is a full column rank $n \times k$ polynomial matrix over $\mathbb{F}$, called the encoder, *u taking values in $\mathbb{F}^k$ is the* information *sequence and v is the* codeword. *Moreover, σ^{-1} denotes the shift $\left(\sigma^{-1}u\right)(\ell) = u(\ell - 1)$, and the subindex* FS *affecting a set of sequences indicates that only its finite support elements are considered.*

The encoders of a code $\mathcal{C}$ are not unique; however they only differ by right multiplication by unimodular matrices over $\mathbb{F}[z]$. An encoder matrix G is called *basic* if it has a polynomial right inverse; from now on we shall only consider basic encoders, and refer to them simply as encoders. The encoder G is called *minimal* if the sum of its column degrees attains the minimal possible value.

We define the *degree* δ of a convolutional code as the sum of the column degrees of one, and hence any, minimal encoder. Note that the list of column degrees (also known as Forney indices) of a minimal encoder is unique up to a permutation. The maximum of the Forney indices is called the *memory* of a code, and is denoted by m. A code $\mathcal{C}$ of rate k/n, degree δ and memory m is said to be an (n, k, δ) code or an (n, k, δ, m) code if the memory is to be specified.

2.2   Periodically Time-Varying Convolutional Codes

In this work we consider convolutional codes $\mathcal{C}$ with P-periodic encoders, i.e.:

$$\mathcal{C} = \left\{ v : v(P\ell + t) = \left(G^t \left(\sigma^{-1} \right) u \right)(P\ell + t); \ t = 0, \dots, P - 1; \atop \ell \in \mathbb{N}_0, u \in \left[\left(\mathbb{F}^k \right)^{\mathbb{N}_0} \right]_{FS} \right\}, \qquad (1)$$

where each $G^t(z)$ is an $n \times k$ time-invariant (basic) encoder. Such codes will be called *P-periodic*.

Inspired by the ideas developed in [1,13] for the case of behaviors, considering the linear map

$$L_p : \left(\mathbb{F}^n \right)^{\mathbb{N}_0} \rightarrow \left(\mathbb{F}^{Pn} \right)^{\mathbb{N}_0}$$

defined by

$$(L_p v)(\ell) = \begin{bmatrix} v(P\ell) \\ v(P\ell + 1) \\ \vdots \\ v(P\ell + P - 1) \end{bmatrix}, \ P \in \mathbb{N}$$

we associate with $\mathcal{C}$ a time-invariant convolutional code $\mathcal{C}^L$, the *lifted* version of $\mathcal{C}$, defined as

$$\mathcal{C}^L = \left\{ \tilde{v} \in \left(\mathbb{F}^{Pn} \right)^{\mathbb{N}_0} : \tilde{v} = L_p v, \ v \in \mathcal{C} \right\}.$$

Note that, since

$$\left(G^t \left(\sigma^{-1} \right) u \right)(P\ell + t) = \left(\left(\sigma^t G^t \left(\sigma^{-1} \right) \right) u \right)(P\ell),$$

the equation in (1) can also be written as

$$\left(\Omega_{P,n} \left(\sigma \right) v \right)(P\ell) = \left(G \left(\sigma, \sigma^{-1} \right) u \right)(P\ell), \ \ell \in \mathbb{N}_0,$$

where for $r \in \mathbb{N}$

$$\Omega_{P,r} \left(\sigma \right) = \begin{bmatrix} I_r \\ \sigma I_r \\ \vdots \\ \sigma^{P-1} I_r \end{bmatrix}$$

is a polynomial matrix operator in the shift σ and

$$
G\left(\sigma, \sigma^{-1}\right) = \begin{bmatrix} G^0\left(\sigma^{-1}\right) \\ \sigma G^1\left(\sigma^{-1}\right) \\ \vdots \\ \sigma^{P-1} G^{P-1}\left(\sigma^{-1}\right) \end{bmatrix}
$$

is a polynomial matrix operator in the shifts σ and σ^{-1}.

Moreover, it is possible to show that the matrix G can be decomposed as

$$
G\left(\sigma, \sigma^{-1}\right) = G^L\left(\sigma^{-P}\right) \Omega_{P,k}\left(\sigma\right)
$$

where

$$
G^L\left(\sigma^{-1}\right) = \left[G^{L_0}\left(\sigma^{-1}\right) \mid G^{L_1}\left(\sigma^{-1}\right) \mid \cdots \mid G^{L_{P-1}}\left(\sigma^{-1}\right) \right]
$$

and the blocks $G^{L_j}\left(\sigma^{-1}\right)$ have size $Pn \times k$, $j = 0, \ldots, P-1$.

Thus, the lifted code can be represented as

$$
\mathcal{C}^L = \left\{ \widetilde{v} : \widetilde{v}(\ell) = (G^L\left(\sigma^{-1}\right) \widetilde{u})(\ell), \ \ell \in \mathbb{N}_0, \ \widetilde{u} \in \left[\left(\mathbb{F}^{kP}\right)^{\mathbb{N}_0} \right]_{FS} \right\},
$$

where $\widetilde{v} = L_P v$ and $\widetilde{u} = L_P u$.

2.3   Distance Properties

In recent years great effort has been dedicated to developing constructions of non-binary convolutional codes having good distance [2,9,14]. However, in contrast to block codes, the theoretical tools for the construction of convolutional codes with good designed distance have not been fully exploited. In fact, most convolutional codes used in practice have been found by systematic computer search and their distance properties must be also computed by full search.

One of our objectives will be the construction of convolutional codes with a large free distance, which is defined as follows.

Definition 2. *The free distance of a convolutional code $\mathcal{C}$ is given by*

$$
d_{free}(\mathcal{C}) = \min \left\{ \sum_{\ell=0}^{\infty} \mathrm{wt}(v(\ell)) : v \in \mathcal{C} \setminus \{0\} \right\},
$$

where wt *denotes the Hamming weight, that is, $\mathrm{wt}(v(\ell))$ corresponds to the number of nonzero components of $v(\ell)$.*

Rosenthal and Smarandache [21] showed that the free distance of a time-invariant (n, k, δ) convolutional code is upper bounded by

$$
d_{free}(\mathcal{C}) \leq (n-k)\left(\left\lfloor \frac{\delta}{k} \right\rfloor + 1 \right) + \delta + 1.
$$

This bound is called the *generalized Singleton bound*. It is well-known [21] that over sufficiently large finite fields, there always exist convolutional codes that achieve this bound for any given set of parameters (n, k, δ).

However, in this paper we will consider instead the Griesmer bound defined in the next theorem. This bound is always less than or equal to the generalized Singleton bound, and can be considerably lower for codes over small fields, whereas it coincides with the Singleton for codes over sufficiently large finite fields.

Theorem 1 ([7,10,19]). *Let (n, k, δ, m) be a 4-tuple of nonnegative integers such that $k < n$, consider $q \in \mathbb{N}$ and $\hat{\mathbb{N}} = \begin{cases} \mathbb{N} & \text{if } km = \delta \\ \mathbb{N}_0 & \text{if } km > \delta \end{cases}$. Let further*

$$GB_q(n, k, \delta, m) = \max \left\{ d' : \sum_{j=0}^{k(m+i)-\delta-1} \left\lceil \frac{d'}{q^j} \right\rceil \leq n(m+i), \ \forall i \in \hat{\mathbb{N}} \right\}.$$

Then, every (n, k, δ, m)-convolutional code $\mathcal{C}$ over the field $\mathbb{F}_q$ is such that $d_{free}(\mathcal{C}) \leq GB_q(n, k, \delta, m)$.

$GB_q(n, k, \delta, m)$ is known as the *Griesmer bound*.

3   State Space Realizations

A state space system

$$\begin{cases} x(\ell + 1) = Ax(\ell) + Bu(\ell) \\ v(\ell) \quad = Cx(\ell) + Du(\ell) \end{cases}, \ l \in \mathbb{N}_0,$$

denoted by (A, B, C, D), where $A \in \mathbb{F}^{\delta \times \delta}, B \in \mathbb{F}^{\delta \times k}, C \in \mathbb{F}^{n \times \delta}$ and $D \in \mathbb{F}^{n \times k}$, is said to be a state space realization of the time-invariant (n, k, δ) convolutional code $\mathcal{C}$ if $\mathcal{C}$ is the set of finite support output sequences v corresponding to finite support input sequences u and zero inicial conditions, i.e., $x(0) = 0$.

Remark 1. This definition implicitly assumes that (A, B, C, D) is a minimal realization of $\mathcal{C}$, i.e., that A has the minimal possible dimension. This implies that A is nilpotent, (A, B) is controllable and (A, C) is observable, i.e., the polynomial matrices $\begin{bmatrix} z^{-1}I - A \mid B \end{bmatrix}$ and $\begin{bmatrix} z^{-1}I - A \\ C \end{bmatrix}$ have, respectively, right and left polynomial inverses (in z^{-1}).

State space realizations for convolutional codes can be obtained as minimal state space realizations of minimal encoders.

The next proposition, adapted from [8, Proposition 2.3], provides a state space realization for a given (not necessarily minimal) encoder.

Proposition 1. *Let $G \in \mathbb{F}^{n \times k}[z]$ be a polynomial matrix with rank k and column degrees $\nu_1, \ldots, \nu_k$. Consider $\bar{\delta} = \sum_{i=1}^{k} \nu_i$. Let G have columns $g_i = \sum_{\ell=0}^{\nu_i} g_{\ell,i} z^\ell$, $i = 1, \ldots, k$ where $g_{\ell,i} \in \mathbb{F}^n$. For $i = 1, \ldots, k$ define the matrices*

$$A_i = \begin{bmatrix} 0 & \cdots & \cdots & 0 \\ 1 & & & \vdots \\ & \ddots & & \vdots \\ & & 1 & 0 \end{bmatrix} \in \mathbb{F}^{\nu_i \times \nu_i}, \; B_i = \begin{bmatrix} 1 \\ 0 \\ \vdots \\ 0 \end{bmatrix} \in \mathbb{F}^{\nu_i}, \; C_i = \begin{bmatrix} g_{1,i} & \cdots & g_{\nu_i,i} \end{bmatrix} \in \mathbb{F}^{n \times \nu_i}.$$

Then a state space realization of G is given by the matrix quadruple $(A, B, C, D) \in \mathbb{F}^{\bar{\delta} \times \bar{\delta}} \times \mathbb{F}^{\bar{\delta} \times k} \times \mathbb{F}^{n \times \bar{\delta}} \times \mathbb{F}^{n \times k}$ where

$$A = \begin{bmatrix} A_1 & & \\ & \ddots & \\ & & A_k \end{bmatrix}, \; B = \begin{bmatrix} B_1 & & \\ & \ddots & \\ & & B_k \end{bmatrix}, \; C = \begin{bmatrix} C_1 & \cdots & C_k \end{bmatrix}, \; D = \begin{bmatrix} g_{0,1} & \cdots & g_{0,k} \end{bmatrix} = G(0).$$

In the case where $\nu_i = 0$ the ith block is missing and in B a zero column occurs.

In this realization (A, B) is controllable, and if G is a minimal encoder, (A, C) is observable.

4   Constructing Periodically Time-Varying Convolutional Codes

In comparison to the literature on time-invariant convolutional codes, there exist few algebraic constructions of time-varying convolutional codes with good properties [11,22]. Here we present a new technique to build time-varying convolutional codes from time-invariant ones. In particular, in this section we focus on constructing 2-periodic codes with optimal free distance. We illustrate our approach by means of an example of a 2-periodic $(3, 2, 2, 1)$ code having larger distance than any $(3, 2, 2, 1)$ time-invariant convolutional code.

We first investigate the problem of finding periodic state space representations of periodic convolutional codes. As shown in Sect. 2.1, using a lifting technique one can transform a time-varying periodic linear system into an equivalent time-invariant one. Following [1], we study the relationship between the periodic state space representations of a given code and the time-invariant state space representations of its lifted version. For the sake of simplicity we assume that the period is $P = 2$. However, whereas in [1] only single-input/single-output systems were considered, here we deal with codes of general rate k/n that are closely related to multi-input/multi-output (MIMO) systems.

Assume that $\Sigma(\cdot) = (A(\cdot), B(\cdot), C(\cdot), D(\cdot))$ is a δ-dimensional state space representation of a code $\mathcal{C}$, as present below:

$$\begin{cases} x(\ell + 1) = A(\ell)x(\ell) + B(\ell)u(\ell) \\ v(\ell) \quad = C(\ell)x(\ell) + D(\ell)u(\ell) \end{cases}, \; l \in \mathbb{N}_0 \tag{2}$$

where $(A(\cdot), B(\cdot), C(\cdot), D(\cdot)) \in \mathbb{F}^{\delta \times \delta} \times \mathbb{F}^{\delta \times k} \times \mathbb{F}^{n \times \delta} \times \mathbb{F}^{n \times k}$ are periodic functions with period 2. Letting

$$w(\ell) = x(2\ell)$$

$$u^L(\ell) = \begin{bmatrix} u(2\ell) \\ u(2\ell + 1) \end{bmatrix}$$

$$v^L(\ell) = \begin{bmatrix} v(2\ell) \\ v(2\ell + 1) \end{bmatrix}$$

we obtain the following time-invariant δ-dimensional state space representation $\Sigma^L = (E, F, H, J)$ for the lifted code $\mathcal{C}^L$:

$$\begin{cases} w(\ell + 1) = Ew(\ell) + Fu^L(\ell) \\ v^L(\ell) = Hw(\ell) + Ju^L(\ell) \end{cases}, \tag{3}$$

with

$$E = A(1)A(0) \qquad F = \begin{bmatrix} A(1)B(0) & B(1) \end{bmatrix}$$

$$H = \begin{bmatrix} C(0) \\ C(1)A(0) \end{bmatrix} \qquad J = \begin{bmatrix} D(0) & 0 \\ C(1)B(0) & D(1) \end{bmatrix}.$$

The representation $\Sigma^L = (E, F, H, J)$ of $\mathcal{C}^L$ is said to be *induced by* the representation $\Sigma(\cdot) = (A(\cdot), B(\cdot), C(\cdot), D(\cdot))$ of $\mathcal{C}$, or equivalently, $\Sigma(\cdot)$ is said to induce Σ^L. Moreover, a time-invariant representation $\Sigma^L = (E, F, H, J)$ of $\mathcal{C}^L$ is called *induced* whenever it is induced by some periodic representation $\Sigma(\cdot)$ of $\mathcal{C}$.

The following proposition is a generalization of [1, Proposition 3.1] (with identical proof) and characterizes induced representations.

Proposition 2. *Let $\mathcal{C}$ be a 2-periodic code and $\mathcal{C}^L$ the lifted code associated to $\mathcal{C}$. Then a δ-dimensional state space representation $\Sigma^L = (E, F, H, J)$ of $\mathcal{C}^L$, with*

$$E \in \mathbb{F}^{\delta \times \delta} \qquad F = \begin{bmatrix} F_1 & F_2 \end{bmatrix} \in \mathbb{F}^{\delta \times 2k}$$

$$H = \begin{bmatrix} H_1 \\ H_2 \end{bmatrix} \in \mathbb{F}^{2n \times \delta} \qquad J = \begin{bmatrix} J_{11} & J_{12} \\ J_{21} & J_{22} \end{bmatrix} \in \mathbb{F}^{2n \times 2k}.$$

is induced if and only if

$$\operatorname{rank} \mathcal{M} = \begin{bmatrix} E & F_1 \\ H_2 & J_{21} \end{bmatrix} \leq \delta.$$

Moreover, in this case, decomposing the matrix $\mathcal{M}$ as

$$\mathcal{M} = \begin{bmatrix} N_1 \\ N_2 \end{bmatrix} \begin{bmatrix} Q_1 & Q_2 \end{bmatrix},$$

the 2-periodic δ-dimensional state space representation of $\mathcal{C}$ that induces Σ^L is $\Sigma(\cdot) = (A(\cdot), B(\cdot), C(\cdot), D(\cdot))$, where

$$A(0) = Q_1 \quad A(1) = N_1 \quad B(0) = Q_2 \quad B(1) = F_2$$
$$C(0) = H_1 \quad C(1) = N_2 \quad D(0) = J_{11} \quad D(1) = J_{22}.$$

Hence, Proposition 2 characterizes the state space realizations of time-invariant convolutional codes from which (2-periodic) time-varying codes can be constructed. In this way one can use the large body of literature and constructions for the time-invariant case in order to build time-varying convolutional codes with good properties. In the next section we illustrate this with an example.

4.1 2-Periodic $(3, 2, 2, 1)$ Convolutional Code with Free Distance 4

It is know from the literature that $(3, 2, 2, 1)$ time-invariant convolutional codes have at most free distance 3, whereas the Griesmer bound for this kind of codes is 4. In this section we construct a 2-periodic $(3, 2, 2, 1)$ convolutional code with free distance 4 based on the construction of a time-invariant code whose state space realization is induced by a 2-periodic realization. This shows that time-varying convolutional codes can attain larger free distance than time-invariant ones.

Example 1. Consider the $(6, 4, 2, 1)$ time-invariant convolutional code, C^L, over $\mathbb{F}_2$ with generator matrix $G = G_1 z + G_0$, where

$$
G_0 = \begin{bmatrix} 1 & 1 & 0 & 0 \\ 0 & 1 & 0 & 0 \\ 1 & 0 & 0 & 0 \\ 1 & 0 & 1 & 0 \\ 1 & 1 & 1 & 1 \\ 0 & 1 & 1 & 0 \end{bmatrix} \quad \text{and} \quad G_1 = \begin{bmatrix} 0 & 0 & 1 & 1 \\ 0 & 0 & 0 & 1 \\ 0 & 0 & 0 & 1 \\ 0 & 0 & 0 & 0 \\ 0 & 0 & 0 & 0 \\ 0 & 0 & 0 & 0 \end{bmatrix} .
$$

As $\begin{bmatrix} G_0 \\ G_1 \end{bmatrix}$ is an encoder of a $(12, 3)$ block code of free distance 4, we conclude that the free distance of C^L is at most 4. Moreover, it can be computed via a program that the free distance of C^L is indeed 4. Since the column degrees of G are $\nu_1 = 0, \nu_2 = 0, \nu_3 = 1, \nu_4 = 1$, by Proposition 1, a state space realization of G is given by $(E, F, H, J) \in \mathbb{F}^{2 \times 2} \times \mathbb{F}^{2 \times 4} \times \mathbb{F}^{6 \times 2} \times \mathbb{F}^{6 \times 4}$ where

$$
E = \begin{bmatrix} 0 & 0 \\ 0 & 0 \end{bmatrix}, \ F = \begin{bmatrix} 0 & 0 & 1 & 0 \\ 0 & 0 & 0 & 1 \end{bmatrix}, \ H = \begin{bmatrix} 1 & 1 \\ 0 & 1 \\ 0 & 1 \\ 0 & 0 \\ 0 & 0 \\ 0 & 0 \end{bmatrix}, \ J = G_0. \tag{4}
$$

Now, the matrix $\mathcal{M}$ defined in Proposition 2 is:

$$
\mathcal{M} = \begin{bmatrix} 0 & 0 & 0 & 0 \\ 0 & 0 & 0 & 0 \\ 0 & 0 & 1 & 0 \\ 0 & 0 & 1 & 1 \\ 0 & 0 & 0 & 1 \end{bmatrix} .
$$

Since rank $\mathcal{M} = 2 \leq \delta$, by this proposition we conclude that realization (4) is induced by a 2-periodic $(3,2,2,1)$ convolutional code, with realization $\Sigma(\cdot) = (A(\cdot), B(\cdot), C(\cdot), D(\cdot))$ where

$$A(0) = \begin{bmatrix} 0 & 0 \\ 0 & 0 \end{bmatrix} \quad A(1) = \begin{bmatrix} 0 & 0 \\ 0 & 0 \end{bmatrix} \quad B(0) = \begin{bmatrix} 1 & 0 \\ 0 & 1 \end{bmatrix} \quad B(1) = \begin{bmatrix} 1 & 0 \\ 0 & 1 \end{bmatrix}$$

$$C(0) = \begin{bmatrix} 1 & 1 \\ 0 & 1 \\ 0 & 1 \end{bmatrix} \quad C(1) = \begin{bmatrix} 1 & 0 \\ 1 & 1 \\ 0 & 1 \end{bmatrix} \quad D(0) = \begin{bmatrix} 1 & 1 \\ 0 & 1 \\ 1 & 0 \end{bmatrix} \quad D(1) = \begin{bmatrix} 1 & 0 \\ 1 & 1 \\ 1 & 0 \end{bmatrix}.$$

This 2-periodic code can also be described as in (1) with $P = 2$, and $G^t(z) = C(t)\left(z^{-1}I - A(t)\right)^{-1} B(t) + D(t)$, for $t = 0, 1$, which yields

$$G^0(z) = \begin{bmatrix} 1 & 1 \\ 0 & 1 \\ 0 & 1 \end{bmatrix} z + \begin{bmatrix} 1 & 1 \\ 0 & 1 \\ 1 & 0 \end{bmatrix} \quad \text{and} \quad G^1(z) = \begin{bmatrix} 1 & 0 \\ 1 & 1 \\ 0 & 1 \end{bmatrix} z + \begin{bmatrix} 1 & 0 \\ 1 & 1 \\ 1 & 0 \end{bmatrix}.$$

This example is equivalent to the one presented by Palazzo in [18].

5   Conclusions

In this paper we have studied the relation between time-invariant and time-varying convolutional codes by means of input-state-output representations. Using a well known lifting technique we have shown how it is possible to transform a given periodically time-varying convolutional code into a time-invariant one. Moreover, we have provided conditions, in terms of input-state-output representations, to transform a time-invariant convolutional code into a time-varying one. Using these ideas, we have illustrated how to construct a 2-periodic $(3,2,2,1)$ convolutional code with optimal free distance from a $(3,2,2,1)$ time-invariant one. This showed that time-varying convolutional codes can attain larger free distance than time-invariant ones. Constructions of periodic convolutional codes of higher periods and with other parameters are currently under investigation.

References

1. Aleixo, J.C., Rocha, P., Willems, J.C.: State space representation of SISO periodic behaviors. In: Proceedings of the 50th IEEE Conference on Decision and Control and European Control Conference (CDC-ECC), Orlando, FL, USA, pp. 1545–1550, 12–15 December 2011
2. Almeida, P., Napp, D., Pinto, R.: A new class of superregular matrices and MDP convolutional codes. Linear Algebra Appl. **439**(7), 2145–2147 (2013)
3. Bocharova, I., Kudryashov, B.: Rational rate punctured convolutional codes for soft-decision Viterbi decoding. IEEE Trans. Inf. Theory **43**(4), 1305–1313 (1997)

4. Climent, J.-J., Herranz, V., Perea, C., Tomás, V.: A systems theory approach to periodically time-varying convolutional codes by means of their invariant equivalent. In: Bras-Amorós, M., Høholdt, T. (eds.) AAECC 2009. LNCS, vol. 5527, pp. 73–82. Springer, Heidelberg (2009). doi:10.1007/978-3-642-02181-7_8

5. Costello, D.: Free distance bounds for convolutional codes. IEEE Trans. Inf. Theory **20**(3), 356–365 (1974)

6. Fekri, F., Sartipi, M., Mersereau, R.M., Schafer, R.W.: Convolutional codes using finite-field wavelets: time-varying codes and more. IEEE Trans. Signal Process. **53**(5), 1881–1896 (2005)

7. Gluesing-Luerssen, H., Schmale, W.: Distance bounds for convolutional codes and some optimal codes, arXiv:math/0305135v1 (2003)

8. Gluesing-Luerssen, H., Schneider, G.: State space realizations and monomial equivalence for convolutional codes. Linear Algebra Appl. **425**, 518–533 (2007)

9. Gluesing-Luerssen, H., Rosenthal, J., Smarandache, R.: Strongly-MDS convolutional codes. IEEE Trans. Inf. Theory **52**(2), 584–598 (2006)

10. Johannesson, R., Zigangirov, K.S.: Fundamentals of Convolutional Coding. IEEE press, New York (1999)

11. Justesen, J.: New convolutional code constructions and a class of asymptotically good time-varying codes. IEEE Trans. Inf. Theory **19**(2), 220–225 (1973)

12. Kuijper, M., Polderman, J.W.: Reed-Solomon list decoding from a system-theoretic perspective. IEEE Trans. Inf. Theory **50**(2), 259–571 (2004)

13. Kuijper, M., Willems, J.C.: A behavioral framework for periodically time-varying systems. In: Proceedings of the 36th IEEE Conference on Decision & Control - CDC 1997, vol. 3, San Diego, California USA, 10–12 December 1997, pp. 2013–2016 (1997)

14. La Guardia, G.: Convolutional codes: techniques of construction. Comput. Appl. Math. **35**(2), 501–517 (2016)

15. Lee, P.J.: There are many good periodically time-varying convolutional codes. IEEE Trans. Inf. Theory **35**(2), 460–463 (1989)

16. Mooser, M.: Some periodic convolutional codes better than any fixed code. IEEE Trans. Inf. Theory **29**(5), 750–751 (1983)

17. Napp, D., Perea, C., Pinto, R.: Input-state-output representations and constructions of finite support 2D convolutional codes. Adv. Math. Commun. **4**(4), 533–545 (2010)

18. Palazzo, R.: A time-varying convolutional encoder better than the best time-invariant encoder. IEEE Trans. Inf. Theory **39**(3), 1109–1110 (1993)

19. Porras, J.M., Curto, J.I.: Classification of convolutional codes. Linear Algebra Appl. **432**(10), 2701–2725 (2010)

20. Rosenthal, J.: Connections between linear systems and convolutional codes. In: Marcus, B., Rosenthal, J. (eds.) Codes, Systems, and Graphical Models, vol. 123, pp. 39–66. Springer, New York (2001). doi:10.1007/978-1-4613-0165-3_2

21. Rosenthal, J., Smarandache, R.: Maximum distance separable convolutional codes. Appl. Algebra Eng. Commun. Comput. **10**, 15–32 (1999)

22. Truhachev, D., Zigangirov, K., Costello, D.: Distance bounds for periodically time-varying and tail-biting LDPC convolutional codes. IEEE Trans. Inf. Theory **56**(9), 4301–4308 (2010)

23. Viterbi, A.J.: Convolutional codes and their performance in communication systems. IEEE Trans. Commun. Technol. **19**(5), 751–772 (1971)

Column Rank Distances of Rank Metric Convolutional Codes

Diego Napp[1], Raquel Pinto[1(✉)], Joachim Rosenthal[2], and Filipa Santana[1]

[1] Department of Mathematics, CIDMA – Center for Research and Development
in Mathematics and Applications, University of Aveiro, Aveiro, Portugal
{diego,raquel,vfssantana}@ua.pt
[2] Department of Mathematics, University of Zurich,
Winterthurstrasse 190, 8057 Zürich, Switzerland
rosenthal@math.uzh.ch

Abstract. In this paper, we deal with the so-called multi-shot network coding, meaning that the network is used several times (shots) to propagate the information. The framework we present is slightly more general than the one which can be found in the literature. We study and introduce the notion of column rank distance of rank metric convolutional codes for any given rate and finite field. Within this new framework we generalize previous results on column distances of Hamming and rank metric convolutional codes [3,8]. This contribution can be considered as a continuation follow-up of the work presented in [10].

1 Introduction

The theory of Random Linear Network Coding has been mainly devoted to non-coherent one-shot network coding, meaning that the random structure of the network is used just once to propagate information. One of the problems in this situation is that in order to increase the error-correcting capabilities of a code, one necessarily needs to increase the field size or the packet size and this might not be optimal or impossible in many applications.

Hence, in these situations one of the solutions proposed is to create dependencies across multiple shots aiming to approach the channel capacity. In fact, it was been recently shown that spreading redundancy among the transmitted codewords (row spaces) at different instances (shots) can improve the error-correction capabilities of the code. These ideas gave rise to the area of *multi-shot network coding*. Although the potential of using multi-shot network coding was already observed in the seminal paper [5], only recently this interesting approach has been investigated [1,7,14,16].

There are basically two ways for constructing multi-shot codes: one using concatenation of codes and other using rank metric convolutional codes. In [14],

D. Napp, R. Pinto, F. Santana—This work was supported by Portuguese funds through the *Center for Research and Development in Mathematics and Applications* (CIDMA), and *The Portuguese Foundation for Science and Technology* (FCT - Fundação para a Ciência e a Tecnologia), within project UID/MAT/04106/2013.

© Springer International Publishing AG 2017
A.I. Barbero et al. (Eds.): ICMCTA 2017, LNCS 10495, pp. 248–256, 2017.
DOI: 10.1007/978-3-319-66278-7_21

a concatenated code was introduced based on a multilevel code construction. In [12], a concatenation scheme was presented using a Hamming metric convolutional code as an outer code and a rank metric code as an inner code. A different type of concatenation was introduced in [7] where the authors use codes that layer both Maximum Sum Rank (MSR) codes and Gabidulin in order to achieve the streaming capacity for the Burst Erasure Channel.

Apart from concatenated codes, another very natural way to spread redundancy across codewords is by means of convolutional codes [2–4,9,13]. Adapting this class of codes to the context of networks brought about the notion of *rank metric convolutional codes* and interestingly there has been little research on these codes, see [1,6–8,16]. The work in [16] was pioneer in this direction by presenting the first class of rank metric convolutional codes together with a decoding algorithm able to deal with errors, erasures and deviations. However, the results were only valid for unit memory convolutional codes and in [1,6–8] (see also the references therein) an interesting and more general class of rank metric convolutional codes was introduced to cope with network streaming applications.

In this paper we continue our work in [10] and propose a framework slightly more general than the existing ones in the literature on rank metric convolutional codes. In the proposed framework, rank metric codes can be defined for all rates and fields. In this setting, an extension of the standard rank metric has been considered to provide the proper measure for the number of rank erasures that a multi-shot network code can tolerate. Here we continue this line of work and investigate the notion of column rank distance of rank metric convolutional codes in this more general setting. We show that the existing results on column distance in both contexts of Hamming [3] and rank metric [8] can be generalized to this more general point of view.

2   Convolutional Codes

Let $\mathbb{F}$ be a finite field and $\mathbb{F}[D]$ be the ring of polynomials with coefficients in $\mathbb{F}$. A *convolutional code* C of rate k/n is an $\mathbb{F}[D]$-submodule of $\mathbb{F}[D]^n$, with rank k. If $G(D) \in \mathbb{F}[D]^{k \times n}$ is a full row rank matrix such that

$$\mathcal{C} = \mathrm{im}_{\mathbb{F}_q[D]} G(D) = \left\{ u(D)G(D) : u(D) \in \mathbb{F}[D]^k \right\},$$

then $G(D)$ is called an *encoder* of C.

Any other encoder $\tilde{G}(D)$ of $\mathcal{C}$ differ from $G(D)$ by left multiplication by a unimodular matrix $U(D) \in \mathbb{F}[D]^{k \times k}$, i.e., $\tilde{G}(D) = U(D)G(D)$. Therefore, if $\mathcal{C}$ admits a left prime convolutional encoder then all its encoders are left prime. Such a code is called observable.

A convolutional code always admits a *minimal* encoder, i.e., in row reduced form[1]. The sum of the row degrees of a minimal encoder attains its minimum

---

[1] A polynomial matrix $G(D) \in \mathbb{F}[D]^{k \times n}$ is in row reduced form if the constant matrix G_{lrc}, called *leading row coefficient matrix*, constituted by the coefficients of the term of degree equal to the row degree, is full row rank.

among all the encoders of $\mathcal{C}$. Such sum is usually denoted by δ and called the *degree* of $\mathcal{C}$. A rate k/n convolutional code $\mathcal{C}$ of degree δ is called an (n, k, δ) convolutional code [9].

The free distance and the column distances of a convolutional code are important measures of the capability of error detection and error correction of the code. The free distance of a convolutional code $\mathcal{C}$ is given by

$$d_{free}(\mathcal{C}) = \min_{v(D) \in \mathcal{C}, v(D) \neq 0} wt\big(v(D)\big),$$

where $wt\big(v(D)\big)$ is the Hamming weight of a polynomial vector

$$v(D) = \sum_{i \in \mathbb{N}_0} v_i D^i \in \mathbb{F}[D]^n,$$

defined as

$$wt\big(v(D)\big) = \sum_{i \in \mathbb{N}_0} wt(v_i),$$

being $wt(v_i)$ the number of the nonzero components of v_i.

Rosenthal and Smarandache [15] showed that the free distance of an (n, k, δ) convolutional code is upper bounded by

$$d_{free}(\mathcal{C}) \leq (n - k)\left(\left\lfloor \frac{\delta}{k} \right\rfloor + 1\right) + \delta + 1.$$

This bound was called the *generalized Singleton bound*. An (n, k, δ) convolutional code whose free distance is equal to the generalized Singleton bound is called *maximum distance separable* (MDS) code [15].

Let us now consider the column distances of an (n, k, δ) convolutional code $\mathcal{C}$. For that we will consider that $\mathcal{C}$ is observable. Observable convolutional codes admit a kernel representation $H(D) \in \mathbb{F}[D]^{(n-k) \times n}$, i.e. such that $\mathcal{C} = \ker H(D)$. Let

$$G(D) = \sum_{j=0}^{\nu} G_j D^j \in \mathbb{F}[D]^{k \times n}, G_i \in \mathbb{F}^{k \times n}, G_\nu \neq 0$$

be an encoder of $\mathcal{C}$ and

$$H(D) = \sum_{j=0}^{\mu} H_j D^j \in \mathbb{F}[D]^{(n-k) \times n}, H_i \in \mathbb{F}^{(n-k) \times n}, H_\mu \neq 0$$

be a parity-check matrix of $\mathcal{C}$. For every $j \in \mathbb{N}_0$, the truncated sliding generator matrices $G_j^c \in \mathbb{F}^{(j+1)k \times (j+1)n}$ and the truncated sliding parity-check matrices $H_j^c \in \mathbb{F}^{(j+1)(n-k) \times (j+1)n}$ are given by

$$G_j^c = \begin{bmatrix} G_0 & G_1 & \cdots & G_j \\ & G_0 & \cdots & G_{j-1} \\ & & \ddots & \vdots \\ & & & G_0 \end{bmatrix}$$

$$H_j^c = \begin{bmatrix} H_0 & & & \\ H_1 & H_0 & & \\ \vdots & \vdots & \ddots & \\ H_j & H_{j-1} & \dots & H_0 \end{bmatrix},$$

respectively, and when $j > \nu$, we let $G_j = 0$ and when $j > \mu, H_j = 0$.

Using the above assumptions the j-th *column distance* of $\mathcal{C}$ is given by

$$
\begin{aligned}
d_j^c &= \min\{wt((v(D))_{|[0,j]}) : v_0 \neq 0\} \\
&= \min\{wt([v_0 v_1 \cdots v_j]) : [v_0 v_1 \cdots v_j] = [u_0 u_1 \cdots u_j] G_j^c, u_i \in \mathbb{F}^k, u_0 \neq 0\} \\
&= \min\{wt(v), v = (v_0, \dots, v_j) \in \mathbb{F}^{(j+1)n}, v(H_j^c)^T = 0, v_0 \neq 0\},
\end{aligned}
$$

where $v(D) = \sum_{i \in \mathbb{N}_0} v_i D^i$ and $(v(D))_{|[0,j]} = \sum_{i=0}^{j} v_i D^i$.

The following results give a bound on the column distances of an (n, k, δ) convolutional code and some properties of these distances.

Proposition 1 ([3, Proposition 2.2]). *Let $\mathcal{C}$ be an (n, k, δ) convolutional code. For every $j \in \mathbb{N}_0$ we have*

$$d_j^c \leq (n - k)(j + 1) + 1.$$

Corollary 2 ([3, Corollary 2.3]). *Let $\mathcal{C}$ be an (n, k, δ) convolutional code. If $d_j^c = (n - k)(j + 1) + 1$ then $d_i^c = (n - k)(i + 1) + 1$, for every $i \leq j$.*

Proposition 3 ([3, Proposition 2.7]). *Let $\mathcal{C}$ be an MDS (n, k, δ) convolutional code with column distances $d_j^c, j \in \mathbb{N}_0$ and free distance d_{free}. Let $M = min\{j \in \mathbb{N}_0, d_j^c = d_{free}\}$. Then,*

$$M \geq \left\lfloor \frac{\delta}{k} \right\rfloor + \left\lceil \frac{\delta}{n - k} \right\rceil.$$

3 Rank Metric Convolutional Codes

In this section we will define rank metric convolutional codes whose codewords are polynomials matrices in $\mathbb{F}[D]^{n \times m}$ and we aim to further explore this more general approach in order to introduce the definition of column rank distances of a rank metric convolutional code and to propose a bound on this important measure.

A *rank metric convolutional code* $\mathcal{C} \subset \mathbb{F}^{n \times m}$ is the image of an homomorphism $\varphi : \mathbb{F}[D]^k \to \mathbb{F}[D]^{n \times m}$. We write $\varphi = \psi \circ \gamma$ as a composition of a monomorphism γ and an isomorphism ψ:

$$
\begin{aligned}
\varphi : \mathbb{F}[D]^k &\xrightarrow{\gamma} \mathbb{F}[D]^{nm} \xrightarrow{\psi} \mathbb{F}[D]^{n \times m} \\
u(D) &\mapsto v(D) = u(D)G(D) \mapsto V(D)
\end{aligned}
\tag{1}
$$

where $G(D) \in \mathbb{F}^{k \times nm}$ is a full row rank polynomial matrix, called *encoder* of $\mathcal{C}$, and let $V(D) = \text{rmat}_{n \times m}(v(D))$, such that $V_{i,j}(D) = v_{mi+j}(D)$, i.e., the rows of $V(D)$ are n consecutive blocks with m elements of $v(D)$.

As for convolutional codes, two encoders of $\mathcal{C}$ differ by left multiplication by a unimodular matrix and therefore $\mathcal{C}$ always admits minimal encoders. The degree δ of a rank metric convolutional code $\mathcal{C}$ is the sum of the row degrees of a minimal encoder of $\mathcal{C}$, i.e. the minimum value of the sum of the row degrees of its encoders. Rank metric convolutional codes with left prime encoders will also be called observable.

A rank metric convolutional code $\mathcal{C}$ of degree δ, defined as in (1), is called an $(n \times m, k, \delta)$-rank metric convolutional code.

When dealing with rank metric codes a different measure of distance must be considered. The *rank weight* of a polynomial matrix $A(D) = \sum_{i \in \mathbb{N}_0} A_i D^i \in \mathbb{F}[D]^{n \times m}$, is given by

$$\text{rwt}(A(D)) = \sum_{i \in \mathbb{N}_0} \text{rank} A_i. \tag{2}$$

If $B(D) = \sum_{i \in \mathbb{N}_0} B_i \in \mathbb{F}[D]^{n \times m}$, we define the *sum rank distance* between $A(D)$ and $B(D)$ as

$$\begin{aligned} d_{\text{SR}}(A(D), B(D)) &= \text{rwt}(A(D) - B(D)) \\ &= \sum_{i \in \mathbb{N}_0} \text{rank}(A_i - B_i). \end{aligned} \tag{3}$$

Lemma 4. *The sum rank distance d_{SR} is a distance in $\mathbb{F}[D]^{n \times m}$.*

Next we will focus on two sum rank distances definitions of a rank metric convolutional code. The sum rank distance defined in [10,11] and the novel notion of column rank distance.

The *sum rank distance* of a rank metric convolutional code $\mathcal{C}$ is defined as

$$\begin{aligned} d_{\text{SR}}(\mathcal{C}) &= \min_{V(D), U(D) \in \mathcal{C}, V(D) \neq U(D)} d_{\text{SR}}(V(D), U(D)) \\ &= \min_{0 \neq V(D) \in \mathcal{C}} \text{rwt}(V(D)). \end{aligned}$$

Next theorem establishes a bound on the sum rank distance of a rank metric convolutional code. Analogously as for the free distance of a convolutional code, this bound is referred as the Singleton bound for rank metric convolutional codes.

Theorem 5 ([10, Theorem 3] [11, Theorem 3]). *Let $\mathcal{C}$ be an $(n \times m, k, \delta)$ rank metric convolutional code. Then the sum rank distance of $\mathcal{C}$ is upper bounded by*

$$d_{SR}(\mathcal{C}) \leq n \left(\left\lfloor \frac{\delta}{k} \right\rfloor + 1 \right) - \left\lceil \frac{k(\lfloor \frac{\delta}{k} \rfloor + 1) - \delta}{m} \right\rceil + 1. \tag{4}$$

An $(n \times m, k, \delta)$ rank metric convolutional code whose sum rank distance attains the Singleton bound is called *Maximum Rank Distance* (MRD).

Let us now restrict to $(n \times m, k, \delta)$ observable codes.

Definition 6. *Let $\mathcal{C}$ be an $(n \times m, k, \delta)$ observable rank metric convolutional code. For $j \in \mathbb{N}_0$ we define the j-th column rank distance of $\mathcal{C}$ as*

$$d_j^{cr} = \min\{\mathrm{rwt}(V(D)_{|[0,j]}) : V(D) \in \mathcal{C} \text{ and } V_0 \neq 0\},$$

where for $V(D) = \sum_{i \in \mathbb{N}_0} V_i D^i$ we define $V(D)_{|[0,j]} = \sum_{i=0}^{j} V_i D^i$.

Theorem 7. *Let $\mathcal{C}$ be an $(n \times m, k, \delta)$ observable rank metric convolutional code. Then the j-th column rank distance of $\mathcal{C}$ is upper bounded by*

$$d_j^{cr} \leq j \left(n - \left\lfloor \frac{k}{m} \right\rfloor \right) + n - \left\lfloor \frac{k-1}{m} \right\rfloor$$

Proof. Let $G(D) = \sum_{i \in \mathbb{N}_0} G_i D^i$ be an encoder of $\mathcal{C}$. Since G_0 is full row rank it admits an invertible $k \times k$ submatrix. We can assume without loss of generality that the $k \times k$ submatrix of G_0 constituted by the first k columns is invertible.

We will prove the theorem by induction on j. For $j = 0$ let $u_0 \in \mathbb{F}^k$ be such that $v_0 = u_0 G_0$ has the first $k-1$ entries equal to zero, i.e., $wt(v_0) \leq nm - k + 1$, and let $V_0 = \mathrm{rmat}_{n \times m}(v_0)$. Then the first $\left\lfloor \frac{k-1}{m} \right\rfloor$ rows of V_0 are equal to zero and therefore $\mathrm{rwt}(V_0) \leq n - \left\lfloor \frac{k-1}{m} \right\rfloor$ and therefore $d_0^{cr} \leq n - \left\lfloor \frac{k-1}{m} \right\rfloor$.

Let us suppose now that $d_j^{cr} \leq j \left(n - \left\lfloor \frac{k}{m} \right\rfloor \right) + n - \left\lfloor \frac{k-1}{m} \right\rfloor$ and let us prove that $d_{j+1}^{cr} \leq (j+1) \left(n - \left\lfloor \frac{k}{m} \right\rfloor \right) + n - \left\lfloor \frac{k-1}{m} \right\rfloor$. Let $u(D) \in \mathbb{F}[D]^k$, $v(D) = u(D)G(D)$ and $V(D) = \mathrm{rmat}_{n \times m}(v(D)) = \sum_{i \in \mathbb{N}_0} V_i D^i \in \mathcal{C}$ be such that $\mathrm{rwt}(V(D)_{|[0,j]}) = d_j^{cr}$. Moreover, since the $k \times k$ submatrix of G_0 constituted by the first k columns is invertible, we can consider u_{j+1} such that $v_{j+1} = u_{j+1}G_0 + u_{j-1}G_1 + \cdots + u_0 G_{j+1}$ has the first k entries equal to zero. Then

$$
\begin{aligned}
d_{j+1}^{cr} &\leq \mathrm{rwt}((V(D))_{|[0,j+1]}) \\
&= d_j^{cr} + \mathrm{rwt}(V_{j+1}) \\
&\leq j \left(n - \left\lfloor \frac{k}{m} \right\rfloor \right) + n - \left\lfloor \frac{k-1}{m} \right\rfloor + n - \left\lfloor \frac{k}{m} \right\rfloor \\
&= (j+1) \left(n - \left\lfloor \frac{k}{m} \right\rfloor \right) + n - \left\lfloor \frac{k-1}{m} \right\rfloor.
\end{aligned}
$$

With a similar reasoning as in the proof of the above theorem we can prove that if the j-th column distance of a rank metric convolutional code achieves the corresponding bound then the same happens for all the i-th column distances for $i < j$.

Theorem 8. *Let $\mathcal{C}$ be an $(n \times m, k, \delta)$ observable rank metric convolutional code. If $d_j^{cr} = j \left(n - \left\lfloor \frac{k}{m} \right\rfloor \right) + n - \left\lfloor \frac{k-1}{m} \right\rfloor$ for some $j \in \mathbb{N}_0$, then $d_i^{cr} = i \left(n - \left\lfloor \frac{k}{m} \right\rfloor \right) + n - \left\lfloor \frac{k-1}{m} \right\rfloor$ for all $i \leq j$.*

Proof. It is enough to prove that $d_j^{cr} = j\left(n - \lfloor\frac{k}{m}\rfloor\right) + n - \lfloor\frac{k-1}{m}\rfloor$ implies that $d_{j-1}^{cr} = (j-1)\left(n - \lfloor\frac{k}{m}\rfloor\right) + n - \lfloor\frac{k-1}{m}\rfloor$. Let us assume that $d_{j-1}^{cr} < (j-1)\left(n - \frac{k}{m}\right) + n - \lfloor\frac{k-1}{m}\rfloor$ and let $u(D) \in \mathbb{F}[D]^k$, $v(D) = u(D)G(D)$ and $V(D) = \text{rmat}_{n\times m}(v(D)) = \sum_{i\in\mathbb{N}_0} V_i D^i \in \mathcal{C}$ be such $\text{rwt}(V(D))_{|[0,j-1]} = d_{j-1}^{cr}$. Let u_j be such that $v_j = u_0 G_j + u_1 G_{j-1} + \cdots + u_{j-1}G_1 + u_j G_0$ has weight $nm-k$. Then $rank(V_j) \le n - \lfloor\frac{k}{m}\rfloor$ and, therefore, $w_{rank}(V(D)_{[0,j]}) < j\left(n - \lfloor\frac{k}{m}\rfloor\right) + n - \lfloor\frac{k-1}{m}\rfloor$. Consequently, $d_j^{cr} < j\left(n - \lfloor\frac{k}{m}\rfloor\right) + n - \lfloor\frac{k-1}{m}\rfloor$

It is obvious that the sequence of column rank distances of the code is non-decreasing. However, there exists an $M \in \mathbb{N}_0$ such that $d_M^{cr} = d_j^{cr}$ for $j > M$ since the column rank distances of a rank convolutional code can not be greater than the sum rank distance of the code. If the code is MRD then M is precisely determined as stated in the next result.

Proposition 9. *Let* $\mathcal{C}$ *be an MRD* $(n \times m, k, \delta)$ *observable rank metric convolutional code with column rank distances* $d_j^{cr}, j \in \mathbb{N}_0$, *and sum rank distance* d_{SR}. *Let* $M = min\{j \in \mathbb{N}_0, d_j^{cr} = d_{SR}\}$. *Then,*

$$M = \left\lceil \frac{n\lfloor\frac{\delta}{k}\rfloor + \left\lfloor\frac{\delta-k\lfloor\frac{\delta}{k}\rfloor}{m}\right\rfloor}{n - \lfloor\frac{k}{m}\rfloor} \right\rceil$$

Proof. Let $\tilde{M} = \frac{n\lfloor\frac{\delta}{k}\rfloor + \left\lfloor\frac{\delta-k\lfloor\frac{\delta}{k}\rfloor}{m}\right\rfloor}{n - \lfloor\frac{k}{m}\rfloor}$. We will consider two cases, when $m \mid k$ and when $m \nmid k$.

Case 1: $m \mid k$. Then

$$\tilde{M}\left(n - \left\lfloor\frac{k}{m}\right\rfloor\right) + n - \left\lfloor\frac{k-1}{m}\right\rfloor = n\left\lfloor\frac{\delta}{k}\right\rfloor + \left\lfloor\frac{\delta - k\lfloor\frac{\delta}{k}\rfloor}{m}\right\rfloor + n - \left\lfloor\frac{k-1}{m}\right\rfloor$$

$$= n\left(\left\lfloor\frac{\delta}{k}\right\rfloor + 1\right) - \frac{k}{m}\left\lfloor\frac{\delta}{k}\right\rfloor + \left\lfloor\frac{\delta}{m}\right\rfloor - \left\lfloor\frac{k-1}{m}\right\rfloor.$$

Then, since $\lfloor\frac{k-1}{m}\rfloor = \frac{k}{m} - 1$, we have that

$$\tilde{M}\left(n - \left\lfloor\frac{k}{m}\right\rfloor\right) + n - \left\lfloor\frac{k-1}{m}\right\rfloor = n\left(\left\lfloor\frac{\delta}{k}\right\rfloor + 1\right) - \frac{k}{m}\left(\left\lfloor\frac{\delta}{k}\right\rfloor + 1\right) + \left\lfloor\frac{\delta}{m}\right\rfloor + 1$$

$$= n\left(\left\lfloor\frac{\delta}{k}\right\rfloor + 1\right) - \left\lceil\frac{k(\lfloor\frac{\delta}{k}\rfloor + 1) - \delta}{m}\right\rceil + 1$$

$$= d_{SR}.$$

Case 2: $m \nmid k$. In this case

$$\tilde{M}\left(n - \frac{k}{m}\right) + n - \left\lfloor \frac{k-1}{m} \right\rfloor = n\left(\left\lfloor \frac{\delta}{k} \right\rfloor + 1\right) + \left\lfloor \frac{\delta - k\left\lfloor \frac{\delta}{k} \right\rfloor}{m} \right\rfloor - \left\lfloor \frac{k-1}{m} \right\rfloor$$

$$= n\left(\left\lfloor \frac{\delta}{k} \right\rfloor + 1\right) - \left\lceil \frac{k\left(\left\lfloor \frac{\delta}{k} \right\rfloor + 1\right) - \delta - k}{m} \right\rceil - \left\lfloor \frac{k-1}{m} \right\rfloor$$

$$= n\left(\left\lfloor \frac{\delta}{k} \right\rfloor + 1\right) - \left(\left\lceil \frac{k\left(\left\lfloor \frac{\delta}{k} \right\rfloor + 1\right) - \delta}{m} \right\rceil - \left\lfloor \frac{k}{m} \right\rfloor\right) - \left\lfloor \frac{k-1}{m} \right\rfloor$$

$$= n\left(\left\lfloor \frac{\delta}{k} \right\rfloor + 1\right) - \left\lceil \frac{k\left(\left\lfloor \frac{\delta}{k} \right\rfloor + 1\right) - \delta}{m} \right\rceil + 1$$

$$= d_{\mathrm{SR}},$$

because $\left\lfloor \frac{k}{m} \right\rfloor - \left\lfloor \frac{k-1}{m} \right\rfloor = 1$.

In both cases $M = \lceil \tilde{M} \rceil$.

References

1. Badr, A., Khisti, A., Tan, W.-T., Apostolopoulos, J.: Layered constructions for low-delay streaming codes. IEEE Trans. Inf. Theor. (2013)
2. Climent, J.J., Napp, D., Perea, C., Pinto, R.: Maximum distance separable 2D convolutional codes. IEEE Trans. Inf. Theor. **62**(2), 669–680 (2016)
3. Gluesing-Luerssen, H., Rosenthal, J., Smarandache, R.: Strongly MDS convolutional codes. IEEE Trans. Inf. Theor. **52**(2), 584–598 (2006)
4. Johannesson, R., Zigangirov, K.S.: Fundamentals of Convolutional Coding. IEEE Press, New York (1999)
5. Kötter, R., Kschischang, F.R.: Coding for errors and erasures in random network coding. IEEE Trans. Inf. Theor. **54**(8), 3579–3591 (2008)
6. Mahmood, R.: Rank metric convolutional codes with applications in network streaming. Master of applied science (2015)
7. Mahmood, R., Badr, A., Khisti, A.: Streaming-codes for multicast over burst erasure channels. IEEE Trans. Inf. Theor. **61**(8), 4181–4208 (2015)
8. Mahmood, R., Badr, A., Khisti, A.: Convolutional codes with maximum column sum rank for network streaming. IEEE Trans. Inf. Theor. **62**(6), 3039–3052 (2016)
9. McEliece, R.J.: The algebraic theory of convolutional codes. In: Pless, V., Huffman, W.C. (eds.) Handbook of Coding Theory, vol. 1, pp. 1065–1138. Elsevier Science Publishers, Amsterdam (1998)
10. Napp, D., Pinto, R., Rosenthal, J., Vettori, P.: Rank metric convolutional codes. In: Proceedings of the 22nd International Symposium on Mathematical Theory of Network and Systems (MTNS), Minnesota (2016)
11. Napp, D., Pinto, R., Rosenthal, J., Vettori, P.: MRD rank metric convolutional codes. In: IEEE International Symposium on Information Theory (ISIT) (2017)
12. Napp, D., Pinto, R., Sidorenko, V.R.: Concatenation of convolutional codes and rank metric codes for multi-shot network coding. Des. Codes Cryptogr
13. Napp, D., Pinto, R., Toste, M.: On MDS convolutional codes over $\mathbb{Z}_{p^r}$. Des. Codes Cryptogr. **83**, 101–114 (2017)
14. Nóbrega, R.W., Uchoa-Filho, B.F.: Multishot codes for network coding using rank-metric codes. In: Wireless Network Coding Conference (WiNC), pp. 1–6. IEEE, June 2010

15. Rosenthal, J., Smarandache, R.: Maximum distance separable convolutional codes. Appl. Algebra Eng. Commun. Comput. **10**(1), 15–32 (1999)
16. Wachter-Zeh, A., Stinner, M., Sidorenko, V.: Convolutional codes in rank metric with application to random network coding. IEEE Trans. Inf. Theor. **61**(6), 3199–3213 (2015)

On Minimality of ISO Representation of Basic 2D Convolutional Codes

Raquel Pinto and Rita Simões[(✉)]

CIDMA – Center for Research and Development in Mathematics and Applications,
Department of Mathematics, University of Aveiro,
Campus Universitário de Santiago, 3810-193 Aveiro, Portugal
{raquel,ritasimoes}@ua.pt

Abstract. In this paper we study the minimality of input-state-output (ISO) representations of basic two-dimensional (2D) convolutional codes. For that we consider the Fornasini-Marchesini ISO representations of such codes. We define the novel property of strongly modally reachable representations and we show that such representations are minimal representations of a basic 2D convolutional code. Moreover, we prove that the dimension of such minimal representations equals the complexity of the code.

1 Introduction

Two-dimensional (2D) convolutional codes are a natural generalization of one-dimensional (1D) convolutional codes. These codes are naturally suitable to deal with data recorded in two dimensions, like pictures, video, data storage, etc. [9,16,19]. 2D/nD convolutional codes were introduced in [7,8,17,18]. In [1,3] the authors introduce the "locally invertible encoders" and the "Two-Dimensional Tail-Biting Convolutional Codes" with the objective of obtaining constructions of 2D convolutional codes with particular decoding properties. Decoding of 2D convolutional codes over the erasure channel was investigated in [4]. In [12] the authors define input-state-output (ISO) representations of 2D convolutional codes, in which the codewords of a code are generated by a 2D linear system.

One of the most important problems studied in the theory of convolutional codes is the minimality of representations of these codes, i.e., the determination of ISO representations with minimal dimension among all ISO representations of the code. Minimality of an ISO representation leads to more efficient practical implementations in terms of the memory space required. This problem is completely solved when we consider 1D convolutional codes [14,15]. However, this seems to be very hard in the 2D case. We address this problem by considering the Fornasini-Marchesini state-space model, originally studied in the theory of 2D

This work was supported by Portuguese funds through the *Center for Research and Development in Mathematics and Applications* (CIDMA), and *The Portuguese Foundation for Science and Technology* (FCT - Fundação para a Ciência e a Tecnologia), within project UID/MAT/04106/2013.

© Springer International Publishing AG 2017
A.I. Barbero et al. (Eds.): ICMCTA 2017, LNCS 10495, pp. 257–271, 2017.
DOI: 10.1007/978-3-319-66278-7_22

linear systems [6]. Unlike the 1D case, it does not exist necessary and sufficient conditions for the minimality of a realization of a 2D polynomial matrix (i.e., a polynomial matrix in two indeterminates), which makes very hard to solve the general problem of minimality of ISO representations of 2D convolutional codes.

In this paper we consider basic 2D convolutional codes, i.e., 2D convolutional codes which are image of a zero left prime 2D polynomial matrix. We introduce the concept of strongly modally 2D linear system and we show that if a 2D basic convolutional code admits a strongly modally reachable ISO representation then this representation is minimal. Moreover, we prove that the dimension of this minimal ISO representation is equal to the complexity of the code.

2   Convolutional Codes

In this section we will introduce 1D and 2D convolutional codes and a representation of these codes by means of a linear system. We start by giving some preliminaries on polynomial matrices in one indeterminate and in two indeterminates that will be important for the definition of these codes.

2.1   Polynomial Matrices

Let $\mathbb{F}$ be a field and let $\overline{\mathbb{F}}$ denote the algebraic closure of $\mathbb{F}$. Denote by $\mathbb{F}[z]$ the ring of polynomials in one indeterminate with coefficients in $\mathbb{F}$, by $\mathbb{F}(z)$ the field of fractions of $\mathbb{F}[z]$ and by $\mathbb{F}[[z]]$ the ring of formal powers series in one indeterminate with coefficients in $\mathbb{F}$.

Definition 1 (Chap. 6, [10]). *A matrix $U(z) \in \mathbb{F}[z]^{n \times k}$, with $n \geq k$ is,*

(a) *unimodular (i.e., it admits a polynomial inverse) if $n = k$ and $\det(U(z)) \in \mathbb{F} \backslash \{0\}$;*
(b) *right prime (rP) if for every factorization*

$$U(z) = \overline{U}(z)T(z),$$

with $\overline{U}(z) \in \mathbb{F}[z]^{n \times k}$ and $T(z) \in \mathbb{F}[z]^{n \times n}$, $T(z)$ is unimodular.

A matrix is *left prime (ℓP)* if its transpose is rP.

The following lemma gives characterizations of right primeness that will be needed later.

Lemma 1 (Chap. 6, [10]). *Let $U(z) \in \mathbb{F}[z]^{n \times k}$, with $n \geq k$. Then the following are equivalent:*

(a) *$U(z)$ is rP;*
(b) *there exists $P(z) \in \mathbb{F}[z]^{n \times (n-k)}$ such that $[U(z)\ P(z)]$ is unimodular;*
(c) *$U(z)$ admits a polynomial left inverse;*
(d) *the $k \times k$ minors of $U(z)$ have no common factor;*
(e) *for all $\hat{v}(z) \in \mathbb{F}(z)^n$, $\hat{v}(z)^T U(z) \in \mathbb{F}[z]^n$ implies that $\hat{v}(z) \in \mathbb{F}[z]^n$;*
(f) *$U(\lambda)$ is full column rank, for all $\lambda \in \overline{\mathbb{F}}$.*

Let us consider now polynomial matrices in two indeterminates. Denote by $\mathbb{F}[z_1, z_2]$ the ring of polynomials in two indeterminates with coefficients in $\mathbb{F}$, by $\mathbb{F}(z_1, z_2)$ the field of fractions of $\mathbb{F}[z_1, z_2]$ and by $\mathbb{F}[[z_1, z_2]]$ the ring of formal powers series in two indeterminates with coefficients in $\mathbb{F}$.

Definition 2 ([7]). *A matrix* $G(z_1, z_2) \in \mathbb{F}[z_1, z_2]^{n \times k}$, *with* $n \geq k$ *is,*

(a) *unimodular (i.e., it admits a polynomial inverse) if* $n = k$ *and* det $(G(z_1, z_2)) \in \mathbb{F} \backslash \{0\}$;

(b) *right factor prime (rFP) if for every factorization*

$$G(z_1, z_2) = \overline{G}(z_1, z_2)T(z_1, z_2),$$

with $\overline{G}(z_1, z_2) \in \mathbb{F}[z_1, z_2]^{n \times k}$ *and* $T(z_1, z_2) \in \mathbb{F}[z_1, z_2]^{k \times k}$, $T(z_1, z_2)$ *is unimodular;*

(c) *right zero prime (rZP) if the ideal generated by the* $k \times k$ *minors of* $G(z_1, z_2)$ *is* $\mathbb{F}[z_1, z_2]$.

A matrix is *left factor prime* (ℓFP)/*left zero prime* (ℓZP) if its transpose is rFP/rZP, respectively. When we consider polynomial matrices in one indeterminate, the notions (b) and (c) of the above definition are equivalent. However this is not the case for polynomial matrices in two indeterminates. In fact, zero primeness implies factor primeness, but the contrary does not happen. The following lemmas give characterizations of right factor primeness and right zero primeness that will be needed later (see [11,13]).

Lemma 2. *Let* $G(z_1, z_2) \in \mathbb{F}[z_1, z_2]^{n \times k}$, *with* $n \geq k$. *Then the following are equivalent:*

(a) $G(z_1, z_2)$ *is right factor prime;*

(b) *for all* $\hat{u}(z_1, z_2) \in \mathbb{F}(z_1, z_2)^k$, $G(z_1, z_2)\hat{u}(z_1, z_2) \in \mathbb{F}[z_1, z_2]^n$ *implies that* $\hat{u}(z_1, z_2) \in \mathbb{F}[z_1, z_2]^k$;

(c) *the* $k \times k$ *minors of* $G(z_1, z_2)$ *have no common factor.*

Lemma 3. *Let* $G(z_1, z_2) \in \mathbb{F}[z_1, z_2]^{n \times k}$, *with* $n \geq k$. *Then the following are equivalent:*

(a) $G(z_1, z_2)$ *is right zero prime;*

(b) $G(z_1, z_2)$ *admits a polynomial left inverse;*

(c) $G(\lambda_1, \lambda_2)$ *is full column rank, for all* $\lambda_1, \lambda_2 \in \overline{\mathbb{F}}$.

It is well known (see [7]) that given a full column rank polynomial matrix $G(z_1, z_2) \in \mathbb{F}[z_1, z_2]^{n \times k}$, there exists a square polynomial matrix $V(z_1, z_2) \in \mathbb{F}[z_1, z_2]^{k \times k}$ and a rFP matrix $\bar{G}(z_1, z_2) \in \mathbb{F}[z_1, z_2]^{n \times k}$ such that

$$G(z_1, z_2) = \bar{G}(z_1, z_2)V(z_1, z_2).$$

The following lemma will be needed in the sequel. Let $H(z_1, z_2) \in \mathbb{F}[z_1, z_2]^{(n-k) \times n}$, $G(z_1, z_2) \in \mathbb{F}[z_1, z_2]^{n \times k}$, $n > k$, c_i the ith column of $H(z_1, z_2)$

and r_j the jth row of $G(z_1, z_2)$. We say that the full size minor of $H(z_1, z_2)$ constituted by the columns $c_{i_1}, \ldots, c_{i_{n-k}}$ and the full size minor of $G(z_1, z_2)$ constituted by the rows $r_{j_1}, \ldots, r_{j_k}$ are corresponding maximal order minors of $H(z_1, z_2)$ and $G(z_1, z_2)$, if

$$\{i_1, \ldots, i_{n-k}\} \cup \{j_1, \ldots, j_k\} = \{1, \ldots, n\}$$

and $\{i_1, \ldots, i_{n-k}\} \cap \{j_1, \ldots, j_k\} = \emptyset$.

Lemma 4 (Proposition A.4., [7]). *Let $H(z_1, z_2) \in \mathbb{F}[z_1, z_2]^{(n-k) \times n}$ and $G(z_1, z_2) \in \mathbb{F}[z_1, z_2]^{n \times k}$ be a ℓFP and a rFP matrices, respectively, such that $H(z_1, z_2)G(z_1, z_2) = 0$. Then the corresponding maximal order minors of $H(z_1, z_2)$ and $G(z_1, z_2)$ are equal, modulo a unit of the ring $\mathbb{F}[z_1, z_2]$.*

2.2   1D Convolutional Codes

A $1D$ *(finite support) convolutional code* $\mathcal{C}$ of rate k/n is a (free) $\mathbb{F}[z]$-submodule of $\mathbb{F}[z]^n$, where k is the rank of $\mathcal{C}$. A full column rank matrix $G(z) \in \mathbb{F}[z]^{n \times k}$ such that

$$\mathcal{C} = \mathrm{Im}_{\mathbb{F}[z]}\, G(z)$$
$$= \left\{ \hat{v}(z) \in \mathbb{F}[z]^n \mid \hat{v}(z) = G(z)\hat{u}(z), \text{ with } \hat{u}(z) \in \mathbb{F}[z]^k \right\},$$

is called an *encoder* of $\mathcal{C}$. The elements of $\mathcal{C}$ are called *codewords*. Two full column rank matrices $G(z), \bar{G}(z) \in \mathbb{F}[z]^{n \times k}$ are equivalent encoders, i.e. they generate the same 1D convolutional code, if and only if $G(z)U(z) = \bar{G}(z)$ for some unimodular matrix $U(z) \in \mathbb{F}[z]^{k \times k}$. We denote the *complexity* (or *degree*) δ of a 1D convolutional code $\mathcal{C}$ as the maximum of the degree of the $k \times k$ minors of any encoder of $\mathcal{C}$ and we say that $\mathcal{C}$ is an (n, k, δ) 1D convolutional code.

Note that the fact that two equivalent encoders differ by unimodular matrices also implies that the primeness properties of the encoders of a code are preserved, i.e., if $\mathcal{C}$ admits a rP encoder then all its encoders are rP. A 1D convolutional code $\mathcal{C}$ that admits a rP encoder is called *basic* (or *noncatastrophic*) [14,15].

Another way of obtaining the codewords of a $1D$ *convolutional code* is by means of a 1D linear system. A 1D linear system, denoted by $\Sigma = (A, B, C, D)$, is given by the updating equations

$$\begin{aligned} x(t+1) &= Ax(t) + Bu(t) \\ y(t) &= Cx(t) + Du(t), \end{aligned} \qquad (1)$$

where $A \in \mathbb{F}^{s \times s}$, $B \in \mathbb{F}^{s \times k}$, $C \in \mathbb{F}^{(n-k) \times s}$, $D \in \mathbb{F}^{(n-k) \times k}$, $s, n, k \in \mathbb{N}$, $n > k$ and with $x(0) = 0$. We say that Σ has dimension s. The vectors $x(t)$, $u(t)$ and $y(t)$ represent the local state, input and output at instant t, respectively.

The input, state and output 1D sequences (trajectories), $\{u(t)\}_{t \in \mathbb{N}}$, $\{x(t)\}_{t \in \mathbb{N}}$, $\{y(t)\}_{t \in \mathbb{N}}$, respectively, can be represented as formal power series:

$$\hat{u}(z) = \sum_{t \in \mathbb{N}} u(t)z^t \in \mathbb{F}[[z]]^k,$$

$$\hat{x}(z) = \sum_{t\in\mathbb{N}} x(t)z^t \in \mathbb{F}[[z]]^\delta,$$

$$\hat{y}(z) = \sum_{t\in\mathbb{N}} y(t)z^t \in \mathbb{F}[[z]]^{n-k}.$$

In the sequel we shall use the sequence and the corresponding series interchangeably.

Since the codewords of a 1D convolutional code have finite support, we will only consider the finite support input-output trajectories $(\hat{u}(z), \hat{y}(z))$ of (1). Moreover, we will restrict to the finite support input-output trajectories with corresponding state trajectory $\hat{x}(z)$ also with finite support, otherwise the system would remain indefinitely excited. The finite support input-output trajectories $(\hat{u}(z), \hat{y}(z))$ with corresponding state $\hat{x}(z)$ also having finite support are called *finite-weight input-output trajectories*. The set of all these trajectories form a 1D convolutional code, as it is stated in the following theorem (see [15]).

Theorem 1. *The set of finite-weight input-output trajectories of (1) is a 1D convolutional code of rate k/n.*

We denote by $\mathcal{C}(A, B, C, D)$ the 1D convolutional code whose codewords are the finite-weight input-output trajectories of the 1D linear system $\Sigma = (A, B, C, D)$. Moreover, Σ is called an *input-state-output (ISO) representation* of $\mathcal{C}(A, B, C, D)$. All the 1D convolutional codes admit (many) ISO representations. Next we will consider some properties of an ISO representation of a 1D convolutional code $\mathcal{C}$ and see how these properties are reflected on $\mathcal{C}$.

Definition 3 (Chap. 6, [10]). *Let $\Sigma = (A, B, C, D)$ be a 1D linear system with dimension s.*

(a) *Σ is reachable if the reachability matrix*

$$\mathcal{R} = \begin{bmatrix} B & AB & A^2B & \cdots & A^{n-1}B \end{bmatrix}$$

is full row rank, or equivalently, if the matrix $\begin{bmatrix} I_s - Az & Bz \end{bmatrix}$ is ℓP.

(b) *Σ is observable if the observability matrix*

$$\mathcal{O} = \begin{bmatrix} C \\ CA \\ \vdots \\ CA^{n-1} \end{bmatrix}$$

is full column rank, or equivalently, if the matrix $\begin{bmatrix} I_s - Az \\ C \end{bmatrix}$ is rP.

Theorem 2 (Lemma 2.1.1., [14]). *Let Σ be a reachable ISO representation of a 1D convolutional code $\mathcal{C}$. Then $\mathcal{C}$ is basic (or noncatastrophic) if and only if Σ is observable.*

An ISO representation of a 1D convolutional code is said to be *minimal* if it has minimal dimension among all the ISO representations of the code. Minimality is an important property in the sense that minimal ISO representations are more efficient because they require less memory space in their implementation. Moreover, such representations have also strong structural properties which can be useful in the construction of good codes or in the implementation of decoding algorithms.

Next theorem gives a characterization of the minimal ISO representations of a 1D convolutional code and shows how these minimal ISO representations are related.

Lemma 5 (Theorem 3.4., [14]). *Let Σ be an ISO representation of an (n, k, δ) 1D convolutional code C. Then Σ is an minimal ISO representation of C if and only if Σ is reachable.*

Moreover, a minimal ISO representation $\Sigma = (A, B, C, D)$ of C has dimension δ and any other minimal ISO representation of C is of the form $\tilde{\Sigma} = (SAS^{-1}, SB, CS^{-1}, D)$, where S is a $\delta \times \delta$ invertible constant matrix.

Note that if Σ is a minimal ISO representation of C then the complexity of C is equal to the dimension of Σ.

We can obtain a minimal ISO representation of an (n, k, δ) 1D convolutional code C from any ISO representation $\Sigma = (A, B, C, D)$ of C, with dimension $s \geq \delta$. For that we consider a $s \times s$ invertible constant matrix S such that

$$SAS^{-1} = \begin{bmatrix} A_{11} & A_{12} \\ 0 & A_{22} \end{bmatrix}, \quad SB = \begin{bmatrix} B_1 \\ 0 \end{bmatrix}, \quad CS^{-1} = \begin{bmatrix} C_1 & C_2 \end{bmatrix}$$

where $A_{11} \in \mathbb{F}^{\delta \times \delta}$, $B_1 \in \mathbb{F}^{\delta \times k}$ and $C_1 \in \mathbb{F}^{(n-k) \times \delta}$ and $\begin{bmatrix} I_\delta - A_{11}z & B_1 \end{bmatrix}$ is ℓP. Such representation is in the Kalman reachability canonical form (see [10]). Then $\Sigma_1 = (A_{11}, B_1, C_1, D)$ is a minimal ISO representation of C (see [15]).

2.3 2D Convolutional Codes

A 2D *(finite support) convolutional code* C of rate k/n is a free $\mathbb{F}[z_1, z_2]$-submodule of $\mathbb{F}[z_1, z_2]^n$, where k is the rank of C. A full column rank matrix $G(z_1, z_2) \in \mathbb{F}[z_1, z_2]^{n \times k}$ whose columns constitute a basis for C, i.e., such that

$$C = \mathrm{Im}_{\mathbb{F}[z_1, z_2]}\, G(z_1, z_2)$$
$$= \left\{ \hat{v}(z_1, z_2) \in \mathbb{F}[z_1, z_2]^n \mid \hat{v}(z_1, z_2) = G(z_1, z_2)\hat{u}(z_1, z_2), \text{ with } \hat{u}(z_1, z_2) \in \mathbb{F}[z_1, z_2]^k \right\},$$

is called an *encoder* of C. The elements of C are called *codewords*. Two full column rank matrices $G(z_1, z_2), \bar{G}(z_1, z_2) \in \mathbb{F}[z_1, z_2]^{n \times k}$ are equivalent encoders if they generate the same 2D convolutional code, i.e., if

$$\mathrm{Im}_{\mathbb{F}[z_1, z_2]}\, G(z_1, z_2) = \mathrm{Im}_{\mathbb{F}[z_1, z_2]}\, \bar{G}(z_1, z_2),$$

which happens if and only if there exists a unimodular matrix $U(z_1, z_2) \in \mathbb{F}[z_1, z_2]^{k \times k}$ such that $G(z_1, z_2)U(z_1, z_2) = \bar{G}(z_1, z_2)$ (see [17]).

Note that the fact that two equivalent encoders differ by unimodular matrices also implies that the primeness properties of the encoders of a code are preserved, i.e., if $\mathcal{C}$ admits a rFP (rZP) encoder then all its encoders are rFP (rZP). A 2D convolutional code $\mathcal{C}$ that admits rFP encoders is called *noncatastrophic* and it is named *basic* if all its encoders are rZP. Finally, we denote the *complexity* δ of a 2D convolutional code $\mathcal{C}$ as the maximum of the degree of the $k \times k$ minors of any encoder of $\mathcal{C}$.

2D convolutional codes can also be represented by a linear system. Unlike the 1D case, there are several state space models of a 2D linear system. In this paper we consider the Fornasini-Marchesini state-space models (see [6]). In this model a *first quarter plane 2D linear system*, denoted by $\Sigma = (A_1, A_2, B_1, B_2, C, D)$, is given by the updating equations

$$x(i+1, j+1) = A_1 x(i, j+1) + A_2 x(i+1, j) + B_1 u(i, j+1) + B_2 u(i+1, j)$$
$$y(i, j) = Cx(i, j) + Du(i, j), \tag{2}$$

where $A_1, A_2 \in \mathbb{F}^{s \times s}$, $B_1, B_2 \in \mathbb{F}^{s \times k}$, $C \in \mathbb{F}^{(n-k) \times s}$, $D \in \mathbb{F}^{(n-k) \times k}$, $s, n, k \in \mathbb{N}$, $n > k$ and with past finite support of the input and of the state (i.e., $u(i, j) = 0$ and $x(i, j) = 0$, where 0 denotes the zero vector of appropriate lenght,t for $i < 0$ or $j < 0$) and zero initial conditions (i.e., $x(0, 0) = 0$). We say that Σ has dimension s. The vectors $x(i, j)$, $u(i, j)$ and $y(i, j)$ represent the local state, input and output at (i, j), respectively.

We will also represent the input, state and output 2D trajectories, $\{u(i, j)\}_{(i,j) \in \mathbb{N}^2}$, $\{x(i, j)\}_{(i,j) \in \mathbb{N}^2}$, $\{y(i, j)\}_{(i,j) \in \mathbb{N}^2}$ as formal power series,

$$\hat{u}(z_1, z_2) = \sum_{(i,j) \in \mathbb{N}^2} u(i, j) z_1^i z_2^j \in \mathbb{F}[[z_1, z_2]]^k,$$

$$\hat{x}(z_1, z_2) = \sum_{(i,j) \in \mathbb{N}^2} x(i, j) z_1^i z_2^j \in \mathbb{F}[[z_1, z_2]]^\delta,$$

$$\hat{y}(z_1, z_2) = \sum_{(i,j) \in \mathbb{N}^2} y(i, j) z_1^i z_2^j \in \mathbb{F}[[z_1, z_2]]^{n-k}.$$

For the same reasons stated for 1D convolutional codes we will restrict ourselves to finite support input-output trajectories $(\hat{u}(z_1, z_2), \hat{y}(z_1, z_2))$ with corresponding state $\hat{x}(z_1, z_2)$ also having finite support, i.e., to the finite-weight input-output trajectories. Next theorem states that the set of these trajectories also constitute a 2D convolutional code.

Theorem 3 (Theorem 1, [12]). *The set of finite-weight input-output trajectories of (2) is a 2D convolutional code of rate k/n.*

Proof. Let us denote by S and S_{io} the set of finite-weight trajectories and the set of finite-weight input-output trajectories of (2), respectively. Then

$$S = \ker_{\mathbb{F}[[z_1, z_2]]} X(z_1, z_2) \cap \mathbb{F}[z_1, z_2]^{n+\delta} = \ker_{\mathbb{F}(z_1, z_2)} X(z_1, z_2) \cap \mathbb{F}[z_1, z_2]^{n+\delta},$$

where
$$X(z_1, z_2) = \begin{bmatrix} I_s - A_1 z_1 - A_2 z_2 & -B_1 z_1 - B_2 z_2 & 0 \\ -C & -D & I_{n-k} \end{bmatrix}.$$

Since $\ker_{\mathbb{F}(z_1, z_2)} X(z_1, z_2)$ has dimension k, there exists an rFP matrix such that

$$\ker_{\mathbb{F}(z_1, z_2)} X(z_1, z_2) = \operatorname{Im}_{\mathbb{F}(z_1, z_2)} \tilde{L}(z_1, z_2),$$

and as $\tilde{L}(z_1, z_2)$ is rFP, we use Lemma 2 to conclude that $S = \operatorname{Im}_{\mathbb{F}[z_1, z_2]} \tilde{L}(z_1, z_2)$. Representing

$$\tilde{L}(z_1, z_2) = \begin{bmatrix} \tilde{L}_1(z_1, z_2) \\ \tilde{L}_2(z_1, z_2) \end{bmatrix},$$

with $\tilde{L}_1(z_1, z_2) \in \mathbb{F}[z_1, z_2]^{\delta \times k}$ and $\tilde{L}_2(z_1, z_2) \in \mathbb{F}[z_1, z_2]^{n \times k}$, it follows that $S_{io} = \operatorname{Im}_{\mathbb{F}[z_1, z_2]} \tilde{L}_2(z_1, z_2)$. Let $F(z_1, z_2) \in \mathbb{F}[z_1, z_2]^{(\delta+n-k) \times (\delta+n-k)}$ be a nonsingular square matrix such that

$$X(z_1, z_2) = F(z_1, z_2) \begin{bmatrix} M_1(z_1, z_2) & M_2(z_1, z_2) & M_3(z_1, z_2) \end{bmatrix},$$

where $M_1(z_1, z_2) \in \mathbb{F}[z_1, z_2]^{(\delta+n-k) \times \delta}$, $M_2(z_1, z_2) \in \mathbb{F}[z_1, z_2]^{(\delta+n-k) \times k}$, $M_3(z_1, z_2) \in \mathbb{F}[z_1, z_2]^{(\delta+n-k) \times (n-k)}$ are such that $\begin{bmatrix} M_1(z_1, z_2) & M_2(z_1, z_2) & M_3(z_1, z_2) \end{bmatrix}$ is ℓFP. Then

$$\begin{bmatrix} M_1(z_1, z_2) & M_2(z_1, z_2) & M_3(z_1, z_2) \end{bmatrix} \begin{bmatrix} \tilde{L}_1(z_1, z_2) \\ \tilde{L}_2(z_1, z_2) \end{bmatrix} = 0.$$

Since $\det \begin{bmatrix} I_\delta - A_1 z_1 - A_2 z_2 & 0 \\ -C & I_{n-k} \end{bmatrix}$ is nonzero, it immediately follow that the we have that $\det \begin{bmatrix} M_1(z_1, z_2) & M_3(z_1, z_2) \end{bmatrix} \neq 0$ and, by Lemma 4, the corresponding maximal order minor of $\tilde{L}_2(z_1, z_2)$ is also nonzero, which implies that $\tilde{L}_2(z_1, z_2)$ is full column rank, and therefore S_{io} is a 2D finite support convolutional code with rate k/n.

We denote by $\mathcal{C}(A_1, A_2, B_1, B_2, C, D)$ the 2D convolutional code whose codewords are the finite-weight input-output trajectories of the 2D linear system $\Sigma = (A_1, A_2, B_1, B_2, C, D)$. Moreover, Σ is called an *input-state-output (ISO) representation* of $\mathcal{C}(A_1, A_2, B_1, B_2, C, D)$ (see [12]).

A 2D convolutional code admits many ISO representations and as happens in the 1D case, properties of the ISO representations reflect on the properties of the code. 2D linear systems as in (2) admit two types of reachability and observability notions stated in the following definition (see [6]).

Definition 4. *Let* $\Sigma = (A_1, A_2, B_1, B_2, C, D)$ *be a 2D linear system with dimension* s.

(a) Σ *is locally reachable if the reachability matrix*

$$\mathcal{R} = [R_1 \ R_2 \ R_3 \ \cdots] \text{ is full row rank,}$$

where R_k represents the block matrix including all columns defined by

$$\left(A_1{}^{i-1}\Delta^j\, A_2\right) B_1 + \left(A_1{}^{i}\Delta^{j-1}\, A_2\right) B_2$$

with $i + j = k$, for $i, j \geq 0$ and

$$A_1{}^{r}\Delta^t\, A_2 = 0, \text{ when either } r \text{ or } t \text{ is negative,}$$

$$A_1{}^{r}\Delta^0\, A_2 = A_1^r, \quad A_1{}^{0}\Delta^t\, A_2 = A_2^t, \text{ for } r, t \geq 0,$$

$$A_1{}^{r}\Delta^t\, A_2 = A_1\left(A_1{}^{r-1}\Delta^t\, A_2\right) + A_2\left(A_1{}^{r}\Delta^{t-1}\, A_2\right), \text{ for } r, t \geq 1.$$

(b) Σ *is modally reachable if the matrix*

$$\begin{bmatrix} I_s - A_1 z_1 - A_2 z_2 & B_1 z_1 + B_2 z_2 \end{bmatrix}$$

is ℓFP.

(c) Σ *is modally observable if the matrix*

$$\begin{bmatrix} I_s - A_1 z_1 - A_2 z_2 \\ C \end{bmatrix}$$

is rFP.

We will not consider the notion of local observability in this paper. For 1D linear systems, the notions (a) and (b) (and the corresponding observability notions) presented in the above definitions are equivalent. Such equivalence is stated in the Definition 3 (see [10]). However, this does not happen in the 2D case. There are systems which are locally reachable (observable) but not modally reachable (observable) and vice-versa (see [6]).

Given an input trajectory $\hat{u}(z_1, z_2)$ with corresponding state $\hat{x}(z_1, z_2)$ and output $\hat{y}(z_1, z_2)$ trajectories obtained from (2), the matrix

$$\hat{r}(z_1, z_2) = \begin{bmatrix} \hat{x}(z_1, z_2) \\ \hat{u}(z_1, z_2) \\ \hat{y}(z_1, z_2) \end{bmatrix}$$

is called an input-state-output trajectory of $\Sigma = (A_1, A_2, B_1, B_2, C, D)$. The set of input-state-output trajectories of Σ is given by

$$\ker_{\mathbb{F}[[z_1, z_2]]} X(z_1, z_2) = \left\{ \hat{r}(z_1, z_2) \in \mathbb{F}[[z_1, z_2]]^{s+n} \mid X(z_1, z_2)\hat{r}(z_1, z_2) = 0 \right\} \quad (3)$$

where

$$X(z_1, z_2) = \begin{bmatrix} I_s - A_1 z_1 - A_2 z_2 & -B_1 z_1 - B_2 z_2 & 0 \\ -C & -D & I_{n-k} \end{bmatrix} \in \mathbb{F}^{(s+n-k)\times(s+n)}. \quad (4)$$

Moreover, there exist polynomial matrices $L(z_1, z_2) \in \mathbb{F}[z_1, z_2]^{s\times k}$ and $G(z_1, z_2) \in \mathbb{F}[z_1, z_2]^{n\times k}$ such that

$$X(z_1, z_2) \begin{bmatrix} L(z_1, z_2) \\ G(z_1, z_2) \end{bmatrix} = 0$$

where $\begin{bmatrix} L(z_1, z_2) \\ G(z_1, z_2) \end{bmatrix}$ is rFP and $G(z_1, z_2)$ is an encoder of $\mathcal{C}(A_1, A_2, B_1, B_2, C, D)$ (see [12]).

The next result gives us a necessary and sufficient condition for modal reachability in terms of the matrix $X(z_1, z_2)$.

Lemma 6 (Lemma III.4, [5]). *Let* $\Sigma = (A_1, A_2, B_1, B_2, C, D)$ *be a 2D linear system and* $X(z_1, z_2)$ *the corresponding matrix defined in (4). Then* Σ *is modally reachable if and only if the matrix* $X(z_1, z_2)$ *is* ℓFP.

If S is an invertible constant matrix, it is said that the 2D linear systems

$$\Sigma = (A_1, A_2, B_1, B_2, C, D)$$

and

$$\tilde{\Sigma} = \left(SA_1S^{-1}, SA_2S^{-1}, SB_1, SB_2, CS^{-1}, D \right)$$

are algebraically equivalent (see [6]). Such systems represent the same code, as stated in the following lemma.

Lemma 7 (Proposition 4, [12]). *Let* $\Sigma = (A_1, A_2, B_1, B_2, C, D)$ *be a 2D linear system with dimension* s *and* S *a* $s \times s$ *invertible constant matrix. Then*

$$\mathcal{C}(A_1, A_2, B_1, B_2, C, D)$$

$$= \mathcal{C} \left(SA_1S^{-1}, SA_2S^{-1}, SB_1, SB_2, CS^{-1}, D \right).$$

An ISO representation of a 2D convolutional code is said to be *minimal* if it has minimal dimension among all the ISO representations of the code. Also in [6], Fornasini and Marchesini generalized the Kalman reachability canonical form for 2D linear systems, considered in the next definition, and showed that every 2D linear system is algebraically equivalent to a system in the Kalman reachability form.

Definition 5 ([6]). *A 2D linear system* $\Sigma = (A_1, A_2, B_1, B_2, C, D)$, *with dimension* s, k *inputs and* $n - k$ *outputs is in the Kalman reachability canonical form if*

$$A_1 = \begin{bmatrix} A_{11}^{(1)} & A_{12}^{(1)} \\ 0 & A_{22}^{(1)} \end{bmatrix}, \quad A_2 = \begin{bmatrix} A_{11}^{(2)} & A_{12}^{(2)} \\ 0 & A_{22}^{(2)} \end{bmatrix}, \quad B_1 = \begin{bmatrix} B_1^{(1)} \\ 0 \end{bmatrix}, \quad B_2 = \begin{bmatrix} B_1^{(2)} \\ 0 \end{bmatrix}, \quad C = [C_1 \ C_2]$$

where $A_{11}^{(1)}, A_{11}^{(2)} \in \mathbb{F}^{\delta \times \delta}$, $B_1^{(1)}, B_1^{(2)} \in \mathbb{F}^{\delta \times k}$, $C_1 \in \mathbb{F}^{(n-k) \times \delta}$, *with* $s \geq \delta$ *and the remaining matrices of suitable dimensions, and* $\Sigma_1 = \left(A_{11}^{(1)}, A_{11}^{(2)}, B_1^{(1)}, B_1^{(2)}, C_1, D \right)$ *is a locally reachable system, which is the largest locally reachable subsystem of* Σ.

Proposition 1 (Proposition 4, [12]). *Let $\Sigma = (A_1, A_2, B_1, B_2, C, D)$ be a ISO representation of a 2D convolutional code $\mathcal{C}$. Let S be an invertible constant matrix such that*

$$\tilde{\Sigma} = (SA_1S^{-1}, SA_2S^{-1}, SB_1, SB_2, CS^{-1}, D)$$

is in the Kalman reachability canonical form and let

$$\tilde{\Sigma}_1 = \left(\tilde{A}_{11}^{(1)}, \tilde{A}_{11}^{(2)}, \tilde{B}_1^{(1)}, \tilde{B}_1^{(2)}, \tilde{C}_1, D\right)$$

be the largest locally reachable subsystem of $\tilde{\Sigma}$. Then $\mathcal{C} = \mathcal{C}\left(\tilde{A}_{11}^{(1)}, \tilde{A}_{11}^{(2)}, \tilde{B}_1^{(1)}, \tilde{B}_1^{(2)}, \tilde{C}_1, D\right)$.

The next result follows immediately.

Corollary 1. *Minimal ISO representations of a 2D convolutional code must be locally reachable.*

However, it does not exist a sufficient condition for minimality of ISO representations of a 2D convolutional code. In fact, minimality of these ISO representations is a hard problem investigated by many authors which has been open for many decades. In the next section we will investigate minimality of ISO representations of basic 2D convolutional codes and we will obtain a sufficient condition for an ISO representation to be minimal.

3   On Minimality of ISO Representations of Basic 2D Convolutional Codes

In this section we only consider basic 2D convolutional codes.

Let $\Sigma = (A_1, A_2, B_1, B_2, C, D)$ be a modally reachable ISO representation of a basic 2D convolutional code $\mathcal{C}$ and let $G(z_1, z_2)$ be an encoder of $\mathcal{C}$. Then

$$X(z_1, z_2)\begin{bmatrix} L(z_1, z_2) \\ G(z_1, z_2) \end{bmatrix} = 0$$

where $X(z_1, z_2)$ is defined in (4) and $L(z_1, z_2)$ is a suitable polynomial matrix. Since $G(z_1, z_2)$ is rZP, so it is $\begin{bmatrix} L(z_1, z_2) \\ G(z_1, z_2) \end{bmatrix}$ and, by Definition 2 and Lemma 4, $X(z_1, z_2)$ must be ℓZP.

Next lemma relates the property of left zero primeness of the matrix $X(z_1, z_2)$ of a 2D linear system Σ with a special type of modal reachability of Σ.

Lemma 8. *Let $A_1, A_2 \in \mathbb{F}^{s \times s}$, $B_1, B_2 \in \mathbb{F}^{s \times k}$, $C \in \mathbb{F}^{(n-k) \times s}$, $D \in \mathbb{F}^{(n-k) \times k}$, $s, n, k \in \mathbb{N}$, $n > k$. Then*

$$\begin{bmatrix} I_s - A_1 z_1 - A_2 z_2 & B_1 z_1 + B_2 z_2 \end{bmatrix}$$

is ℓZP if and only if the corresponding matrix $X(z_1, z_2)$ defined in (4) is ℓZP.

Proof. Suppose that the matrix $\begin{bmatrix} I_s - A_1 z_1 - A_2 z_2 & B_1 z_1 + B_2 z_2 \end{bmatrix}$ is ℓZP; then there exist $U_1(z_1, z_2) \in \mathbb{F}[z_1, z_2]^{s \times s}$ and $U_2(z_1, z_2) \in \mathbb{F}[z_1, z_2]^{k \times s}$ such that

$$\begin{bmatrix} I_s - A_1 z_1 - A_2 z_2 & B_1 z_1 + B_2 z_2 \end{bmatrix} \begin{bmatrix} U_1(z_1, z_2) \\ U_2(z_1, z_2) \end{bmatrix} = I_s$$

So

$$\begin{bmatrix} I_s - A_1 z_1 - A_2 z_2 & -B_1 z_1 - B_2 z_2 & 0 \\ -C & -D & I_{n-k} \end{bmatrix} \begin{bmatrix} U_1(z_1, z_2) & 0 \\ -U_2(z_1, z_2) & 0 \\ CU_1(z_1, z_2) - DU_2(z_1, z_2) & I_{n-k} \end{bmatrix} = I_{s+n-k}$$

and therefore $X(z_1, z_2)$ is ℓZP. The other implication follows trivially.

This means that ISO representations $\Sigma = (A_1, A_2, B_1, B_2, C, D)$ with dimension s of 2D basic convolutional codes which are modally reachable are such that $\begin{bmatrix} I_s - A_1 z_1 - A_2 z_2 & B_1 z_1 + B_2 z_2 \end{bmatrix}$ is ℓZP. This property will very important throughout the paper. So we propose the next definition.

Definition 6. *Let* $\Sigma = (A_1, A_2, B_1, B_2, C, D)$ *be a 2D linear system with dimension* s. Σ *is said to be strongly modally reachable if* $\begin{bmatrix} I_s - A_1 z_1 - A_2 z_2 & B_1 z_1 + B_2 z_2 \end{bmatrix}$ *is* ℓZP.

It is obvious that strongly modally reachable systems are also modally reachable, but the converse is not true.

Next we will consider the projections of a 2D convolutional code $\mathcal{C}$ onto the two semi-axis $\{\ell e_i | \ell \in \mathbb{N}\}$, for $i = 1, 2$, with $e_1 = (1, 0)$ and $e_2 = (0, 1)$, respectively:

$$\mathcal{C}_1 = \text{proj}_{z_1} \mathcal{C} = \{\hat{v}(z_1, 0) : \hat{v}(z_1, z_2) \in \mathcal{C}\}$$

and

$$\mathcal{C}_2 = \text{proj}_{z_2} \mathcal{C} = \{\hat{v}(0, z_2) : \hat{v}(z_1, z_2) \in \mathcal{C}\}.$$

$\mathcal{C}_1$ and $\mathcal{C}_2$ are 1D convolutional codes (see [15]). Moreover, if $G(z_1, z_2) \in \mathbb{F}[z_1, z_2]^{n \times k}$ is an encoder of $\mathcal{C}$ then

$$\mathcal{C}_1 = \text{Im}_{\mathbb{F}[z_1]} G(z_1, 0) \qquad \text{and} \qquad \mathcal{C}_2 = \text{Im}_{\mathbb{F}[z_2]} G(0, z_2)$$

Note that $G(z_1, 0)$ or $G(0, z_2)$ may not have full column rank and therefore they may not be encoders of $\mathcal{C}_1$ and $\mathcal{C}_2$, respectively. Furthermore, the noncatastrophicity of $\mathcal{C}$ does not imply the noncatastrophicity of $\mathcal{C}_1$ and $\mathcal{C}_2$ (see [12]). However, if $\mathcal{C}$ is basic then $\mathcal{C}_1$ and $\mathcal{C}_2$ are basic. In fact, if $G(z_1, z_2)$ is rZP then there exists $Y(z_1, z_2) \in \mathbb{F}[z_1, z_2]^{k \times n}$ such that

$$Y(z_1, z_2)G(z_1, z_2) = I_k \Rightarrow Y(z_1, 0)G(z_1, 0) = I_k$$

i.e., $G(z_1, 0)$ is rP and then $\mathcal{C}_1$ is basic and $G(z_1, 0)$ is its encoder. Analogously, we prove that $\mathcal{C}_2$ is basic and $G(0, z_2)$ is its encoder. Moreover, if $\mathcal{C}$ has rate k/n, $\mathcal{C}_1$ and $\mathcal{C}_2$ also have rate k/n.

Furthermore, let $\Sigma = (A_1, A_2, B_1, B_2, C, D)$ be an ISO representation of a 2D convolutional code C and consider the restriction of a trajectory of $\{x(i,j), u(i,j), y(i,j)\}_{(i,j) \in \mathbb{N}^2}$ of Σ to the semi-axis $\{\ell e_1 | \ell \in \mathbb{N}\}$, i.e., $\{x(i,0), u(i,0), y(i,0)\}_{i \in \mathbb{N}}$. By the zero initial condition and the past finite support property of the input and state, we have that

$$x(i+1, 0) = A_1 x(i, 0) + B_1 u(i, 0)$$
$$y(i, 0) = C x(i, 0) + D u(i, 0)$$

with $x(0,0) = 0$. This means that the 1D linear system $\Sigma_1 = (A_1, B_1, C, D)$ generates the restrictions to the semi-axes $\{\ell e_1 | \ell \in \mathbb{N}\}$, of all trajectories of Σ, i.e., Σ_1 is an ISO representation of C_1, and analogously, $\Sigma_2 = (A_2, B_2, C, D)$ is an ISO representation of C_2.

Theorem 4. *Let* $\Sigma = (A_1, A_2, B_1, B_2, C, D)$ *be a strongly modally reachable ISO representation of a 2D convolutional code with dimension* s. *Then* Σ *is a minimal ISO representation.*

Proof. By the previous lemma, $X(z_1, z_2)$ is ℓZP and therefore $X(z_1, 0)$ and $X(0, z_2)$ are ℓP. In fact, if $X(z_1, z_2)$ is ℓZP then there exists $\widetilde{X}(z_1, z_2) \in \mathbb{F}[z_1, z_2]^{(s+n) \times (s+n-k)}$ such that $X(z_1, z_2)\widetilde{X}(z_1, z_2) = I_{s+n-k}$ which implies that

$$X(z_1, 0)\widetilde{X}(z_1, 0) = I_{s+n-k} \qquad \text{and} \qquad X(0, z_2)\widetilde{X}(0, z_2) = I_{s+n-k}$$

which means that

$$X(z_1, 0) = \begin{bmatrix} I_s - A_1 z_1 & -B_1 z_1 & 0 \\ -C & -D & I_{n-k} \end{bmatrix} \text{ and } X(0, z_2) = \begin{bmatrix} I_s - A_2 z_2 & -B_2 z_2 & 0 \\ -C & -D & I_{n-k} \end{bmatrix}$$

are ℓP.

Then $\begin{bmatrix} I_s - A_1 z & B_1 z \end{bmatrix}$ and $\begin{bmatrix} I_s - A_2 z & B_2 z \end{bmatrix}$ are ℓP and therefore, by Definition 3, $\Sigma_1 = (A_1, B_1, C, D)$ and $\Sigma_2 = (A_2, B_2, C, D)$ are reachable. Thus, by Lemma 5, Σ_i is a minimal ISO representation of $C_i = C(A_i, B_i, C, D)$, for $i = 1, 2$.

Now suppose that Σ is not a minimal ISO representation of C. Then there exists $\widetilde{\Sigma} = \left(\widetilde{A}_1, \widetilde{A}_2, \widetilde{B}_1, \widetilde{B}_2, \widetilde{C}, \widetilde{D}\right)$ a minimal ISO representation of C with dimension $\widetilde{s} < s$. Then, for $i = 1, 2$, $\widetilde{\Sigma}_i = \left(\widetilde{A}_i, \widetilde{B}_i, \widetilde{C}, \widetilde{D}\right)$ is an ISO representation of C_i with smaller dimension than Σ_i, which contradicts the fact that Σ_i is a minimal ISO representation of C_i. Then Σ is a minimal ISO representation of C.

The next results follow immediately. It shows that if C is a basic 2D convolutional code with a strongly modally reachable ISO representation $\Sigma = (A_1, A_2, B_1, B_2, C, D)$, then the complexity of C is equal to the dimension of a minimal ISO representation of C.

Corollary 2. *Let $\mathcal{C}$ be a basic 2D convolutional code of rate k/n with a strongly modally reachable ISO representation Σ of dimension s. Then $\mathcal{C}$ has complexity s. Moreover, the projections of $\mathcal{C}$ onto the semi-axes $\{\ell e_1 | \ell \in \mathbb{N}\}$ and $\{\ell e_2 | \ell \in \mathbb{N}\}$, respectively have rate k/n and complexity s.*

Proof. Let us assume that $\Sigma = (A_1, A_2, B_1, B_2, C, D)$ is a strongly modally reachable ISO representation of $\mathcal{C}$ with dimension s. By the above theorem Σ is a minimal ISO representation of $\mathcal{C}$ and $\Sigma_1 = (A_1, B_1, C, D)$ and $\Sigma_2 = (A_2, B_2, C, D)$ are also minimal ISO representations of $\mathcal{C}_1$ and $\mathcal{C}_2$, respectively, with dimension δ. Then, by Lemma 5, $\mathcal{C}_1$ and $\mathcal{C}_2$ have complexity s.

Let $G(z_1, z_2)$ be an encoder of $\mathcal{C}$ and $L(z_1, z_2)$ a suitable polynomial matrix such that

$$X(z_1, z_2) \begin{bmatrix} L(z_1, z_2) \\ G(z_1, z_2) \end{bmatrix} = 0,$$

where $X(z_1, z_2)$ is defined in (4). Since the full size minors of $X(z_1, z_2)$ have degree smaller or equal than s, it follows from Lemma 4 that also the full size minors of $G(z_1, z_2)$ have degree less or equal than s. On the other hand, as $\mathcal{C}_1$ has complexity s, $G(z_1, 0)$ is an encoder of $\mathcal{C}_1$ that has one full size minor of degree s and therefore $G(z_1, z_2)$ has one full size minor of degree greater or equal than s. Consequently, the greatest degree of the full size minors of $G(z_1, z_2)$ is s and therefore $\mathcal{C}$ has complexity s.

Corollary 3. *Let Σ be a strongly modally reachable 2D linear system. Then Σ is locally reachable.*

Proof. It follows from Corollary 1.

4   Conclusion

In this paper we have investigated the minimality of ISO representations of basic 2D convolutional codes. We have showed that if a basic 2D convolutional code admits a strongly modally reachable ISO representation then this ISO representation is minimal with dimension equal to the complexity of the code. This result is a natural generalization of the characterization of minimal ISO representations of basic (or noncatastrophic) 1D convolutional codes. We believe that all basic 2D convolutional codes admit a strongly modally reachable ISO representation and that, as happens in the 1D case, all minimal ISO representations of a basic 2D convolutional code are algebraically equivalent. We will investigate this problem in the future. For that we will make use of the so-called first order representations of a code.

References

1. Alfandary, L., Raphaeli, D.: Ball codes - Two-dimensional tail-biting convolutional codes. In: Proceedings 2010 IEEE Global Communications Conference (GLOBE-COM 2010), Miami, FL (2010)

2. Benedetto, S., Divsalar, D., Montorsi, G., Pollara, F.: Serial concatenation of inter-leaved codes: performance analysis, design, and iterative decoding. IEEE Trans. Inf. Theory **44**(3), 909–926 (1998)
3. Charoenlarpnopparut, C.: Applications of Grobner bases to the structural description and realization of multidimensional convolutional code. Sci. Asia **35**, 95–105 (2009)
4. Climent, J.-J., Napp, D., Pinto, R., Simões, R.: Decoding of 2D convolutional codes over the erasure channel. Adv. Math. Commun. **10**(1), 179–193 (2016)
5. Climent, J.-J., Napp, D., Pinto, R., Simões, R.: Series concatenation of 2D convolutional codes. In: Proceedings IEEE 9th International Workshop on Multidimensional (nD) Systems (nDS). Vila Real, Portugal (2015)
6. Fornasini, E., Marchesini, G.: Structure and properties of two-dimensional systems. In: Tzafestas, S.G. (ed.) Multidimensional Systems, Techniques and Applications. Electrical and Computer Engineering, vol. 29, pp. 37–88 (1986)
7. Fornasini, E., Valcher, M.E.: Algebraic aspects of two-dimensional convolutional codes. IEEE Trans. Inf. Theory **40**(4), 1068–1082 (1994)
8. Gluesing-Luersen, H., Rosenthal, J., Weiner, P.A.: Duality between mutidimensinal convolutional codes and systems. In: Colonius, F., Helmke, U., Wirth, F., Pratzel-Wolters, D. (eds.) Advances in Mathematical Systems Theory, A Volume in Honor of Diedrich Hinrichsen, pp. 135–150, Birkhauser, Boston (2000)
9. Justesen, J., Forchhammer, S.: Two Dimensional Information Theory and Coding: With Applications to Graphics Data and High-Density Storage Media. Cambridge University Press, Cambridge (2010)
10. Kailath, T.: Linear Systems. Prentice-Hall, Englewood Cliffs (1980)
11. Lévy, B.C.: 2d-polynomial and rational matrices and their applications for the modelling of 2-d dynamical systems. Ph.D. dissertation, Department of Electrical Engineering, Stanford University, Stanford, CA (1981)
12. Napp, D., Perea, C., Pinto, R.: Input-state-output representations and constructions of finite support 2D convolutional codes. Adv. Math. Commun. **4**(4), 533–545 (2010)
13. Rocha, P.: Structure and representation of 2-d systems. Ph.D. dissertation, University of Groningen, Groningen, The Netherlands (1990)
14. Rosenthal, J., Schumacher, J.M., York, E.V.: On behaviors and convolutional codes. IEEE Trans. Inf. Theory **42**(6), 1881–1891 (1996)
15. Rosenthal, J., York, E.V.: BCH convolutional codes. IEEE Trans. Inf. Theory **45**(6), 1833–1844 (1999)
16. Singh, J., Singh, M.L.: A new family of two-dimensional codes for optical CDMA systems. Optik Int. J. Light Electr. Opt. **120**(18), 959–962 (2009)
17. Valcher, M.E., Fornasini, E.: On 2D finite support convolutional codes: an algebraic approach. Multidimension. Syst. Signal Process. **5**, 231–243 (1994)
18. Weiner, P.A.: Multidimensional convolutional codes. Ph.D. dissertation, Department of Mathematics, University of Notre Dame, Indiana, USA (1998)
19. Zhou, X.-L., Hu, Y.: Multilength two-dimensional codes for optical CDMA systems. Optoelectron. Lett. **1**(3), 232–234 (2005)

A New Construction of Minimum Distance Robust Codes

Hila Rabii and Osnat Keren[✉]

Faculty of Engineering, Bar-Ilan University, Ramat Gan, Israel
hila.rabii@live.biu.ac.il, osnat.keren@biu.ac.il

Abstract. Robust codes are codes that can detect any nonzero error with nonzero probability. This property makes them useful in protecting hardware systems from fault injection attacks which cause arbitrary number of bit flips. There are very few high rate robust codes, non of them has minimum distance greater than two. Therefore, robust codes with error correction capability are derived by concatenation of linear codes with high rate robust codes. This paper presents a new construction of non-linear robust codes with error correction capability. The codes are built upon linear codes; however, the redundant symbols that were originally allocated to increase the minimum distance of the code, are modified to provide both correction capability and robustness. Consequently, the codes are more effective and have higher rate than concatenated codes of the same error masking probability.

Keywords: Fault injection attacks · Security oriented codes · Robust · Nonlinear · Error correction

1 Introduction

The need to secure hardware systems and memories against side channel attacks (SCA) has become more apparent as the attacker capabilities have strengthened. One of the most powerful side-channel attacks is the fault injection attack [1] in which an attacker injects a fault into the device to change its behavior and manipulate its functioning. An efficient and effective approach to protect hardware against this kind of attack is use of error detecting codes which can detect errors of arbitrary multiplicity (Hamming weight) [2,3]. These codes are termed 'security-oriented codes'.

Codes from classic coding theory are designed to address the problem of the *reliability* of information transmitted over a noisy channel or stored in storage media. In classic coding theory, the errors are assumed to be random and the probability that the channel will introduce an error is relatively small. Thus, the minimum distance of the code plays a crucial role in the code's effectiveness. A different class of problems addresses the *security* of information transmitted

This research was supported by the ISRAEL SCIENCE FOUNDATION (grant No. 923/16).

Á.I. Barbero et al. (Eds.): ICMCTA 2017, LNCS 10495, pp. 272–282, 2017.
DOI: 10.1007/978-3-319-66278-7_23

over a noisy channel or stored in storage media. In this class of problems it is assumed that errors are injected by an attacker and therefore can be of any multiplicity. The effectiveness of *security oriented codes* is then measured in terms of the ability to detect **arbitrary** errors with nonzero probability. A *robust code* is a security oriented code that can detect **any** nonzero error.

Security oriented codes can have a deterministic encoding or random encoding [4–7]. The effectiveness of codes with random-encoding depends on the entropy of the random portion. In practice, it is difficult and expensive to implement in hardware a true (i.e., maximal entropy) random number generator nor to protect it from fault injection attacks which could neutralized it. This fact makes codes with deterministic encoding an attractive alternative. Indeed, in some cases, when properly designed, codes with deterministic encoding can be more effective than random codes of the same rate [8]. In this paper we consider robust codes with deterministic encoding.

As far as we know, there are two basic high rate binary systematic robust codes, the Quadratic-Sum (QS) code [9] and the Punctured-Cubic (PC) code [10,11]. All other systematic robust codes (e.g., the codes in [12]) employ them as ground codes. The QS code is an optimum robust code for the case where $k = 2sr$ and q is any power of a prime. However, it does not have any correction capabilities since its minimum distance equals one. The PC code is a close to optimum robust code for any $1 < r \leq k$ where q is a power of two. As shown in [11], also this code does not have correction capabilities.

Furthermore, there are some minimum distance partially robust codes, such as the Vasil'ev code, the Phelps code, the One Switching Code and the generalized cubic code [9,13]. These codes have $2 \leq d \leq 5$; however, they are not robust.

To date, the only way to provide reliability and security is by concatenation of codes. Namely, use of a linear code with correction capability and a nonlinear code to obtain a robust concatenated code [14]. Overall, in existing coding schemes $r_{nl} + r_l$ redundancy symbols are required in order to provide both reliability and security.

This paper presents a new construction of q-ary codes, which utilizes the r_l symbols which were originally allocated to provide reliability, to provide robustness (i.e., security). The codes are q-ary codes robust code with $q = 2^m$, m odd, and minimum distance d.

The paper is organized as follows. In Sect. 2, basic definitions and theoretical background are presented. Section 3 introduces the new construction and Sect. 4 concludes the paper.

2   Effectiveness Criterion for Robust Codes

Notations: Double stroke capital letters are used to denote an algebraic structure; e.g., $\mathbb{F}_q$ is a finite field with q elements. Calligraphic capital letters are used to denote codebooks; e.g., $\mathcal{C}$, where $|\mathcal{C}|$ is the number of codewords in $\mathcal{C}$. The operators $\oplus$ and $\ominus$ denote addition and subtraction in a finite field, respectively.

A code $\mathcal{C}$ of size $|\mathcal{C}|$ and length n is a subset of $\mathbb{F}_q^n$. A systematic code is a code of the form $\mathcal{C} = \{c|c = (x, w(x)), x \in \mathbb{F}_q^k, w \in \mathbb{F}_q^r\}$, where x contains k information symbols and w contains r redundant symbols, $n = k+r$. Systematic codes are widely used in hardware systems since they have a simple, low cost hardware implementation, and the information can be processed by the receiving block while its correctness is being checked in parallel.

A fault in a system may manifests itself as an error at the output of the circuit or at the output of a communication channel. It is convenient to model the correct output as a codeword $c \in \mathcal{C}$ and the injected fault as an additive error $e = (e_x, e_w) \in \mathbb{F}_q^n$, where e_x and e_w are the errors in the information and redundancy portions, respectively. The Hamming weight of the error vector is termed error multiplicity. Random errors are characterized by small multiplicity since the probability of a bit flip is smaller than 0.5, whereas injected errors are assumed to have an arbitrary multiplicity. An error vector e is detected by a codeword $c \in \mathcal{C}$ if $c \oplus e \notin \mathcal{C}$. Similarly, an error e is undetected (masked) by a codeword $c \in \mathcal{C}$ if $c \oplus e \in \mathcal{C}$.

Definition 1. *The masking probability of an error e, $Q(e)$, is the probability that an additive error e is not detected by the code, that is,*

$$Q(e) = \sum_{c \in \mathcal{C}} Pr(c)\delta_{\mathcal{C}}(c \oplus e), \tag{1}$$

where $Pr(c)$ is the probability that the codeword c was used and $\delta_{\mathcal{C}}$ is the characteristic function of the code $\mathcal{C}$,

$$\delta_{\mathcal{C}}(c) = \begin{cases} 1 & c \in \mathcal{C} \\ 0 & c \notin \mathcal{C}. \end{cases}$$

The number of codewords that mask an error e is denoted by $R(e)$ and equals

$$R(e) = \sum_{v \in \mathbb{F}_q^n} \delta_{\mathcal{C}}(v)\delta_{\mathcal{C}}(v \oplus e).$$

Hence, if the codewords are uniformly distributed, Eq. 1 becomes

$$Q(e) = \frac{R(e)}{R(0)}. \tag{2}$$

Note that R is called the autocorrelation function of the code.

Robust codes detect all nonzero errors with some nonzero probability. Formally,

Definition 2 (Robust codes). *[15] A code $\mathcal{C}$ is robust if $Q(e) < 1$ for any nonzero error e.*

The detection kernel of a code is denoted by K_d and consists of all the error vectors that are never detected; that is, $K_d = \{e|Q(e) = 1\}$. The detection kernel

K_d is a linear subspace and its dimension is denoted by ω_d. Robust codes have $|K_d| = 1$, that is, K_d contains only the all-zero word. Partially robust codes are codes whose detection kernel is of size $1 < |K_d| < |\mathcal{C}|$. If $\mathcal{C}$ is a linear code $K_d = \mathcal{C}$ and hence linear codes are not robust and cannot be used for security.

For any error vector $e \in GF(2^n)$, one of the following may occur:

1. The error vector e will always be detected. That is, there exists no $c \in \mathcal{C}$ such that $c + e \in \mathcal{C}$, and therefore, $Q(e) = 0$.
2. The error vector e will never be detected; i.e., the error vector will always be masked. That is, $c + e \in \mathcal{C}$, for every $c \in \mathcal{C}$, and therefore, $Q(e) = 1$. This group of errors is called the kernel of the code and is denoted by K_d.
3. The error vector e will be detected with nonzero probability $0 < Q(e) < 1$. That is, there exists at least one $c \in \mathcal{C}$ that satisfies $c + e \in \mathcal{C}$; in addition, there is at least one $\tilde{c} \in \mathcal{C}$ that does not solve it, this $\tilde{c}$ allows to detect the presence of the error.

These three possible scenarios are depicted in Fig. 1.

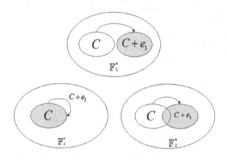

Fig. 1. The error e_1 is always detected, the error e_2 is never detected and the error e_3 is detected with some probability.

Denote by $wt(e)$ the Hamming weight of e, then,

Property 1. [16] A code is of a minimum distance d if and only if $Q(e) = 0$ for all the nonzero error vectors e having $0 < wt(e) < d$.

Definition 3 (Error masking probability of a code). *The error masking probability of the code $\mathcal{C}$ is defined as*

$$\overline{Q} = \max_{e \neq K_d} Q(e).$$

In the binary case, for given k and r the error masking probability $\overline{Q}$ is lower bounded by $\overline{Q} \geq \max\{2^{-r}, 2^{-k+1}\}$ [16]. A code that satisfies equality in the lower bound on $\overline{Q}$ is called *optimum*.

The following theorem sets a lower bound on the error masking probability of a systematic code of minimum distance d.

Property 2. Consider a systematic code $\mathcal{C}$ with minimum distance d over $\mathbb{F}_q$, $q = p^m$ and p is a prime number. The error masking probability of the code is lower bounded by

$$\overline{Q} \geq \max \left\{ \frac{q^k - q^{w_d}}{q^n - q^r - q^{w_d} + d\binom{r}{d}(q-1)^d - 1}, q^{-k} \right\}. \tag{3}$$

Proof. The maximal correlation value is greater than or equal to the average one. Thus, the error masking probability of the code is lower bounded by

$$\overline{Q} = \frac{\max_{e \notin K_d} R(e)}{R(0)} \geq \frac{R_{avr}}{R(0)}$$

where the average autocorrelation R_{avr} is computed over all the error vectors of interest.

The sum of the autocorrelation values of a code at positions that correspond to error vectors that are not in the detection kernel of the code equals

$$\sum_{e \notin K_d} R(e) = \sum_{e \notin K_d} \sum_{v \in \mathbb{F}_q^n} \delta_\mathcal{C}(v)\delta_\mathcal{C}(v \oplus e) =$$

$$\sum_{v \in \mathbb{F}_q^n} \delta_\mathcal{C}(v) \sum_{e \notin K_d} \delta_\mathcal{C}(v \oplus e) = q^k(q^k - q^{w_d}). \tag{4}$$

Thus, for a general code we have

$$\max_{e \notin K_d}(R(e)) \geq \frac{q^k(q^k - q^{w_d})}{q^n - q^{w_d}}.$$

However, in a systematic code, all the $q^r - 1$ nonzero errors of the form $e_x = 0$ and $e_w \neq 0$ are always detected. Moreover, the code has a minimum distance d and hence nonzero errors with Hamming weight smaller than d are always detected; There are $\sum_{j=1}^{d-1} \binom{n}{j}(q-1)^j$ errors of this type. Therefore, at most $q^n - (q^{w_d} + q^r + \sum_{j=0}^{d-1}(\binom{n}{j} - \binom{r}{j})(q-1)^j)$ error vectors that are neither always detected nor always masked contribute to the sum in Eq. 4. Consequently,

$$\max_{e \notin K_d}(R(e)) \geq \frac{q^k(q^k - q^{w_d})}{q^n - (q^{w_d} + q^r + \sum_{j=0}^{d-1}(\binom{n}{j} - \binom{r}{j})(q-1)^j)},$$

and hence,

$$\overline{Q} \geq \frac{(q^k - q^{w_d})}{q^n - q^{w_d} - q^r + d\binom{r}{d}(q-1)^d - 1}.$$

In Addition, since for there exist an error, say $e = c_1 \ominus c_2$, $c_1, c_2 \in \mathcal{C}$, that is masked by at least one codeword (c_2) then $\overline{Q} \geq 1/q^k$. Note that if $q = 2^m$ then e is masked by at least two codewords c_1 and c_2, hence, $\overline{Q} \geq 2/q^k$.

In robust codes, $w_d = 0$, hence asymptotically, $\overline{Q} \geq q^{-r}\frac{1}{1-\epsilon}$ where $\epsilon \to 0$ for $k/n \to 1$.

3   Construction of Robust Codes with Distance

In this section we present a construction of a robust code with distance. In order to simplify the description and the analysis, the cubic function $(f(x) = x^3)$ is employed. Nevertheless, any perfect nonlinear function can be used as well.

Construction 1. *Let m be an odd integer and $q = 2^m$. Let $G = (I|A)$ be a generator matrix of a systematic linear q-ary code C with minimum distance d_L where $A = \{a_{ij}\}_{i,j=1}^{k,r}, a_{ij} \in \mathbb{F}_{2^m}$. Let $x = (x_1, x_2, \ldots, x_k)$ where $x_i \in \mathbb{F}_{2^m}$ for $1 \le i \le k$. Code $\tilde{C}$ is defined as follows,*

$$\tilde{C} = \{(x, w) : x \in \mathbb{F}_{2^m}^k, w \in \mathbb{F}_{2^m}^r\}$$

where $w = (w_1, w_2, \ldots, w_r)$, $w_j \in \mathbb{F}_{2^m}$ for $1 \le j \le r$, and

$$w_j = \sum_{i=1}^{k} a_{ij} x_i^3$$

Property 3. Let m be an odd integer and $q = 2^m$, then the minimum distance of code $\tilde{C}$ is d_L.

The correctness of this property follows from the fact that C and $\tilde{C}$ are equivalent codes. Namely, since for odd values of m the cubic function is invertible, it is possible to obtain one code from the other by a permutation of the alphabet in each column.

Theorem 1. *Let m be an odd integer and $q = 2^m$. The code $\tilde{C}$ is a robust code with $\overline{Q} \le (\frac{2}{q})$. All the nonzero errors of Hamming weight (wt) less than d_L are always detected.*

Errors $e = (e_x, e_w)$ with $wt(e_x) \ge d_L$ are masked with probability $Q(e) \le (\frac{2}{q})^{d_L-1}$, and errors having $0 < wt(e_x) \le d_L - 1$ are masked with probability $Q(e) \le (\frac{2}{q})^{wt(e_x)}$. There are at most

$$\binom{k}{wt(e_x)} (q-1)^{wt(e_x)} \min\left\{ \left(\frac{q}{2}\right)^{wt(e_x)}, q^r \right\}$$

errors $e = (e_x, e_w)$, $wt(e_x) > 0$, that are being masked with a probability $Q(e) > 0$; the remaining errors of this form are always detected.

Proof. Let $c = (x, w)$ be a codeword and let $e = (e_x, e_w)$ be an error vector, where $e_x = (e_{x_1}, \ldots e_{x_k}) \in \mathbb{F}_{2^m}^k$ and $e_w = (e_{w_1}, \ldots e_{w_r}) \in \mathbb{F}_{2^m}^r$. Any nonzero error of Hamming weight smaller than d_L is detected because of the minimum distance of the code. Additionally, since the code is systematic, an error of the form $e = (0, e_w)$ is always detected.

An error $e = (e_x \neq 0, e_w)$ is masked by a codeword c iff $(x \oplus e_x, w \oplus e_w) \in \mathcal{C}$. The r error masking equations of the code are then

$$\sum_{i=1}^{k} a_{ij}(x_i \oplus e_{x_i})^3 = e_{w_j} \oplus \sum_{i=1}^{k} a_{ij}x_i^3, \quad 1 \leq j \leq r. \tag{5}$$

In other words, the error is masked by the j'th redundancy check, $1 \leq j \leq r$, if

$$\sum_{i=1}^{k} a_{ij}(e_{x_i}x_i^2 \oplus e_{x_i}^2 x_i \oplus e_{x_i}^3) = e_{w_j}. \tag{6}$$

This set of error masking equations can be represented in a matrix form as:

$$A^T z = e_w \tag{7}$$

where

$$z = (z_1, z_2, \ldots, z_k)^T \in \mathbb{F}_{2^m}^k,$$

and

$$z_i = (e_{x_i}x_i^2 \oplus e_{x_i}^2 x_i \oplus e_{x_i}^3) \in \mathbb{F}_{2^m}, \quad 1 \leq i \leq k.$$

This quadratic equation in x_i over $\mathbb{F}_q$ has two distinct solutions if the trace of $y_i = \frac{z_i}{e_{x_i}^3} + 1$ equals zero, that is

$$Tr(y_i) = \sum_{j=0}^{m-1} y_i^{2^j} = 0$$

Otherwise, it has no solutions in $\mathbb{F}_{2^m}$.

Recall that the check matrix of the code is of the form $H = (-A^T | I)$. Since the code has minimum distance d_L, any subset of $d_L - 1$ columns of A^T are linearly independent.

Denote by S the support of e_x, and denote by α the weight of the error in the information part, i.e., $\alpha = wt(e_x) = |S|$. There are two cases:

– CASE I: $0 < \alpha \leq d_L - 1$:
  In this case, also the size of support of z equals α. Any set of $d_L - 1$ columns of A^T, and in particular the ones defined by the support of z, are linearly independent. Hence, depending on the value of e_w, Eq. 7 is either satisfied by a single vector z of the required form, or, it has no such solution. (Although Eq. 7 has q^{k-r} solutions, not all of them have $z_i = 0$ for all $i \notin S$).
  Consequently, if for a given e_x, e_w there exists a z which satisfies Eq. 7, has $z_i = 0$ for $i \notin S$, and has $Tr(y_i) = 0$ for $i \in S$, the error e is masked by codewords in $\mathcal{C}$. Otherwise, it is always detected.
  There are at most $R(e)$ codewords that mask e, where

$$R(e) = \underbrace{1}_{\substack{\text{single solution} \\ \text{to Eq. 7}}} \cdot \underbrace{2^\alpha}_{\substack{x_i, i \in S \\ \text{takes two values}}} \cdot \underbrace{q^{k-\alpha}}_{\substack{x_i, i \notin S \\ \text{takes } q \text{ values}}} \cdot$$

Hence, the error masking probability of e is upper bounded by

$$Q(e) \leq \left(\frac{2}{q}\right)^{\alpha}.$$

The number of error vectors that are masked with this probability is at most

$$\binom{k}{\alpha}(q-1)^{\alpha}\left(\frac{q}{2}\right)^{\alpha}.$$

because e_x can take $\binom{k}{\alpha}(q-1)^{\alpha}$ distinct values, and because that for a given e_x there are only α (out of r) degrees of freedom in choosing the e_w vector. However, if one (or more) z_i for $i \in S$ will not satisfy the trace restriction $(Tr(y_i) = 0)$, the error will be always detected. Therefore, e_w can take $\left(\frac{q}{2}\right)^{\alpha} \cdot 1$ values.

- CASE II: $\alpha \geq d_L$:
Since there are at least $d_L - 1$ elements in the support of e_x, at least $d_L - 1$ equations out of the r equations in z_i are linearly independent. Therefore, there are at most $q^{\alpha - (d_L - 1)}$ solutions $\mathbb{F}_{2^m}^k$ to Eq. 7 for which $z_i = 0$ for all $i \notin S$. Let Z be this set of solutions to Eq. 7. Z is a coset in $\mathbb{F}_{2^m}^k$, hence, for each i, $1 \leq i \leq k$, either $z_i = 0$ in all $z \in Z$, or z_i takes each value in $\mathbb{F}_{2^m}$ exactly $|Z|/q$ times.
Recall that z_i, $i \in S$, defines a quadratic equation in x_i over $\mathbb{F}_q$ and if $Tr(y_i) = 0$ this equation has two solution $x_i \in \mathbb{F}_{2^m}$. Since the trace of half of the elements in $\mathbb{F}_{2^m}$ equals zero, and the trace of the other half equals one. Thus, only half of the quadratic equations associated with z_i, $i \in S$ have solutions. Namely, at most $\left(\frac{q}{2}\right)^{\alpha - (d_L - 1)}$ vectors in $\mathbb{F}_{2^m}^k$ are associated with x's that mask the error e. Consequently,

$$R(e) \leq \left(\frac{q}{2}\right)^{\alpha - (d_L - 1)} \cdot 2^{\alpha} \cdot q^{k-\alpha},$$

and the error masking probability of e is upper bounded by

$$Q(e) \leq \left(\frac{2}{q}\right)^{d_L - 1}.$$

Using the same arguments as before, the number of error vectors that are masked with this probability is at most

$$\binom{k}{\alpha}(q-1)^{\alpha} \min\left\{\left(\frac{q}{2}\right)^{\alpha}, q^r\right\}.$$

In conclusion, the code is robust and its error masking probability is $\overline{Q} \leq \left(\frac{2}{q}\right)$. Note that asymptotically the percentage of error vectors having this masking probability is relatively small.

In the following examples a linear shortened Reed-Solomon code and a robust code derived from it are introduced and the autocorrelation functions of these two codes are calculated.

Example 1. Let $\mathcal{C}$ be a linear shortened Reed-Solomon code $[3,3,5]_q$ where $q = 2^3$ with the following generator matrix:

$$G = \begin{bmatrix} 1 & 0 & 0 & 1 & 1 \\ 0 & 1 & 0 & 4 & 2 \\ 0 & 0 & 1 & 3 & 2 \end{bmatrix}.$$

The autocorrelation function of code $\mathcal{C}$ is shown in Fig. 2. The x-axis is the decimal value of the error vector e when referred to as a number in base two; the y-axis is the autocorrelation value $R(e)$. The number of nonzero error vectors is $q^n - 1 = (2^3)^5 - 1 = 32767$. Out of them, 511 errors are never detected, $R(e) = 512$; all other nonzero errors are always detected, $R(e) = 0$. Note that since the code is a linear code, all the $8^3 - 1$ undetected errors are codewords. Clearly, this code is not a robust code and cannot provide security.

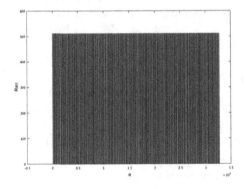

Fig. 2. Autocorrelation values for linear code $\mathcal{C}$. 512 errors have $R(e) = 512$ and all other errors have $R(e) = 0$.

Example 2. Let $\tilde{\mathcal{C}}$ be the code constructed according to Construction 1 with the generator matrix from the previous example. The autocorrelation function of code $\tilde{\mathcal{C}}$ is shown in Fig. 3. Out of 32768 errors, 84 errors have $R(e) = 128$,the all-zero vector has $R(e) = 512$ and all the other errors have smaller autocorrelation values. Note that since $\tilde{\mathcal{C}}$ inherits the minimum distance of the linear code $\mathcal{C}$, all the error vectors of Hamming weight less or equal to three are always detected; i.e., $R(e) = 0$.

The code $\tilde{\mathcal{C}}$ is robust, its error masking probability is $\overline{Q} = \frac{128}{512} = 2^{-2} = 0.25$. As expected, the number of errors with error masking probability of $\frac{2}{q}$ is 84 errors which is smaller than the upper bound $\binom{5}{1}(8-1)^1(8/2)^1 = 140$ errors.

It is interesting to note that these codes are more effective and have higher rate than concatenated codes of the same error masking probability and the same distance; Let $\mathcal{C}_L[n_L, k_L, d_L]$ be a linear q-ary code. Let $\mathcal{C}_{conc}$ be the code constructed from $\mathcal{C}_L$ and a q-ary robust code $\mathcal{C}_{nl}(n_{nl} = n_L + 1, k_{nl} = n_L)$ by

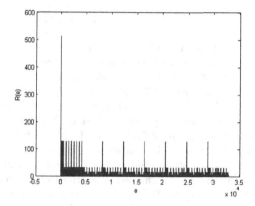

Fig. 3. Autocorrelation values for code $\tilde{\mathcal{C}}$. 10752 errors have $R(e) = 8$, 5376 errors have $R(e) = 16$, 2464 errors have $R(e) = 32$, 84 errors have $R(e) = 128$ and the all-zero error vector has $R(e) = 512$.

concatenation, and let $\tilde{\mathcal{C}}$ be the code constructed according to Construction 1. $\mathcal{C}_{conc}$ is a code of dimension k_L, length $n_L + 1$, distance d_L and $\bar{Q} = 2/q$. Whereas $\tilde{\mathcal{C}}$ is a code with the same dimension, distance and $\bar{Q}$, but of length n_{nl}. Moreover, let $e = (e_x, e_w)$ be an error of Type III, i.e., an error which is detected by some of the codewords and is masked by the others. The concatenated code $\mathcal{C}_{conc}$ masks all the errors of Type III with probability $2/q$, whereas $\tilde{\mathcal{C}}$ masks such an error with probability $(2/q)^{wt(e_x)}$. That is, on average, $\tilde{\mathcal{C}}$ provides a smaller error masking probability than the concatenated code.

4   Summary

Security oriented codes are designed to cope with fault injection attacks. This paper presents a new class of q-ary security oriented codes with minimum distance d for $q = 2^m$, m odd. The codes provide reliability for $d \geq 3$, and provide security as they can detect any arbitrary error with probability greater than $1 - (2/q)$. The proposed construction can be generalized to other q-ary alphabets by use of nonlinear functions other than the cubic function.

References

1. Biham, E., Shamir, A.: Differential fault analysis of secret key cryptosystems. In: Kaliski, B.S. (ed.) CRYPTO 1997. LNCS, vol. 1294, pp. 513–525. Springer, Heidelberg (1997). doi:10.1007/BFb0052259
2. Tomashevich, V., Neumeier, Y., Kumar, R., Keren, O., Polian, I.: Protecting cryptographic hardware against malicious attacks by nonlinear robust codes. In: 2014 IEEE International Symposium on Defect and Fault Tolerance in VLSI and Nanotechnology Systems (DFT), pp. 40–45. IEEE (2014)

3. Verbauwhede, I.M. (ed.): Secure Integrated Circuits and Systems. Springer, New York (2010)
4. Cramer, R., Dodis, Y., Fehr, S., Padró, C., Wichs, D.: Detection of algebraic manipulation with applications to robust secret sharing and fuzzy extractors. In: Smart, N. (ed.) EUROCRYPT 2008. LNCS, vol. 4965, pp. 471–488. Springer, Heidelberg (2008). doi:10.1007/978-3-540-78967-3_27
5. Wang, Z., Karpovsky, M.: Algebraic manipulation detection codes and their applications for design of secure cryptographic devices. In: 2011 IEEE 17th International on On-Line Testing Symposium (IOLTS), pp. 234–239. IEEE (2011)
6. Ngo, X.T., Bhasin, S., Danger, J., Guilley, S., Najm, Z.: Linear complementary dual code improvement to strengthen encoded circuit against hardware Trojan Horses. In: IEEE International Symposium on Hardware Oriented Security and Trust, HOST 2015, Washington, DC, USA, 5–7 May 2015, pp. 82–87 (2015)
7. Dziembowski, S., Pietrzak, K., Wichs, D.: Non-malleable codes. Cryptology ePrint Archive, Report 2009/608 (2009). http://eprint.iacr.org/2009/608
8. Keren, O., Karpovsky, M.: Relations between the entropy of a source and the error masking probability for security-oriented codes. IEEE Trans. Commun. **63**(1), 206–214 (2015)
9. Karpovsky, M.G., Kulikowski, K.J., Wang, Z.: Robust error detection in communication and computational channels. In: International Workshop on Spectral Methods and Multirate Signal Processing, SMMSP 2007, Citeseer (2007)
10. Admaty, N., Litsyn, S., Keren, O.: Puncturing, expurgating and expanding the q-ary BCH based robust codes. In: 2012 IEEE 27th Convention of Electrical Electronics Engineers in Israel (IEEEI), pp. 1–5, November 2012
11. Neumeier, Y., Keren, O.: Robust generalized punctured cubic codes. IEEE Trans. Inf. Theory **60**(5), 2813–2822 (2014)
12. Rabii, H., Neumeier, Y., Keren, O.: Low complexity high rate robust codes derived from the quadratic-sum code. In: 12th International Workshop on Boolean Problems (2016)
13. Engelberg, S., Keren, O.: A comment on the karpovsky-taubin code. IEEE Trans. Inf. Theory **57**(12), 8007–8010 (2011)
14. Neumeier, Y., Keren, O.: A new efficiency criterion for security oriented error correcting codes. In: 2014 19th IEEE European Test Symposium (ETS), pp. 1–6. IEEE (2014)
15. Wang, Z., Karpovsky, M.G., Kulikowski, K.J.: Replacing linear hamming codes by robust nonlinear codes results in a reliability improvement of memories. In: IEEE/IFIP International Conference on Dependable Systems & Networks, DSN 2009, pp. 514–523. IEEE (2009)
16. Wang, Z., Karpovsky, M., Kulikowski, K.J.: Design of memories with concurrent error detection and correction by nonlinear SEC-DED codes. J. Electron. Test. **26**(5), 559–580 (2010)

Constructions and Bounds for Batch Codes with Small Parameters

Eldho K. Thomas$^{(\boxtimes)}$ and Vitaly Skachek$^{(\boxtimes)}$

Institute of Computer Science, University of Tartu, Tartu, Estonia
{eldho.thomas,vitaly.skachek}@ut.ee

Abstract. Linear batch codes and codes for private information retrieval (PIR) with a query size t and a restricted size r of the reconstruction sets are studied. New bounds on the parameters of such codes are derived for small values of t or r by providing corresponding constructions. By building on the ideas of Cadambe and Mazumdar, a new bound in a recursive form is derived for batch codes and PIR codes.

Keywords: PIR codes · Batch codes · Private information retrieval · Locally repairable codes · Distributed data storage

1 Introduction

Batch codes are proposed in [10] for load balancing in the distributed server systems. They can be broadly classified as linear batch codes and combinatorial batch codes. A particular version of the former is known as switch codes and were mainly studied in [5,18,19] in the context of network switches. Some works on linear batch codes and on combinatorial batch codes can be found in [6,11,21] and in [1–3], respectively.

Locally repairable codes (LRC codes), or codes with locality, which are deeply studied in [4,8,9,12,13], share lots of similarities with batch codes, and therefore many of the properties of these two code families are expected to be related to each other. In [21], new upper bounds on the parameters of batch codes based on the classical Singleton bound which do not depend on the size of the underlying alphabet are derived. Batch codes turn out to be a special case of private information retrieval (PIR) codes [7]. Indeed, PIR codes support only queries of type $(x_i, x_i, \ldots, x_i), 1 \le i \le k$, whereas batch codes support queries of a more general form $(x_{i_1}, x_{i_2}, \ldots, x_{i_t})$, possibly for different indices $i_1, i_2, \ldots, i_t$. It follows that batch codes can be used as PIR codes. In [17], batch codes with unrestricted size of reconstruction sets are considered and some bounds on the optimal length of batch and PIR codes for a given batch size and dimension are proposed.

In this work, we construct new families of batch codes with restricted size r of reconstruction sets. We also generalize the existing bounds on the dimension of LRC codes proposed in [12] to batch codes using the connections between the two families. This paper is organized as follows. In Sect. 3, we propose an

© Springer International Publishing AG 2017
A.I. Barbero et al. (Eds.): ICMCTA 2017, LNCS 10495, pp. 283–295, 2017.
DOI: 10.1007/978-3-319-66278-7_24

optimal construction of batch codes with $t = 2, r \geq 2$, and a construction with $t \geq 3, r = 2$. In Sect. 4.1, we present constructions of PIR codes with arbitrary t and r for $r|k$. In Sect. 5, we derive a new upper bound on the dimension k of batch codes.

2   Notations and Related Works

We start with introducing some notations. Denote by $\mathbb{N}$ the set of nonnegative integers. For $n \in \mathbb{N}$ we denote $[n] = \{1, 2, \ldots, n\}$. A $k \times k$ identity matrix will be denoted by $\mathbf{I}_k$, all-one column vector by $\mathbf{1}$. We use $\mathbf{0}$ to denote an all-zero column vector and a zero matrix. The right dimensions will be clear from the context. Let $\mathbf{x}$ be a vector of length n indiced by $[n]$. Take $S \subseteq [n]$. Then $\mathbf{x}_S$ stands for a sub-vector of $\mathbf{x}$ indiced by S. If $\mathbf{A}$ is a matrix, then $\mathbf{A}^{[i]}$ denotes the i-th column in $\mathbf{A}$.

Let $\mathcal{Q}$ be a finite alphabet. Consider an information vector $\mathbf{x} = (x_1, x_2, \ldots, x_k) \in \mathcal{Q}^k$. The code is a set of vectors $\{\mathbf{y} = \mathcal{C}(\mathbf{x}) \mid \mathbf{x} \in \mathcal{Q}^k\} \subseteq \mathcal{Q}^n$, where $\mathcal{C} : \mathcal{Q}^k \rightarrow \mathcal{Q}^n$ is a bijective mapping, and $n \in \mathbb{N}$. By slightly abusing the notation, $\mathcal{C}$ will also be used to denote the above code.

In this work, we study (primitive, multiset) batch codes with restricted size of the recovery sets, as they are defined in [21] (see also [17]).

Definition 1. *An (n, k, r, t) batch code $\mathcal{C}$ over a finite alphabet $\mathcal{Q}$ is defined by an encoding mapping $\mathcal{C} : \mathcal{Q}^k \rightarrow \mathcal{Q}^n$, and a decoding mapping $\mathcal{D} : \mathcal{Q}^n \times [k]^t \rightarrow \mathcal{Q}^t$, such that*

1. For any $\mathbf{x} \in \mathcal{Q}^k$ and a multiset $(i_1, i_2, \cdots, i_t) \subseteq [k]^t$,

$$\mathcal{D}(\mathbf{y} = \mathcal{C}(\mathbf{x}), i_1, i_2, \cdots, i_t) = (x_{i_1}, x_{i_2}, \cdots, x_{i_t}) \ .$$

2. The symbols in the query $(x_{i_1}, x_{i_2}, \cdots, x_{i_t})$ can be reconstructed from t respective disjoint recovery sets of symbols of $\mathbf{y}$ of size at most r each (the symbol x_{i_ℓ} is reconstructed from the ℓ-th recovery set for each ℓ, $1 \leq \ell \leq t$).

If the alphabet $\mathcal{Q}$ is a finite field, and the associated encoding mapping $\mathcal{C} : \mathcal{Q}^k \rightarrow \mathcal{Q}^n$ is linear over $\mathcal{Q}$, the corresponding code is termed linear. In that case, for each fixed $(i_1, i_2, \cdots, i_t) \in [k]^t$, the corresponding decoding mapping $\mathcal{D}$ from $\mathcal{Q}^n$ to $\mathcal{Q}^t$ is linear over $\mathcal{Q}$ too. Additionally, if the encoding mapping $\mathcal{C} : \mathbf{x} \mapsto \mathbf{y}$ is such that $\mathbf{x}$ is a sub-vector of $\mathbf{y}$, then the corresponding code is called *systematic*.

This setup has first appeared in [11]. It was shown therein that the minimum distance d of a batch code satisfies $d \geq t$. It is worth mentioning that batch codes are closely related to locally repairable codes, which have been extensively studied in the context of the distributed data storage. The main difference between them is that in batch codes we are interested in the reconstruction of information symbols in $\mathbf{x}$, while in locally repairable codes we are interested in the recovery of coded symbols in $\mathbf{y}$.

The following property of batch codes is stated as Corollary III.2 in [21].

Lemma 1. *Let C be a linear (n,k,r,t) batch code over Q, and $\mathbf{x} \in Q^k$, whose encoding is $\mathbf{y} \in C$. Let $R_1, R_2, \cdots, R_t \subseteq [n]$ be t disjoint recovery sets for the coordinate x_i. Then, there exist indices $a_2 \in R_2$, $a_3 \in R_3$, $\cdots$, $a_t \in R_t$, such that if we fix the values of all coordinates of $\mathbf{y}$ indexed by the sets $R_1, R_2 \backslash \{a_2\}, R_3 \backslash \{a_3\}, \cdots, R_t \backslash \{a_t\}$, then the values of the coordinates of $\mathbf{y}$ indexed by $\{a_2, a_3, \cdots, a_t\}$ are uniquely determined.*

There is a number of bounds on the parameters of batch codes in the literature, but it is often difficult to make a comparison due to slight variations in the models and assumptions made. Thus, it is proven in [21] that for a linear (n,k,r,t) batch code over Q,

$$n \geq k + d + \max_{1 \leq \beta \leq t} \left\{ (\beta - 1) \left(\left\lceil \frac{k}{r\beta - \beta + 1} \right\rceil - 1 \right) \right\} - 1 . \tag{1}$$

In particular, when the code is systematic, the bound can be tighten a bit, as follows:

$$n \geq k + d + \max_{2 \leq \beta \leq t} \left\{ (\beta - 1) \left(\left\lceil \frac{k}{r\beta - \beta - r + 2} \right\rceil - 1 \right) \right\} - 1 . \tag{2}$$

If the queries in Definition 1 are restricted to $i_1 = i_2 = \cdots = i_t$, then the corresponding code is called an (n,k,r,t) code for private information retrieval (PIR) [7], or simply (n,k,r,t) PIR code. In particular, all batch codes are PIR codes with the corresponding parameters. It should be mentioned that the proofs of (1) and (2) in [21] work in analogous way for the PIR codes too, and therefore these two bounds hold for general and systematic (n,k,r,t) PIR codes, respectively.

Systematic linear (n,k,r,t) PIR codes can be viewed as LRC codes with *locality of information symbols* and availability [13]. A number of bounds on the parameters of the latest family were derived in [13], and in subsequent works. Specifically, when re-written for systematic batch code setting, the following bound holds:

$$d + k + \left\lceil \frac{(t-1)(k-1)+1}{(t-1)(r-1)+1} \right\rceil - 2 \leq n. \tag{3}$$

A comparison between (3) and (2) is not always straightforward, in particular due to the minimization term in (2). However, with the aid of a computer, we verified that the bound (3) gives equal or slightly higher values of n compared to (2) for small values of d, k, r and t. Hereafter, we employ the bound (3) in the analysis of the optimal values of n. However, this bound is not always tight, especially for small alphabets.

Binary simplex codes of length $n = 2^m - 1$ are shown to be optimal batch codes with parameters $k = m$, $t = 2^{m-1} - 2$ and $r = 2$ (for any $m \in \mathbb{N}$) [18], yet those codes exist only for very specific parameters.

The LRC codes were extensively studied in the last years. Thus, in [16], a lower bound on the length n is presented for an LRC with locality *of all symbols* and availability. It should be noted that codes with locality (and availability) of all symbols are a special case of codes with locality (and availability) of information symbols, and therefore the bounds derived for the former family are not directly applicable to the latter family.

It is shown in [16] that it is possible to construct a t-fold power of the binary $(r+1, r)$ single parity check code in order to obtain an LRC with availability (termed *direct product code*) for specific parameters. An algebraic construction of binary LRC codes in [20] further improves on the rate of the direct product code. However, in general, the resulting construction is non-systematic, and thus it is not straightforward how to derive an analogous result for batch/PIR codes, see [15, Example 5]. It would be interesting to extend those techniques to batch/PIR codes, but that is left out of the scope of this paper.

In Sect. 4.1, we present constructions of binary PIR codes for arbitrary r and t achieving rate $\frac{r}{r+t-1}$ for $k \geq r^2$ similar to their counterparts in [20]. For $t \in \{2, 3, 4\}$ these codes are batch codes. The achieved rate is close to optimal, especially for small values of t and r.

A special case of (n, k, r, t) batch and PIR codes, where the size of the recovery sets r is not restricted (for example, it can be assumed that $r = n$), is studied in [17]. Let $\mathcal{B}(k, t)$ be the shortest length n of any systematic linear batch code with unrestricted size of the recovery set, and $\mathcal{P}(k, t)$ be the shortest length n of any linear systematic PIR code with unrestricted size of the recovery set. Then the optimal redundancy of batch and PIR codes, respectively, is defined as $\gamma_{\mathcal{B}}(k, t) = \mathcal{B}(k, t) - k$ and $\gamma_{\mathcal{P}}(k, t) = \mathcal{P}(k, t) - k$.

Proposition 1. *[17] It holds $\mathcal{B}(k, t) = \mathcal{P}(k, t)$ for $1 \leq t \leq 4$, and $\gamma_{\mathcal{B}}(k, t) \leq \gamma_{\mathcal{P}}(k, t) + 2\lceil \log(k) \rceil \cdot \gamma_{\mathcal{P}}(k/2, t-2)$ for $5 \leq t \leq 7$.*

Hereafter, we denote the optimal length of a linear *systematic* batch code and PIR code with the size of the reconstruction sets r as $\mathcal{B}(k, r, t)$ and $\mathcal{P}(k, r, t)$, respectively.

3 Batch Codes with $r = 2$ or $t = 2$

3.1 Optimal Batch Codes with $r \geq 2$ and $t = 2$

In this and subsequent sections, we construct (n, k, r, t) batch codes for specific values of t and r. To this end, consider an (n, k, r, t) systematic batch code with $r \geq 2$ and $t = 2$. Then, by using $d \geq t$, from the bound in (3), we have

$$n \geq \left\lceil \frac{k}{r} \right\rceil + k . \tag{4}$$

The construction of codes attaining this bound, for $t = 2$ and $r = 2$, is presented in [21, Example 2]. In the sequel, we generalize that construction to other values of t and r.

First, we show that the bound (4) is optimal for $t = 2$ and any $r \geq 2$. We achieve that by constructing corresponding (n, k, r, t) batch codes.

Take $\mathbf{G}$ to be a $k \times n$ binary systematic generator matrix of a code $\mathcal{C}$ defined as follows:

$$
\mathbf{G} = \begin{pmatrix}
\mathbf{I}_r & \mathbf{0} & \cdots & \mathbf{0} & \mathbf{0} & 1\,0 & \cdots\,0\,0 \\
\mathbf{0} & \mathbf{I}_r & \cdots & \mathbf{0} & \mathbf{0} & 0\,1 & \cdots\,0\,0 \\
\vdots & \vdots & \ddots & \vdots & \vdots & \vdots\ \vdots & \ddots\ \vdots\ \vdots \\
\mathbf{0} & \mathbf{0} & \cdots & \mathbf{I}_r & \mathbf{0} & 0\,0 & \cdots\,1\,0 \\
\mathbf{0} & \mathbf{0} & \cdots & \mathbf{0} & \mathbf{I}_s & 0\,0 & \cdots\,0\,1
\end{pmatrix}, \tag{5}
$$

where $s = k \bmod r$, and recall that $\mathbf{1}$ denotes all-one column vector.

It is easy to see that $\mathcal{C}$ supports any query of size $t = 2$. If $r|k$, then $n = \frac{k}{r}(r+1)$, which satisfies the lower bound (4) with equality. When $r \nmid k$, we have $s = k - \lfloor \frac{k}{r} \rfloor r$, and

$$
n = \left\lfloor \frac{k}{r} \right\rfloor (r+1) + s + 1 = \left\lceil \frac{k}{r} \right\rceil + k ,
$$

which also satisfies (4) with equality.

Since $\mathcal{B}(k,r,t) \geq \mathcal{P}(k,r,t)$, we can summarize the result as in the following proposition.

Proposition 2. *For any k, $t = 2$ and $r \geq 2$,*

$$
\mathcal{B}(k,r,t) = \mathcal{P}(k,r,t) = \left\lceil \frac{k}{r} \right\rceil + k .
$$

Corollary 1. *For any $k \geq 1$, $t = 2$ and $r \geq 2$, the optimal length of a non-systematic batch and PIR code, $\mathcal{B}_n(k,r,t)$ and $\mathcal{P}_n(k,r,t)$, respectively, satisfies*

$$
\left\lceil \frac{k}{r} \right\rceil + k \geq \mathcal{B}_n(k,r,t) \geq \mathcal{P}_n(k,r,t) \geq \left\lceil \frac{k}{2r-1} \right\rceil + k .
$$

The right-most inequality is obtained from the bound (1) by substituting $d \geq t = 2$ and $\beta = 2$.

3.2   Batch Codes and PIR Codes with $t \geq 2$ and $r = 2$

In this section, we propose a construction of (systematic) binary PIR codes such that

$$
2 \leq t \leq \max \left\{ \left\lceil \frac{k}{r} \right\rceil, r \right\} + 2 \tag{6}
$$

and $r = 2$. We achieve this by using a generator matrix with columns of weight at most 2. The constructed codes are batch codes for $t = 2, 3, 4$. In particular, for $t = 2$ and $r = 2$, the construction is identical to the one in the previous section.

Let $\mathbf{G}$ be a binary generator matrix defined as $\mathbf{G} = [\,\mathbf{I}_k \,|\, \mathbf{A}\,]$. In the sequel we describe how to construct the sub-matrix $\mathbf{A}$.

When k is even or $t-1$ is even, the sub-matrix $\mathbf{A}$ has all its columns of weight 2 and all its rows of weight $t-1$, such that there is no 1-square pattern. In other words, for any i_1, i_2, j_1, j_2, $i_1 \neq i_2$, $j_1 \neq j_2$, at least one of the entries A_{i_1,j_1}, A_{i_1,j_2}, A_{i_2,j_1} and A_{i_2,j_2} in $\mathbf{A}$ is zero.

The total number of columns in $\mathbf{G}$ is $n = k + (t-1)\frac{k}{2}$. In particular, for $r = 2$ and $t = 2$, this bound coincides with (4). We remark that the above construction requires some modification when $k = 2$ and 3. This is due to the fact that it is impossible to avoid 1-squares if the number of rows in $\mathbf{G}$ is two or three (for $k = 2$, $t = 3, 4$, the corresponding values of n are 5 and 7, respectively, and for $k = 3$, $t = 3, 4$, the values of n are 6 and 9, respectively). Moreover, the right-hand inequality in (6) is required in order to fit $t-2$ ones per row, while avoiding 1-squares.

When k is odd and $t-1$ is odd, then $\mathbf{A}$ has all (except the last) columns of weight 2, and the last column of weight 1. All its rows are of weight $t-1$, and there is no 1-square pattern in $\mathbf{A}$. Therefore the total number of columns in $\mathbf{G}$ is

$$n = k + \left\lceil (t-1) \cdot \frac{k}{2} \right\rceil.$$

Proposition 3. *The code $\mathcal{C}$ defined by the above generator matrix $\mathbf{G}$ for $t = 3$ supports any query of the form (x_i, x_j, x_ℓ) with recovery sets of size at most 2, $i, j, \ell \in [k]$.*

Proposition 4. *The code $\mathcal{C}$ defined by the above generator matrix $\mathbf{G}$ for $t = 4$ supports any query of the form (x_i, x_j, x_ℓ, x_h) with recovery sets of size at most 2, $i, j, \ell, h \in [k]$.*

Proposition 5. *The code $\mathcal{C}$ defined by the above generator matrix $\mathbf{G}$ for general $t \geq 5$ supports any query of size t of the form $(x_i, x_i, \cdots, x_i)$ with recovery sets of size at most 2.*

Proposition 6. *For $r = 2$ and $3 \leq t \leq \max\{\lceil \frac{k}{r} \rceil, r\} + 2$,*

$$k + t - 2 + \left\lceil \frac{t-1)(k-1) + 1}{t} \right\rceil \leq \mathcal{P}(k, r, t) \leq k + \left\lceil (t-1) \cdot \frac{k}{2} \right\rceil.$$

The left-most inequality in Proposition 6 is obtained from (3) by substituting $r = 2$.

Proposition 7. *For $r = 2$ and $t \in \{3, 4\}$,*

$$k + t - 2 + \left\lceil \frac{t-1)(k-1) + 1}{t} \right\rceil \leq \mathcal{B}(k, r, t) \leq k + \left\lceil (t-1) \cdot \frac{k}{2} \right\rceil.$$

The following examples illustrate the above constructions.

Example 1. Consider a binary (n, k, r, t) batch (PIR) code $\mathcal{C}$ with $k = 5$, $t = 3$ and $r = 2$. From Proposition 7, we have $9 \leq \mathcal{B}(5, 2, 3) \leq 10$. We construct a batch (PIR) code $\mathcal{C}$ of length 10 with the above parameters using the following 5×10 generator matrix:

$$\mathbf{G} = \begin{pmatrix} 1\,0\,0\,0\,0 & 1\,0\,0\,1\,0 \\ 0\,1\,0\,0\,0 & 0\,1\,0\,0\,1 \\ 0\,0\,1\,0\,0 & 0\,0\,1\,1\,0 \\ 0\,0\,0\,1\,0 & 0\,0\,1\,0\,1 \\ 0\,0\,0\,0\,1 & 1\,1\,0\,0\,0 \end{pmatrix}.$$

Example 2. Take a binary (n, k, r, t) batch (PIR) code $\mathcal{C}$ with $k = 5$, $t = 4$ and $r = 2$. From Proposition 7, we have $11 \leq \mathcal{B}(5, 2, 4) \leq 13$. We construct a batch (PIR) code $\mathcal{C}$ of length 13 with the above parameters using the following 5×13 generator matrix:

$$\mathbf{G} = \begin{pmatrix} 1\,0\,0\,0\,0 & 1\,0\,0\,1\,0\,1\,0\,0 \\ 0\,1\,0\,0\,0 & 0\,1\,0\,0\,1\,0\,1\,0 \\ 0\,0\,1\,0\,0 & 0\,0\,1\,1\,0\,0\,1\,0 \\ 0\,0\,0\,1\,0 & 0\,0\,1\,0\,1\,1\,0\,0 \\ 0\,0\,0\,0\,1 & 1\,1\,0\,0\,0\,0\,0\,1 \end{pmatrix}.$$

4   PIR Codes for Arbitrary $t > 2$ and $r > 2$

4.1   Case $r \mid k$

In this section, by using a generator matrix with columns of weight at most r, we generalize the construction in Sect. 3 to systematic PIR codes with arbitrary parameters $2 < t \leq \max\{\frac{k}{r}, r\} + 2$ and $r > 2$, $r \mid k$. The corresponding upper bounds are implied by the construction.

Let $\mathbf{G}$ be a generator matrix of the form $\mathbf{G} = [\mathbf{I}_k \mid \mathbf{A} \mid \mathbf{B}]$, where $\mathbf{I}_k$ is the systematic part. Here, $\mathbf{A}$ is a $k \times \frac{k}{r}$ matrix with the jth column of the form $\mathbf{A}^{[j]} = (a_{1,j}, a_{2,j}, \cdots, a_{k,j})^T$, $1 \leq j \leq \frac{k}{r}$, where $a_{i,j} = 1$ if $(j - 1)r + 1 \leq i \leq jr$, and $a_{i,j} = 0$ otherwise. The matrix $\mathbf{B}$ is defined as follows. Each column has weight $\min\{r, k/r\}$, every row in $\mathbf{B}$ has weight $t - 2$, and there are no 1-squares in $[\mathbf{A} \mid \mathbf{B}]$. That is, $\mathbf{B}$ is constructed in such a way that all rows of $\mathbf{G}$ have weight t, columns have weight at most r, and there are no 1-squares in $[\mathbf{A} \mid \mathbf{B}]$. The absence of 1-squares is instrumental in finding disjoint recovery sets for all information symbols.

- **Case 1:** $\frac{k}{r} < r$. In that case we choose $\mathbf{B}$ to be a $k \times (t - 2)r$ matrix defined as per the rules above. Then,

$$n = k + \frac{k}{r} + (t - 2)r = (r + 1)\frac{k}{r} + (t - 2)r \,,$$

and the code rate is

$$\frac{k}{n} = \frac{k}{(r + 1)\frac{k}{r} + (t - 2)r} < \frac{k}{(r + 1)\frac{k}{r} + (t - 2)\frac{k}{r}} = \frac{r}{r + t - 1} \,,$$

where the inequality is due to $\frac{k}{r} < r$.

- **Case 2:** $\frac{k}{r} \geq r$. In this case, we can choose $\mathbf{B}$ as a $k \times (t-2)\frac{k}{r}$ matrix as defined above. To this end,

$$n = (r+1)\frac{k}{r} + (t-2)\frac{k}{r},$$

and the rate of the code is

$$\frac{k}{n} = \frac{k}{(r+1)\frac{k}{r} + (t-2)\frac{k}{r}} = \frac{r}{r+t-1}.$$

By a suitable choice of k, it is always possible to construct a PIR code (or batch code if $t \in \{2,3,4\}$) achieving the rate $\frac{r}{r+t-1}$, which is close to the optimal rate given in [16]. Note that the condition $t \leq \max\{\frac{k}{r}, r\} + 2$ is necessary to make sure that no 1-squares are generated.

Proposition 8. *For $r > 2$ and $2 < t \leq \zeta + 2$ with $r|k$,*

$$\mathcal{P}(k, r, t) \leq (r+1)\frac{k}{r} + (t-2)\zeta$$

where $\zeta = \max\{\frac{k}{r}, r\}$.

Proposition 9. *Let $t = 3$. The code $\mathcal{C}$ defined by the generator matrix $\mathbf{G}$ supports any query of the form (x_i, x_j, x_ℓ) with recovery sets of size at most r, $i, j, \ell \in [k]$. Therefore it is a batch code.*

Proposition 10. *Let $t = 4$. The code $\mathcal{C}$ defined by the generator matrix $\mathbf{G}$ supports any query of the form (x_i, x_j, x_ℓ, x_h) with recovery sets of size at most r, for $i, j, \ell, h \in [k]$. Therefore it is a batch code.*

Proposition 11. *Let $t \geq 5$. The code $\mathcal{C}$ defined by the above generator matrix $\mathbf{G}$ supports any query of size t of the form $(x_i, x_i, \cdots, x_i)$ with recovery sets of size at most r. Therefore it is a PIR code.*

Example 3. Let $k = 8$, $r = 4$, $t = 3$ so that $k/r = 2$. Then the following generator matrix $\mathbf{G}$ generates a batch code of length $n = 14$.

$$\mathbf{G} = \left(\begin{array}{cc|cc|c} \mathbf{I}_4 & \mathbf{0} & 1 & 0 & \mathbf{I}_4 \\ \mathbf{0} & \mathbf{I}_4 & 0 & 1 & \mathbf{I}_4 \end{array} \right).$$

Example 4. Let $k = 12$, $r = 3$, $t = 5$ so that $k/r = 4$. Then the following generator matrix $\mathbf{G}$ generates a PIR code of length $n = 12 + 4 + 3 \cdot 4 = 28$.

$$
\mathbf{G} = \left(
\begin{array}{cccc|cccc|cccc}
 & & & & & & & & 1\,1\,1\,0 & 0\,0\,0\,0 & 0\,0\,0\,0 \\
\mathbf{I}_3\ 0\ 0\ 0 & & 1\,0\,0\,0 & & 0\,0\,0\,0 & 1\,1\,1\,0 & 0\,0\,0\,0 \\
 & & & & 0\,0\,0\,0 & 0\,0\,0\,0 & 1\,1\,1\,0 \\
\hline
 & & & & 0\,1\,0\,0 & 0\,1\,0\,0 & 0\,1\,0\,0 \\
0\ \mathbf{I}_3\ 0\ 0 & & 0\,1\,0\,0 & & 0\,0\,1\,0 & 0\,0\,1\,0 & 0\,0\,1\,0 \\
 & & & & 0\,0\,0\,1 & 0\,0\,0\,1 & 0\,0\,0\,1 \\
\hline
 & & & & 1\,0\,0\,0 & 0\,1\,0\,0 & 0\,0\,0\,1 \\
0\ 0\ \mathbf{I}_3\ 0 & & 0\,0\,1\,0 & & 0\,0\,0\,1 & 1\,0\,0\,0 & 0\,1\,0\,0 \\
 & & & & 0\,1\,0\,0 & 0\,0\,0\,1 & 1\,0\,0\,0 \\
\hline
 & & & & 0\,0\,0\,1 & 0\,0\,1\,0 & 1\,0\,0\,0 \\
0\ 0\ 0\ \mathbf{I}_3 & & 0\,0\,0\,1 & & 0\,0\,1\,0 & 0\,0\,0\,1 & 0\,0\,0\,1 \\
 & & & & 1\,0\,0\,0 & 1\,0\,0\,0 & 0\,0\,1\,0 \\
\end{array}
\right).
$$

The rate of the above code is

$$
\frac{k}{n} = \frac{12}{28} = \frac{3}{7} = \frac{r}{r+t-1}.
$$

This rate is greater than the rate of the direct product code in [16].

4.2 Case $t = 3$ and $r \nmid k$

In the sequel, we extend the results in the previous subsection towards the case where $t = 3$ and $r \nmid k$. Let $\mathbf{G}$ be the generator matrix of the block form $\mathbf{G} = [\mathbf{I}_k \,|\, \mathbf{A} \,|\, \mathbf{B} \,|\, \mathbf{C}]$.

$\mathbf{A}$ is a $k \times \lfloor \frac{k}{r} \rfloor$ matrix with the jth column of the form $\mathbf{A}^{[j]} = (a_{1,j}, a_{2,j}, \cdots, a_{k,j})^T$, $1 \le j \le \lfloor \frac{k}{r} \rfloor$, where $a_{i,j} = 1$ if $(j-1)r + 1 \le i \le jr$, and $a_{i,j} = 0$ otherwise.

Denote $s = k \mod r$. The matrix $\mathbf{B}$ is a $k \times (s+1)$ block matrix, defined as $\mathbf{B} = \left[\mathbf{B}_1^T | \mathbf{B}_2^T \right]^T$, where $\mathbf{B}_1$ is $(k-s) \times (s+1)$ and $\mathbf{B}_2$ is an $s \times (s+1)$ matrix $[\mathbf{1} \,|\, \mathbf{I}_s]$.

We take $\tau \triangleq \min\left\{ r-s, \lfloor \frac{k}{r} \rfloor \right\}$. The first column of $\mathbf{B}_1$, $\mathbf{B}_1^{[1]}$, has τ entries 1, each entry appears in a different block of rows $[(j-1)r + 1, jr]$ for $j = 1, 2, \cdots, \lfloor \frac{k}{r} \rfloor$. We take also $\eta \triangleq \min\left\{ r-1, \lfloor \frac{k}{r} \rfloor \right\}$. The columns $\mathbf{B}_1^{[2]}, \mathbf{B}_1^{[3]}, \cdots, \mathbf{B}_1^{[s+1]}$, all have η entries 1, each entry appears in a different block of rows. Additionally, every row in $\mathbf{B}_1$ contains at most one non-zero entry.

We observe that every column in $\mathbf{B}$ has at most r ones. Denote $\gamma \triangleq \min\left\{ r, \lfloor \frac{k}{r} \rfloor \right\}$. The matrix $\mathbf{C}$ is constructed according to the following rules:

- Each column in $\mathbf{C}$ has γ ones (except possibly for the last column);
- The last s rows in $\mathbf{C}$ are zeros;
- Each row in $[\mathbf{A} \,|\, \mathbf{B} \,|\, \mathbf{C}]$ has two ones;
- There are no 1-squares in $[\mathbf{B} \,|\, \mathbf{C}]$.

Next, we estimate the total number of columns in $\mathbf{G}$. The number of ones in the first $k - s$ positions of $\mathbf{B}^{[1]}$ is τ. The number of ones in the first $k - s$ positions of each of $\mathbf{B}^{[2]}, \mathbf{B}^{[3]}, \cdots, \mathbf{B}^{[s+1]}$ is η. Since there are two ones in each of the first $k - s$ rows of $[\,\mathbf{A}\,|\,\mathbf{B}\,|\,\mathbf{C}\,]$, the total number of columns in $\mathbf{G}$ is

$$n = (r+1)\left\lfloor \frac{k}{r} \right\rfloor + 2s + 1 + \left\lceil \frac{(k-s) - \tau - \eta \cdot s}{\gamma} \right\rceil .$$

Proposition 12. *For $t = 3$ and $r \geq 3$,*

$$k+1+\left\lceil \frac{2k-1}{2r-1} \right\rceil \leq \mathcal{B}(k,r,t) \leq \begin{cases} (r+1)\frac{k}{r} + \zeta & \text{if } r|k \\ (r+1)\left\lfloor \frac{k}{r} \right\rfloor + 2s + 1 + \left\lceil \frac{(k-s)-\tau-\eta \cdot s}{\gamma} \right\rceil & \text{if } r \nmid k \end{cases}$$

where $\zeta = \max\{\frac{k}{r}, r\}$.

Proposition 13. *The code $\mathcal{C}$ defined in this section by the generator matrix $\mathbf{G}$ supports any query of the form (x_i, x_j, x_ℓ) with recovery sets of size at most r, $i, j, \ell \in [k]$.*

Example 5. Let $k = 11$, $r = 3$, $t = 3$ so that $\lfloor k/r \rfloor = 3$ and $s = 2$. Then the following generator matrix $\mathbf{G}$ generates a batch code of length $n = 19$.

$$\mathbf{G} = \begin{pmatrix} \mathbf{I}_3\ 0\ 0\ 0 & 1\ 0\ 0 & 1\ 0\ 0 & 0\ 0 \\ & & 0\ 0\ 1 & 0\ 0 \\ & & 0\ 0\ 0 & 1\ 0 \\ 0\ \mathbf{I}_3\ 0\ 0 & 0\ 1\ 0 & 0\ 1\ 0 & 0\ 0 \\ & & 0\ 0\ 1 & 0\ 0 \\ & & 0\ 0\ 0 & 1\ 0 \\ 0\ 0\ \mathbf{I}_3\ 0 & 0\ 0\ 1 & 0\ 1\ 0 & 0\ 0 \\ & & 0\ 0\ 0 & 0\ 1 \\ & & 0\ 0\ 0 & 1\ 0 \\ 0\ 0\ 0\ \mathbf{I}_2 & 0\ 0\ 0 & 1 \quad \mathbf{I}_2 & 0\ 0 \end{pmatrix} .$$

We summarize the bounds on $\mathcal{B}(k,r,t)$ and $\mathcal{P}(k,r,t)$ for $t \in \{2, 3, 4\}$ and for $r \geq 2$ in Tables 1 and 2.

Table 1. Lower bounds for $\mathcal{B}(k,r,t)$ and $\mathcal{P}(k,r,t)$

	$t = 2$	$t = 3$	$t = 4$
$r \geq 2$	$k + \lceil k/r \rceil$	$k + 1 + \left\lceil \frac{2k-1}{2r-1} \right\rceil$	$k + 2 + \left\lceil \frac{3k-2}{3r-2} \right\rceil$

Table 2. Upper bounds for $\mathcal{B}(k,r,t)$ and $\mathcal{P}(k,r,t)$

	$t = 2$	$t = 3$		$t = 4$
$r = 2, k > 3$	$k + \lceil k/2 \rceil$	$2k$		$k + \lceil 3k/2 \rceil$
$r \geq 3$	$k + \lceil \frac{k}{r} \rceil$	$\begin{cases} (r+1)\frac{k}{r} + \zeta \\ (r+1)\lfloor \frac{k}{r} \rfloor + 2s + 1 + \left\lceil \frac{(k-s)-\tau-\eta \cdot s}{\gamma} \right\rceil \end{cases}$	$\begin{matrix} \text{if } r|k \\ \text{if } r \nmid k \end{matrix}$	$(r+1)\frac{k}{r} + 2\zeta$, if $r|k$

5 Bounds on the Dimension of a Batch Code

Let $k_q^{\text{opt}}(n, d)$ denote the largest possible dimension of a linear code of length n and minimum distance d, for a given alphabet $\mathcal{Q}$ of size q. More formally, by following on the notations in [4], denote:

$$k_q^{\text{opt}}(n, d) \triangleq \max \frac{\log |\mathcal{C}|}{\log q} \, ,$$

where the maximum is taken over all possible linear codes $\mathcal{C}$ of length n with minimum distance d.

Let $\mathcal{I} \subseteq [n]$ be a set of coordinates. Define

$$H(\mathcal{I}) = \frac{\log |\{\mathbf{x}_\mathcal{I} : \mathbf{x} \in \mathcal{C}\}|}{\log q} \, .$$

The following result appears as Lemma 2 in [4].

Lemma 2. *Consider an $[n, k, d]$ code over $\mathcal{Q}$ where there exists a set $\mathcal{I} \subseteq [n]$ such that $H(\mathcal{I}) \leq m$. Then, there exists an $[n - |\mathcal{I}|, (k - m)^+, d]$ code over $\mathcal{Q}$, where the $+$ symbol denotes that the dimension is at least $k - m$.*

Cadambe and Mazumdar show in [4] that, for any r-locally recoverable (n, k, d) code over the alphabet $\mathcal{Q}$, it holds:

$$k \leq \min_t \left[tr + k_q^{\text{opt}}(n - t(r + 1), d) \right] \, . \tag{7}$$

By using similar techniques, in the sequel we show a bound on k for a linear (n, k, r, t) batch code. We restrict our discussion to linear codes only.

Proposition 14. *Let $\mathcal{C}$ be a linear (n, k, r, t) batch code over an alphabet $\mathcal{Q}$ of size q with minimum distance d, and $n - tr \geq d$. Then,*

$$k \leq tr - (t - 1) + k_q^{\text{opt}}(n - tr, d) \, . \tag{8}$$

Proof. Since $\mathcal{C}$ is an (n, k, r, t) batch code, for a query $(x_i, \ldots, x_i)$ of size t, for some $i \in [k]$, there exist t disjoint recovery sets $R_1, \ldots, R_t$ with $|R_j| \leq r$ for all $j \in [t]$.

Denote $\mathcal{I} \triangleq R_1 \cup R_2 \cup \cdots \cup R_t$. Clearly, $|\mathcal{I}| \leq tr$. Take an arbitrary word $\mathbf{y} \in \mathcal{C}$. By Lemma 1, there exist indices $a_2 \in R_2, a_3 \in R_3, \cdots, a_t \in R_t$ such that if we fix the values of all coordinates of $\mathbf{y}$ indexed by the sets $R_1, R_2 \backslash \{a_2\}, R_3 \backslash \{a_3\}, \ldots, R_t \backslash \{a_t\}$, then the values of the coordinates of $\mathbf{y}$ indexed by $S \triangleq \{a_2, a_3, \ldots, a_t\}$ are uniquely determined. It follows that there is one-to-one mapping between $\mathbf{y}_{\mathcal{I} \backslash S}$ and $\mathbf{y}_\mathcal{I}$. Therefore, $H(\mathcal{I}) \leq tr - (t - 1)$.

The main statement now follows by applying Lemma 2.

We remark that the condition $n - tr \geq d$ in Proposition 14 is necessary for the existence of a code of length $n - tr$ and minimum distance d. However, it should be noted that any (n, k, r, t) batch code is a (n, k, r, β) batch code for $1 \leq \beta \leq t$. Thus, if t is too large to satisfy this condition, we can take a smaller value β instead. The following result follows.

Corollary 2. *Let $\mathcal{C}$ be a linear (n, k, r, t) batch code over an alphabet $\mathcal{Q}$ of size q with minimum distance d. Then,*

$$k \leq \min_{1 \leq \beta \leq t} \left\{ \beta r - (\beta - 1) + k_q^{\mathsf{opt}}(n - \beta r, d) \right\} . \tag{9}$$

Comparison with the Bounds in the Literature

Example 6. The asymptotic versions of the classical Singleton and the sphere-packing bounds for a code over alphabet $\mathcal{Q}$ of size q are $R \leq 1 - \delta + o(1)$ and $R \leq 1 - h_q(\delta/2) + o(1)$, respectively, where $R = k/n$, $\delta = d/n$, and $h_q(\cdot)$ denotes the q-ary entropy function [14, Chap. 4].

The asymptotic versions of the bounds (1) and (9) can be rewritten as (when ignoring the $o(1)$ term, for any specific value of β):

$$R \leq 1 - \delta - \frac{\beta - 1}{n} \left(\left\lceil \frac{k}{\beta r - \beta + 1} \right\rceil - 1 \right) , \tag{10}$$

and

$$R \leq \frac{\beta(r - 1)}{n} + R_q^{\mathsf{opt}}(n - \beta r, d) , \tag{11}$$

respectively, where $R_q^{\mathsf{opt}}(n, d)$ denotes the maximum rate of any code of length n and minimum distance d over $\mathcal{Q}$.

For $n - d \gg tr \geq \beta r$, the bound (10) becomes $R \leq (1 - \delta) \cdot \left(1 - \frac{\beta - 1}{\beta r} \right)$. For comparison, when we use the sphere-packing bound for $R_q^{\mathsf{opt}}(n, d)$, then (11) can be rewritten as

$$R \leq \frac{\beta(r - 1)}{n} + 1 - h_q(\delta/2) \approx 1 - h_q(\delta/2) .$$

Therefore, for large blocklength, $n - d \gg tr$, and for a range of δ and r, the bound (9) is tighter than (1) (for any $\beta \geq 1$).

Acknowledgment. The work of E. Thomas and V. Skachek is supported by the Estonian Research Council under the grants PUT405 and IUT2-1. The authors wish to thank Mart Simisker for valuable comments on the earliest version of this work.

References

1. Bhattacharya, S., Ruj, S., Roy, B.: Combinatorial batch codes: a lower bound and optimal constructions. Adv. Math. Commun. **6**(2), 165–174 (2012)
2. Brualdi, R.A., Kiernan, K., Meyer, S.A., Schroeder, M.W.: Combinatorial batch codes and transversal matroids. Adv. Math. Commun. **4**(3), 419–431 (2010)
3. Bujtas, C., Tuza, Z.: Batch codes and their applications. Electron. Notes Discrete Math. **38**, 201–206 (2011)
4. Cadambe, V., Mazumdar, A.: Bounds on the size of locally recoverable codes. IEEE Trans. Inf. Theory **61**(11), 5787–5794 (2015)

5. Chee, Y.M., Gao, F., Teo, S.T.H., Zhang, H.: Combinatorial systematic switch codes. In: Proceedings of the IEEE International Symposium on Information Theory (ISIT), Hong Kong, China, pp. 241–245 (2015)
6. Dimakis, A.G., Gal, A., Rawat, A.S., Song, Z.: Batch codes through dense graphs without short cycles, arXiv:1410.2920, October 2014
7. Fazeli, A., Vardy, A., Yaakobi, E.: PIR with low storage overhead: coding instead of replication, arXiv:1505.06241, May 2015
8. Forbes, M., Yekhanin, S.: On the locality of codeword sysmbols in non-linear codes. Discrete Math. **324**, 78–84 (2014)
9. Gopalan, P., Huang, C., Simitci, H., Yekhanin, S.: On the locality of codeword symbols. IEEE Trans. Inform. Theory **58**(11), 6925–6934 (2012)
10. Ishai, Y., Kushilevitz, E., Ostrovsky, R., Sahai, A.: Batch codes and their applications. In: Proceeding of the 36th ACM Symposium on Theory of Computing (STOC), Chicago, June 2004
11. Lipmaa, H., Skachek, V.: Linear batch codes. In: Proceedings of the 4th International Castle Meeting on Coding Theory and Applications, Palmela, Portugal, September 2014
12. Rawat, A.S., Mazumdar, A., Vishwanath, S.: Cooperative local repair in distributed storage. EURASIP J. Adv. Sign. Process. **1**, 107 (2015)
13. Rawat, A.S., Papailiopoulos, D.S., Dimakis, A.G., Vishwanath, S.: Locality and availability in distributed storage. IEEE Trans. Inform. Theory **62**(8), 4481–4493 (2016)
14. Roth, R.M.: Introduction to Coding Theory. Cambridge University Press, Cambridge (2006)
15. Skachek, V.: Batch and PIR Codes and Their Connections to Locally Repairable Codes, arXiv:1611.09914, November 2016
16. Tamo, I., Barg, A.: Bounds on Locally Recoverable Codes with Multiple Recovering Sets (2014). arXiv:1402.0916v1
17. Vardy, A., Yaakobi, E.: Constructions of batch codes with near-optimal redundancy. In: Proceedings of the IEEE International Symposium on Information Theory (ISIT), Barcelona (2016)
18. Wang, Z., Kiah, H.M., Cassuto, Y.: Optimal binary switch codes with small query size. In: Proceeding of the IEEE International Symposium on Information Theory (ISIT), Hong Kong, pp. 636–640, June 2015
19. Wang, Z., Shaked, O., Cassuto, Y., Bruck, J.: Codes for network switches. In: Proceeding of the IEEE International Symposium on Information Theory (ISIT), Istanbul (2013)
20. Wang, A., Zhang, Z.: Achieving Arbitrary Locality and Availability in Binary Codes, arXiv:1501.04264v1 (January 2015)
21. Zhang, H., Skachek, V.: Bounds for batch codes with restricted query size. In: Proceeding of the IEEE International Symposium on Information Theory (ISIT), Barcelona, pp. 1192–1196 (2016)

Author Index

Printed in the United States
By Bookmasters